Multivariate Analysis
Part 2
Classification, covariance structures and repeated measurements

KENDALL'S ADVANCED THEORY OF STATISTICS
and
KENDALL'S LIBRARY OF STATISTICS

The development of statistical theory in the past fifty years is faithfully reflected in the history of the late Sir Maurice Kendall's volumes THE ADVANCED THEORY OF STATISTICS. The ADVANCED THEORY began life as a two volume work (Volume 1, 1943; Volume 2, 1946) and grew steadily, as a single authored work, until the late fifties. At that point, Alan Stuart became co-author and the ADVANCED THEORY was rewritten in three volumes. When Keith Ord joined in the early eighties, Volume 3 became the largest and plans were developed to expand, yet again, to a four-volume work. Even so, it became evident that there were gaps in the coverage and that it was becoming increasingly difficult to provide timely updates to all volumes, so a new strategy was devised.

In future, the ADVANCED THEORY will be in the form of three core volumes together with a series of related monographs called KENDALL'S LIBRARY OF STATISTICS. The three volumes of ADVANCED THEORY will be:

1 Distribution Theory
2A Classical Inference and Relationships
2B Bayesian Inference (a new companion volume by Anthony O'Hagan)

KENDALL'S LIBRARY OF STATISTICS will encompass the areas previously appearing in the old Volume 3, such as sample surveys, design of experiments, multivariate analysis and time series as well as non-parametrics and log-linear models, previously covered to some extent in Volume 2. In the preface to the first edition of the ADVANCED THEORY Kendall declared that his aim was 'to develop a systematic treatment of [statistical] theory as it exists at the present time' while ensuring that the work remained 'a book on statistics, not on statistical mathematics'. These aims continue to hold true for KENDALL'S LIBRARY OF STATISTICS and the flexibility of the monograph format will enable the series to maintain comprehensive coverage over the whole of modern statistics.

Published volumes:

1. MULTIVARIATE ANALYSIS Part 1 Distributions, Ordination and Inference, W J Krzanowski (University of Exeter) and F H C Marriott (University of Oxford) 1994

2. MULTIVARIATE ANALYSIS Part 2 Classification, Covariance Structures and Repeated Measurements, W J Krzanowski (University of Exeter) and F H C Marriott (University of Oxford) 1995

3. MULTILEVEL STATISTICAL MODELS Second Edition, Harvey Goldstein (University of London) 1995

MULTIVARIATE ANALYSIS
Part 2
Classification, covariance structures and repeated measurements

W J Krzanowski

Professor of Statistics, Department of Mathematical Statistics and Operational Research, University of Exeter, Exeter, UK

and

F H C Marriott

Department of Statistics, University of Oxford, Oxford, UK

A member of the Hodder Headline Group

LONDON

Co-published in the United States of America by
Oxford University Press Inc., New York

First published in Great Britain by Arnold 1995
a member of the Hodder Headline Group
338 Euston Road, London NW1 3BH

Co-published in the United States of America by
Oxford University Press Inc.,
198 Madison Avenue, New York, NY10016

British Library Cataloguing in Publication Data
A catalogue record for this book is available from the British Library

Library of Congress Cataloging-in-Publication Data
A catalog record for this book is available from the Library of Congress

ISBN 0 340 59325 3

4 5 6 7 8 9 10

Typeset in 10/11 Times by Paston Press Ltd, Loddon, Norfolk
Printed and bound in Great Britain by Bookcraft, Bath.

Contents

Preface

This is the second and concluding part of our survey of multivariate analysis. For consistency and ease of cross-referencing we have maintained our previous style and notation, and have continued the sequential numbering of chapters from where we left off in Part 1. Where relevant we have also made explicit that a particular reference is to material in Part 1.

While the first part dealt with what might be termed the 'basic' multivariate ideas and techniques, the present volume is concerned with the more specialised models and methods that follow on from the basic theory. The first two chapters focus on classification, Chapter 9 dealing with discrimination ('supervised learning' in pattern-recognition terminology) and Chapter 10 with cluster analysis ('unsupervised learning'). Both of these topics have a long history, and the two chapters cover the newer ideas in neural networks and image analysis as well as the more traditional methodology. The two following chapters are then concerned with the modelling and analysis of covariance structures. Chapter 11 focuses mainly on underlying models, with a discussion of path models, graphical models and partial least squares, while Chapter 12 surveys all the techniques that rely on the concept of latent variables. These techniques include factor analysis and various manifestations of latent structure analysis, many of which have been developed in the social and behavioural sciences and so may not be familiar to all statisticians. Chapter 13 then extends the linear model ideas outlined in Chapter 7 (Part 1) to cater for correlated observations, and Chapter 14 covers a number of miscellaneous items such as distances between populations, analysis of high-dimensional and functional data, and the treatment of directional or angular data. Finally, Chapter 15 includes a summary overview, some consideration of strategic aspects of multivariate analysis, and brief speculation about future directions.

It almost goes without saying that the span of multivariate analysis is now so enormous that a single work giving full technical details is no longer a viable proposition. Indeed, there are now many texts on the market that are much weightier than ours but which are devoted to only a small aspect of the subject. Our aim has therefore been somewhat more modest: to provide an overview that conveys the scope and flavour of the modern subject in a way that is not only accessible to a variety of readers but that also enables the interested reader to find any necessary detail both easily and quickly. We hope that we have succeeded in this aim, but as always will welcome any comments and constructive suggestions for improvement.

All acknowledgements cited in the Preface to Part 1 hold equally for the present volume. We would also like to add our thanks to Dr V G S Vasdekis for his comments on Chapter 13.

W J Krzanowski, Exeter — F H C Marriott, Oxford

9

Discriminant Analysis

Overview of concepts

9.1 The objectives of discriminant analysis can be stated very simply. Given the existence of several distinct groups of individuals, and a sample of observations from each group, can we find functions of these observations that will distinguish the groups and enable future unidentified individuals to be classified to their correct group? Such questions arise in many diverse application areas. For example: an archaeologist may wish to distinguish between a number of bronze-age populations from each of which some skeletons are known to have come, and also to determine to which (if any) of these populations some unidentified skeletons should be ascribed; hospital patients suffering from jaundice can be treated either by medication or surgery, the registry has available all the records of previous patients known to have belonged to each of these groups, and an efficient method of patient diagnosis is required; and seismologists taking recordings at a particular station wish to distinguish between an earthquake and an underground nuclear test, on the basis of signals recorded on previous events whose provenance was subsequently identified.

Strictly speaking there are two distinct objectives here, those of *discrimination* and *classification* respectively, and each may, in principle, give rise to its own function of observations. Group separation is achieved by means of a *discriminant function,* while identification of future individuals is handled through a *classification rule* (or *allocation rule*). Unsurprisingly, however, it turns out very often that the best function for group separation also provides the best allocation rule for future individuals, so the terms 'discriminant function' and 'allocation rule' are generally used interchangeably. We shall follow this practice, and will only specifically distinguish the two separate objectives when a significant practical difference arises between them. Otherwise, both will be subsumed under the generic title of *discriminant analysis.*

Example 9.1
Table 9.1 gives the set of bivariate observations extracted by Seber (1984) from Lubischew (1962) on specimens of three species of male flea beetles. The two variables measured were maximal width of aedeagus in the forepart (x_1) and front angle of aedeagus (x_2); there were 21 specimens of *Chaetocnema concinna*, 31 specimens of *Chaetocnema heikertingeri* and 22 specimens of *Chaetocnema heptapotamica*. The scatter diagram for these data is shown in Fig. 9.1. It is evident that the three species are well separated. Many different boundaries between species (discriminant functions) could be drawn in on the diagram; a future individual with a pair of values (x_1, x_2) could then be plotted on this diagram, and these boundaries could be used to allocate it to one of the species. See Example 9.3 for further discussion.

Table 9.1 Two variables observed on samples of three species of male flea beetles, reproduced with permission from Seber (1984)

C. concinna		*C. heikertingeri*		*C. heptapotamica*	
x_1	x_2	x_1	x_2	x_1	x_2
150	15	120	14	145	8
147	13	123	16	140	11
144	14	130	14	140	11
144	16	131	16	131	10
153	13	116	16	139	11
140	15	122	15	139	10
151	14	127	15	136	12
143	14	132	16	129	11
144	14	125	14	140	10
142	15	119	13	137	9
141	13	122	13	141	11
150	15	120	15	138	9
148	13	119	14	143	9
154	15	123	15	142	11
147	14	125	15	144	10
137	14	125	14	138	10
134	15	129	14	140	10
157	14	130	13	130	9
149	13	129	13	137	11
147	13	122	12	137	10
148	14	129	15	136	9
		124	15	140	10
		120	13		
		119	16		
		119	14		
		133	13		
		121	15		
		128	14		
		129	14		
		124	13		
		129	14		

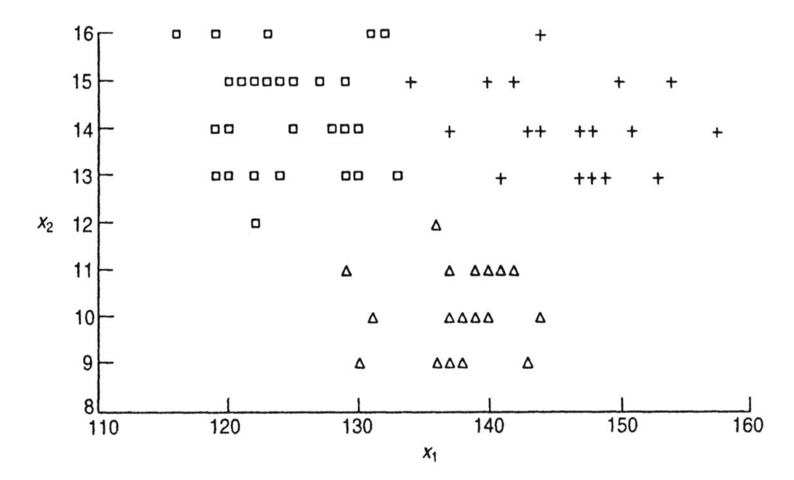

Figure 9.1 Scatter diagram of data from Table 9.1. *C. concinna* individuals denoted +, *C. heikertingeri* individuals denoted □, and *C. heptapotamica* individuals denoted △. Reproduced with permission from Seber (1984).

9.2 There are many different ways of approaching the construction of a discriminant function or allocation rule in a given situation. A traditional statistical approach to the problem is to postulate a probability model for each of the populations under consideration, specify an objective function to be optimised, derive the best population discriminant function or allocation rule, and then estimate this function from the available sample data. Specifying the probability models parametrically and conducting standard estimation of these parameters yields a *parametric* approach to the problem. Alternatively, one might specify a parametric form directly for the discriminant function or allocation rule, and then estimate the parameters of the function from the available data. Here the parameters occur only in the function and not in any probability models, so such an approach is sometimes termed *semiparametric*. The third possibility is to take a fully *non-parametric* approach. This encompasses many different options, ranging from non-parametric estimation of density functions for substitution into population discriminant functions through to construction of discriminant functions from purely data-based characteristics such as distances or neighbourhoods in the sample space.

9.3 Given this plethora of available functions, the practitioner will want some means of objectively assessing performance in order to select the best possible rule in any given practical application. Such an assessment can be provided by the *error rates* associated with each rule. These are the probabilities that the given rule will misclassify future individuals from each population under study. They implicitly provide a measure of 'overlap' of the populations in the data space. Whereas such probabilities can often be calculated theoretically under specific assumptions about the population structures, practical problems arise in trying to *estimate* the error rates in any given application. Once again, many different approaches have been proposed for such estimation.

Another aspect of considerable practical interest in discriminant analysis is that of selecting the best variables (or *features*) on which to conduct the analysis. Intuition suggests that discrimination should always improve whenever new features are added to a data set, because combining the new features with the existing ones can but emphasize group separation. However, many features will be ones that contribute very little to group separation and will thus provide only very marginal improvement. Including them in any discriminant function increases the number of parameters to be estimated, and introduces extra complications in error rate estimation. Consequently, adding such features to an allocation rule may even produce an apparently *worse* performance of the rule than if they had been left out. Such considerations, and those of economy of future data collection, mean that investigators are often interested in choosing just a small subset of features on which to base their analysis. Various criteria have been proposed for effecting such choice of features, including ones based on estimated error rates, and this then introduces the possibility of over-optimism in assessment of performance of the consequent allocation rules.

9.4 The above are all general problems arising in discriminant analysis, whatever the form of the available data. In recent years, further problems have arisen because of a number of special forms of data, in particular those showing strong spatial or temporal dependence or those in which very many observations are taken on each sample individual by some automatic recording device, and special methods have been developed to handle these problems.

The present chapter is devoted to a survey of these topics. The above introduction has already hinted at the range of complicating features of the subject, and it should be said at the outset that this is a very heavily researched area. Moreover, this research has not been restricted just to the traditional statistical sources, as discrimination is also of central interest in such studies as pattern recognition, artificial intelligence and neural networks. Independent research in these areas has led inevitably to divergences in notation and terminology, as well as to an enormous bibliography. A survey such as the present one thus has obvious limitations, and the reader will need to supplement it by plenty of reference to source material. For a single comprehensive text on the subject, however, the book by McLachlan (1992) is highly recommended; much of this chapter draws on the material contained therein.

Empirical discriminant functions

9.5 The foundations of the subject were laid in the now-classic study of differences between various types of Iris plants by Fisher (1936), who sought the linear function $\boldsymbol{a}'\boldsymbol{x}$ of p variables $\boldsymbol{x}' = (x_1, x_2, \ldots, x_p)$ that would best separate samples of *Iris versicolor* and *Iris setosa*. Fisher argued that the coefficients $\boldsymbol{a}' = (a_1, a_2, \ldots, a_p)$ in this function should be given by the values maximising the ratio of squared group mean difference to within-group variance. If there are n_i observations $\boldsymbol{x}_{i1}, \ldots, \boldsymbol{x}_{in_i}$ in the ith group for $i = 1, 2,$

then the coefficients are obtained by maximising the criterion

$$V = (\boldsymbol{a}'\bar{\boldsymbol{x}}_1 - \boldsymbol{a}'\bar{\boldsymbol{x}}_2)^2/\boldsymbol{a}'\boldsymbol{W}\boldsymbol{a}, \tag{9.1}$$

where $\bar{\boldsymbol{x}}_1 = \frac{1}{n_1}\sum_{i=1}^{n_1} \boldsymbol{x}_{1i}$, $\bar{\boldsymbol{x}}_2 = \frac{1}{n_2}\sum_{i=1}^{n_2} \boldsymbol{x}_{2i}$ are the two sample means of $\boldsymbol{x}$ and $\boldsymbol{W} = \frac{1}{n_1+n_2-2}\sum_{i=1}^{2}\sum_{j=1}^{n_i}(\boldsymbol{x}_{ij} - \bar{\boldsymbol{x}}_i)(\boldsymbol{x}_{ij} - \bar{\boldsymbol{x}}_i)'$ is the within-group pooled covariance matrix. Straightforward differentiation of V with respect to $\boldsymbol{a}$ yields the coefficients in *Fisher's Linear Discriminant Function* (LDF) as

$$\boldsymbol{a} = c\boldsymbol{W}^{-1}(\bar{\boldsymbol{x}}_1 - \bar{\boldsymbol{x}}_2) \tag{9.2}$$

for arbitrary constant multiplying factor c (generally taken as 1).

In the same paper, Fisher mentioned that the coefficients (9.2) could be equivalently obtained by regressing a dummy variable y taking value $y_1 = n_2/(n_1 + n_2)$ for each observation in group 1 and value $y_2 = -n_1/(n_1 + n_2)$ for each observation in group 2, on the p variables $\boldsymbol{x}$. Further consideration of this equivalence was given by Fisher (1938); see also Hand (1981) for a formal proof. Whereas the original derivation yields a *discriminant* function, the regression approach can be viewed as one of *classification* (since regression analysis affords the possibility of predicting the y value and hence the group membership of a future individual $\boldsymbol{x}$). Thus, we have here a demonstration of the feature mentioned in **9.2**, that the two different objectives are served by the same function; the individual $\boldsymbol{x}$ can be allocated to group 1 if its predicted y value is closer to y_1 than to y_2, and otherwise to group 2.

Example 9.2

Table 9.2 gives the *Iris* data, used by Fisher (1936) and reproduced from Kendall *et al* (1983); the measurements were x_1 = sepal length, x_2 = sepal width, x_3 = petal length, x_4 = petal width (all in centimetres), and there were 50 plants from each of the two species. The mean vectors are

$$\bar{\boldsymbol{x}}_1' = (5.006, 3.428, 1.462, 0.246)$$

for *setosa* and

$$\bar{\boldsymbol{x}}_2' = (5.936, 2.770, 4.260, 1.326)$$

for *versicolor*. The pooled covariance matrix in the two groups is

$$\boldsymbol{W} = \begin{pmatrix} 0.1953 & 0.0922 & 0.0997 & 0.0331 \\ 0.0922 & 0.1211 & 0.0472 & 0.0253 \\ 0.0997 & 0.0472 & 0.1255 & 0.0396 \\ 0.0331 & 0.0253 & 0.0396 & 0.0251 \end{pmatrix}$$

with inverse

$$\boldsymbol{W}^{-1} = \begin{pmatrix} 11.6342 & -6.5530 & -7.9984 & 3.8844 \\ -6.5530 & 14.2368 & 3.2743 & -10.8539 \\ -7.9984 & 3.2743 & 21.4975 & -26.6582 \\ 3.8844 & -10.8539 & -26.6582 & 87.6661 \end{pmatrix}$$

Thus, from (9.2), Fisher's linear discriminant function for these data is given by $y = 3.0528x_1 + 18.0230x_2 - 21.7662x_3 - 30.8442x_4$. Rescaling so

Table 9.2 Fisher's *Iris* data

Iris setosa				*Iris versicolor*			
x_1	x_2	x_3	x_4	x_1	x_2	x_3	x_4
5.1	3.5	1.4	0.2	7.0	3.2	4.7	1.4
4.9	3.0	1.4	0.2	6.4	3.2	4.5	1.5
4.7	3.2	1.3	0.2	6.9	3.1	4.9	1.5
4.6	3.1	1.5	0.2	5.5	2.3	4.0	1.3
5.0	3.6	1.4	0.2	6.5	2.8	4.6	1.5
5.4	3.9	1.7	0.4	5.7	2.8	4.5	1.3
4.6	3.4	1.4	0.3	6.3	3.3	4.7	1.6
5.0	3.4	1.5	0.2	4.9	2.4	3.3	1.0
4.4	2.9	1.4	0.2	6.6	2.9	4.6	1.3
4.9	3.1	1.5	0.1	5.2	2.7	3.9	1.4
5.4	3.7	1.5	0.2	5.0	2.0	3.5	1.0
4.8	3.4	1.6	0.2	5.9	3.0	4.2	1.5
4.8	3.0	1.4	0.1	6.0	2.2	4.0	1.0
4.3	3.0	1.1	0.1	6.1	2.9	4.7	1.4
5.8	4.0	1.2	0.2	5.6	2.9	3.6	1.3
5.7	4.4	1.5	0.4	6.7	3.1	4.4	1.4
5.4	3.9	1.3	0.4	5.6	3.0	4.5	1.5
5.1	3.5	1.4	0.3	5.8	2.7	4.1	1.0
5.7	3.8	1.7	0.3	6.2	2.2	4.5	1.5
5.1	3.8	1.5	0.3	5.6	2.5	3.9	1.1
5.4	3.4	1.7	0.2	5.9	3.2	4.8	1.8
5.1	3.7	1.5	0.4	6.1	2.8	4.0	1.3
4.6	3.6	1.0	0.2	6.3	2.5	4.9	1.5
5.1	3.3	1.7	0.5	6.1	2.8	4.7	1.2
4.8	3.4	1.9	0.2	6.4	2.9	4.3	1.3
5.0	3.0	1.6	0.2	6.6	3.0	4.4	1.4
5.0	3.4	1.6	0.4	6.8	2.8	4.8	1.4
5.2	3.5	1.5	0.2	6.7	3.0	5.0	1.7
5.2	3.4	1.4	0.2	6.0	2.9	4.5	1.5
4.7	3.2	1.6	0.2	5.7	2.6	3.5	1.0
4.8	3.1	1.6	0.2	5.5	2.4	3.8	1.1
5.4	3.4	1.5	0.4	5.5	2.4	3.7	1.0
5.2	4.1	1.5	0.1	5.8	2.7	3.9	1.2
5.5	4.2	1.4	0.2	6.0	2.7	5.1	1.6
4.9	3.1	1.5	0.2	5.4	3.0	4.5	1.5
5.0	3.2	1.2	0.2	6.0	3.4	4.5	1.6
5.5	3.5	1.3	0.2	6.7	3.1	4.7	1.5
4.9	3.6	1.4	0.1	6.3	2.3	4.4	1.3
4.4	3.0	1.3	0.2	5.6	3.0	4.1	1.3
5.1	3.4	1.5	0.2	5.5	2.5	4.0	1.3
5.0	3.5	1.3	0.3	5.5	2.6	4.4	1.2
4.5	2.3	1.3	0.3	6.1	3.0	4.6	1.4
4.4	3.2	1.3	0.2	5.8	2.6	4.0	1.2
5.0	3.5	1.6	0.6	5.0	2.3	3.3	1.0
5.1	3.8	1.9	0.4	5.6	2.7	4.2	1.3
4.8	3.0	1.4	0.3	5.7	3.0	4.2	1.2
5.1	3.8	1.6	0.2	5.7	2.9	4.2	1.3
4.6	3.2	1.4	0.2	6.2	2.9	4.3	1.3
5.3	3.7	1.5	0.2	5.1	2.5	3.0	1.1
5.0	3.3	1.4	0.2	5.7	2.8	4.1	1.3

that the coefficient of x_1 is 1, we obtain $y' = x_1 + 5.9037x_2 - 7.1299x_3 - 10.1036x_4$, the function quoted by Fisher (1936). The mean of y' for the *setosa* sample is thus 12.3345, and that for the *versicolor* sample is -21.4815. The mid-point between these two values is -4.574, so we allocate a new individual $\boldsymbol{x}$ to *setosa* or *versicolor* according to whether its y' value lies above or below -4.574.

9.6 In fact, there were three types of Iris plants in the study, a sample of *Iris virginica* plants also being available. Fisher (1936) therefore went on to consider discrimination between all three groups. However, additional structure could be imposed on the data because it had been conjectured that *Iris versicolor* was a polyploid hybrid of the other two species in specified proportions. Fisher used these proportions directly in forming the coefficients of a linear function segregating the three types. His solution cannot therefore be viewed as a true generalisation of the two-group discriminant function to the case of three or more arbitrarily defined groups. Such a generalisation was subsequently provided by Rao (1948) and Bryan (1951).

Both these authors took the natural route of extending the criterion (9.1) to the case of $g > 2$ groups, and sought the linear function $\boldsymbol{a}'\boldsymbol{x}$ of the p variables $\boldsymbol{x}' = (x_1, x_2, \ldots, x_p)$ that maximises the ratio of between-group to within-group variance. This criterion can be written

$$V = \boldsymbol{a}'\boldsymbol{B}\boldsymbol{a}/\boldsymbol{a}'\boldsymbol{W}\boldsymbol{a} \tag{9.3}$$

where $\boldsymbol{W}$ is the within-group matrix as before, but now $\boldsymbol{B}$ is the *between-group* covariance matrix. The problem is that this latter quantity can be defined in several different ways. The more usually adopted one is the *weighted* matrix considered by Bryan, $\boldsymbol{B}_w = \frac{1}{n-g}\sum_{i=1}^{g} n_i(\bar{\boldsymbol{x}}_i - \bar{\boldsymbol{x}})(\bar{\boldsymbol{x}}_i - \bar{\boldsymbol{x}})'$, where n_i is the sample size in group i and $n = \sum_{i=1}^{g} n_i$; the alternative, considered by Rao, is the *unweighted* matrix $\boldsymbol{B}_u = \frac{1}{n-g}\sum_{i=1}^{g}(\bar{\boldsymbol{x}}_i - \bar{\boldsymbol{x}})(\bar{\boldsymbol{x}}_i - \bar{\boldsymbol{x}})'$.

The main features of the analysis are the same whichever matrix is used, but their interpretations differ according to the choice. Maximisation of (9.3) leads to the solution of the generalised eigenequation $(\boldsymbol{B} - \lambda\boldsymbol{W})\boldsymbol{a} = \boldsymbol{0}$, so there is, in general, more than one set of coefficients $\boldsymbol{a}$. In fact there are $s = \min(p, g-1)$ of them, $\boldsymbol{a}_1, \ldots, \boldsymbol{a}_s$, and the corresponding linear functions $\boldsymbol{a}_i'\boldsymbol{x}$ are the *canonical variates* already discussed in Part 1, **4.11**. Their use, as described in that section, is in identifying a low-dimensional space in which to plot the sample individuals and, more particularly, the group means, so as to highlight the between-group differences. The recovered configuration will depend on which of the two between-group matrices is used: using the weighted matrix $\boldsymbol{B}_w$ produces the configuration that maximises the between-group to within-group scatter in the chosen dimensionality, while using the unweighted matrix $\boldsymbol{B}_u$ produces the configuration that maximises the total Mahalanobis squared distance between all pairs of groups in the chosen dimensionality. The former property is perhaps the more useful one for highlighting group differences and so this is the one usually adopted, but Ashton *et al* (1957) point out the merits of the latter in descriptive analysis while Gower (1966) shows that the latter solution can also be obtained by a metric scaling of the matrix of Mahalanobis distances between all pairs of groups (see Part 1, **5.8**). Of course, both solutions will be the same if sample sizes are equal in all groups.

Just as the two-group linear discriminant function can be derived by regression of a dummy variable y on the observed variables $\boldsymbol{x}$, so the (weighted) solution above can be obtained by canonical correlation analysis between the observed variables $\boldsymbol{x}$ and a dummy set of design variables $\boldsymbol{y}$ defining the groups (Part 1, **4.11**). Again, the prime motivation of the derivation of the technique is discriminant analysis, as the recovered configurations *describe* differences between groups, but the same functions can be used for classification by superimposing the unclassified individuals on the diagram and allocating each to the 'nearest' group mean point. It is often felt useful to show sampling variation on the diagrams, by means of either $100\alpha\%$ tolerance regions for individuals (Ashton *et al*, 1957) or $100\alpha\%$ confidence regions for population means (Mardia *et al*, 1979, p344). The former are traditionally shown as circles (on two-dimensional canonical diagrams) of squared radii $\chi^2_{2,\alpha}$ and the latter as circles of squared radii $\chi^2_{2,\alpha}/n_i$, where $\chi^2_{2,\alpha}$ is the $100\alpha\%$ point of the chi-squared distribution on two degrees of freedom and n_i is the appropriate group sample size. More recently, the correctness of circles for such diagrams has been questioned because of the sampling variability present in the canonical axes, and various approximating ellipses have been proposed by Krzanowski and Radley (1989), Krzanowski (1989), and Schott (1990). Coverage properties of the various regions are investigated in the simulation study of Ringrose and Krzanowski (1991).

Some generalisations of criterion (9.3) have been provided by Campbell (1984) and Aladjem (1991a,b).

Decision theoretic approach

9.7 The above discriminant functions all have an empirical basis. The first theoretically founded approach was provided by Welch (1939), who considered classification in general and treated it as a decision problem. He fixed attention on the case of just two populations, π_1 and π_2, and viewed the individual $\boldsymbol{x}' = (x_1, \ldots, x_p)$ that is to be classified as a random vector having probability density functions $f_1(\boldsymbol{x}), f_2(\boldsymbol{x})$ in each of these populations. An allocation rule can then be defined by partitioning the p-dimensional sample space into disjoint regions and assigning $\boldsymbol{x}$ to a particular population according to the region into which it falls. If the classification is a *forced* one (i.e. the individual *must* be allocated to one of the two populations, without the possibility of a deferred decision), then we require a partition into two regions R_1, R_2 and allocate $\boldsymbol{x}$ to π_i if it falls in region R_i $(i = 1, 2)$.

Once these regions have been defined, the probability that an individual comes from π_j and is allocated to π_i is given by $p(i|j) = \int_{R_i} f_j(\boldsymbol{x})d\boldsymbol{x}$. Probabilities of a correct decision are thus $p(1|1)$ and $p(2|2)$, while $p(1|2)$ and $p(2|1)$ denote the probabilities of the two types of error. The consequences of each type of error may not be the same, however. For example, it is less serious to classify a 'medical' patient as a 'surgical' case (since, apart from some danger and discomfort during the operation, the mistake will be quickly discovered and the correct treatment easily applied) than it is to make the converse mistake (as delaying surgery might be life-threatening). Thus, it is important to be able to quantify these differential consequences, and this can be done (in theory) by specifying two *costs* due to misclassification: $c(1|2)$ is the cost due to allocating a π_2 individual to π_1, and $c(2|1)$ is the cost due to the reverse mistake. Furthermore, the rate of occurrence of individuals may differ between the two populations so, in general, we should allow *prior probabilities* q_1, q_2 (with $q_1 + q_2 = 1$) of drawing an individual from π_1, π_2 respectively. For example, if it is desired to discriminate very young children who have cystic fibrosis from 'normal' cases, prior experience tells us that only about 1 child in every 2500 is born in Great Britain with the disease while the other 2499 are 'normal'. Thus, in the absence of any other diagnostic information, a randomly chosen individual has an overwhelming probability of being 'normal'. On the other hand, classification of a cystic fibrosis case as 'normal' carries very severe penalties and these two facets must be balanced in the analysis.

9.8 Welch (1939) tackled the problem by looking for the best partition R_1, R_2, as defined by expected behaviour of all future classifications. Given the quantities above, the probability of drawing and misclassifying an individual from π_i is $q_i p(j|i)$ $(j \neq i)$. Hence, the expected cost due to misclassifying future individuals is given by

$$C = q_1 p(2|1)c(2|1) + q_2 p(1|2)c(1|2) \tag{9.4}$$

which in terms of the regions R_1, R_2 is

$$C = q_1 c(2|1) \int_{R_2} f_1(\boldsymbol{x})d\boldsymbol{x} + q_2 c(1|2) \int_{R_1} f_2(\boldsymbol{x})d\boldsymbol{x} \tag{9.5}$$

The optimal classification rule will thus be given by the partition R_1, R_2 that minimises this expected cost, and it is easy to show (see, for example, Krzanowski, 1988a, p335) that this optimal partition puts into R_1 all those vectors $\boldsymbol{x}$ for which

$$f_1(\boldsymbol{x})/f_2(\boldsymbol{x}) \geqslant q_2 c(1|2)/q_1 c(2|1) \tag{9.6}$$

and into R_2 all remaining vectors.

If it is difficult to quantify the costs due to misclassification, common practice is to set $c(1|2) = c(2|1)$, in which case the right-hand side of (9.6) becomes q_2/q_1 and the allocation rule is the intuitively reasonable one of assigning $\boldsymbol{x}$ to the population that has the greater posterior probability $\Pr(\pi_i|\boldsymbol{x})$ in the Bayesian sense. Furthermore, if prior probabilities are unknown then we can additionally set $q_1 = q_2$, the right-hand side of (9.6) becomes unity and we assign $\boldsymbol{x}$ to the population in which it has the greater 'probability' $f_i(\boldsymbol{x})$. In all cases the allocation rule is of the form: allocate $\boldsymbol{x}$ to π_1 if $f_1(\boldsymbol{x})/f_2(\boldsymbol{x}) > k$, for some suitable constant k, and otherwise to π_2 (cf. the Neyman–Pearson lemma for hypothesis testing).

A rule such as (9.6), obtained by minimising C, is known in decision theory as a *Bayes* rule. For a general discussion of concepts in decision theory see, for example, DeGroot (1970). Anderson (1984, Chapter 6) establishes optimal properties of (9.6) in terms of admissibility, and also derives *ab initio* optimality of the special case when prior probabilities are unknown. Rao (1948) discusses the Bayes rule in more detail, and extends the solution to allow a region of doubt R_3, in which the decision to classify is deferred. Other criteria for optimisation can be used in place of C, leading to other rules. The most common alternative is the *minimax* rule, which is the rule that minimises the maximum probability of misclassification. This rule was first considered by von Mises (1945). Seber (1984, p286) shows that it is also of the form: allocate $\boldsymbol{x}$ to π_1 if $f_1(\boldsymbol{x})/f_2(\boldsymbol{x}) > k$, for some suitable constant k, and otherwise to π_2, but this time with constraints on k.

9.9 The above concepts extend readily to the case of classifying $\boldsymbol{x}$ to one of $g(> 2)$ populations $\pi_1, \ldots, \pi_g$. Let the prior probabilities that $\boldsymbol{x}$ comes from each of the g populations now be $q_1, \ldots, q_g$, and let the density functions of $\boldsymbol{x}$ in each of these populations be $f_1(\boldsymbol{x}), \ldots, f_g(\boldsymbol{x})$. We now seek a partition of the sample space into g regions $R_1, \ldots, R_g$ so as to allocate $\boldsymbol{x}$ to π_i if it falls in region R_i, for $i = 1, \ldots, g$. Given such a partition, the probability of misclassifying an individual from π_i into π_j is $p(j|i) = \int_{R_j} f_i(\boldsymbol{x})d\boldsymbol{x}$. Denoting the cost of this misclassification by $c(j|i)$, the total expected cost incurred with these regions is

$$C = \sum_{i=1}^{g} q_i \left\{ \sum_{j \neq i} c(j|i)p(j|i) \right\} \tag{9.7}$$

The Bayes allocation rule is given by the partition minimising C, and it can easily be shown (Anderson, 1984, p224) that this rule assigns $\boldsymbol{x}$ to π_k if

$$\sum_{i \neq k} q_i c(k|i) f_i(\boldsymbol{x}) < \sum_{i \neq j} q_i c(j|i) f_i(\boldsymbol{x}), \quad j = 1, \ldots, g; j \neq k \tag{9.8}$$

Considerable simplification occurs if all costs $c(j|i)$ are assumed to be equal, as substitution of the common value in (9.8) and subtraction of $\sum_{i \neq j,k} q_i f_i(\boldsymbol{x})$ from both sides results in assigning $\boldsymbol{x}$ to π_k if

$$q_j f_j(\boldsymbol{x}) < q_k f_k(\boldsymbol{x}), \quad j \neq k \tag{9.9}$$

i.e. to the population for which its posterior probability is highest. Anderson (1984, Chapter 6) again discusses the Bayes rule when prior probabilities are unobtainable; however, (9.9) is the basis of most applications in practice.

Parametric allocation rules

9.10 The above general theory leads directly to the derivation of *parametric* discriminant functions. Such a function can be obtained in principle for any particular situation, on the assumption of appropriate forms for the densities $f_i(\boldsymbol{x})$ in the various populations. In practice, nearly all applications of the theory have been to the case of multivariate normal density functions, as the discriminant functions that result have been found to be reasonably robust to departures from normality of data. More recently, a division of data into the three broad categories, continuous, categorical and mixed, has been found to be useful, and discriminant functions have been derived in the first case using multivariate normal densities, in the second case using multinomial densities, and in the third case using conditional Gaussian densities (Part 1, **2.25**). Each of these three cases is thus now treated in turn; we give the details for $g = 2$ populations, and indicate extensions to $g > 2$ as appropriate.

9.11 If all elements of $\boldsymbol{x}$ are continuous, then we can assume their joint distribution to be multivariate normal with means $\boldsymbol{\mu}_1$, $\boldsymbol{\mu}_2$ and dispersion matrices $\boldsymbol{\Sigma}_1$, $\boldsymbol{\Sigma}_2$ in π_1, π_2 respectively. Substituting the two density functions in $f_1(\boldsymbol{x})/f_2(\boldsymbol{x}) > k$ and simplifying leads to the rule: allocate $\boldsymbol{x}$ to π_1 if

$$\tfrac{1}{2}\log(|\boldsymbol{\Sigma}_2|/|\boldsymbol{\Sigma}_1|) - \tfrac{1}{2}[\boldsymbol{x}'(\boldsymbol{\Sigma}_1^{-1} - \boldsymbol{\Sigma}_2^{-1})\boldsymbol{x} - 2\boldsymbol{x}'(\boldsymbol{\Sigma}_1^{-1}\boldsymbol{\mu}_1 - \boldsymbol{\Sigma}_2^{-1}\boldsymbol{\mu}_2) + \boldsymbol{\mu}_1'\boldsymbol{\Sigma}_1^{-1}\boldsymbol{\mu}_1 - \boldsymbol{\mu}_2'\boldsymbol{\Sigma}_2^{-1}\boldsymbol{\mu}_2] > \log k,$$

and otherwise to π_2. Note the presence of quadratic terms in elements of $\boldsymbol{x}$ in this rule; it is therefore known as the *quadratic discriminant function.* Often, however, it is appropriate to make the simplifying assumption that $\boldsymbol{\Sigma}_1 = \boldsymbol{\Sigma}_2 = \boldsymbol{\Sigma}$ (cf. the assumptions made in the general linear model, Part 1, Chapter 7), in which case the rule becomes: allocate $\boldsymbol{x}$ to π_1 if

$$(\boldsymbol{\mu}_1 - \boldsymbol{\mu}_2)'\boldsymbol{\Sigma}^{-1}[\boldsymbol{x} - \frac{1}{2}(\boldsymbol{\mu}_1 + \boldsymbol{\mu}_2)] > \log k, \tag{9.10}$$

and otherwise to π_2. Assumption of common dispersion matrices has removed the quadratic terms and produced a *linear discriminant function* again.

When $g > 2$ it is usual to make the common dispersion matrix assumption, so that $\boldsymbol{x}$ has a $N(\boldsymbol{\mu}_i, \boldsymbol{\Sigma})$ distribution in π_i for $i = 1, \dots, g$, and also to assume equal costs due to misclassification. Application of (9.8) to these normal densities produces regions of classification $R_i = \{\boldsymbol{x}|u_{ik}(\boldsymbol{x}) > \log(q_k/q_i), k = 1, \dots, g; k \neq i\}$ where $u_{ik}(\boldsymbol{x})$ is the ratio of densities in π_i and π_k, and so is given by the left-hand side of (9.10) but with $\boldsymbol{\mu}_i$ and $\boldsymbol{\mu}_k$ replacing $\boldsymbol{\mu}_1$ and $\boldsymbol{\mu}_2$ respectively. More directly, the logarithm of q_i times the density in π_i is

$$\log q_i - \frac{p}{2}\log(2\pi) - \frac{1}{2}\log|\boldsymbol{\Sigma}| - \frac{1}{2}\Delta^2(\boldsymbol{x}, \pi_i), \tag{9.11}$$

where $\Delta^2(\boldsymbol{x}, \pi_i) = (\boldsymbol{x} - \boldsymbol{\mu}_i)'\boldsymbol{\Sigma}^{-1}(\boldsymbol{x} - \boldsymbol{\mu}_i)$ is the squared Mahalanobis distance between $\boldsymbol{x}$ and π_i. Thus, $\boldsymbol{x}$ is allocated to the population for which its *discriminant score* (9.11) is highest. Note that in the case of equal prior probabilities, $q_i = 1/g\ \forall i$, this rule reduces to the allocation of $\boldsymbol{x}$ to the population π_i from which its distance $\Delta^2(\boldsymbol{x}, \pi_i)$ is least (i.e. to which it is 'nearest').

Non-normal models for continuous variables are comparatively rare in discriminant analysis, but allocation rules have been derived for some general families of distributions, such as the linear exponential family (Day, 1969a) and the elliptically symmetric family (Cooper, 1963; Glick, 1976), and for some specific distributions including the θ-generalised normal (Chhikara and Odell, 1973; Goodman and Kotz, 1973), the exponential (Bhattacharya and Das Gupta, 1964; Basu and Gupta, 1974), the inverse normal (Folks and Chhikara, 1978; Chhikara and Folks, 1989) and the multivariate t (Sutradhar, 1990).

9.12 If all elements of $\boldsymbol{x}$ are categorical, then an allocation rule can be derived very simply in general terms. Suppose that x_i has s_i categories ($i = 1, \dots, p$), and let $s = \prod_{i=1}^{p} s_i$. Then $\boldsymbol{x}$ can be treated as defining an s-state multinomial variable, z say, which has probability p_{im} of falling in state m when it comes from population π_i $(i = 1, 2;\ m = 1, , s)$. Thus, if $\boldsymbol{x}$ is identified as belonging to state m, then the optimal allocation rule (9.6) is to classify it to π_1 if

$$p_{1m}/p_{2m} > k \tag{9.12}$$

where $k = q_2c(1|2)/q_1c(2|1)$ as before. Given the discrete nature of the data, we must explicitly allow the case that $p_{1m}/p_{2m} = k$ for each m, so that the appropriate alternatives are to allocate $\boldsymbol{x}$ to π_2 if $p_{1m}/p_{2m} < k$ and to make a randomised decision if $p_{1m}/p_{2m} = k$.

Extension to the g–group case is immediate. If costs due to misclassification are assumed equal, then rule (9.9) becomes: allocate $\boldsymbol{x}$ to that population for which q_ip_{im} is greatest $(i = 1, \ldots, g)$. If, additionally, prior probabilities are all equal then $\boldsymbol{x}$ is allocated to the population for which p_{im} is greatest.

This last case again has a 'distance' interpretation, similar to the continuous variable situation of **9.11** above. The distance between two multinomial populations, π_i and π_j say, has been considered by a number of authors (Bhattacharyya, 1946; Matusita, 1956; Rao, 1973 among others; see also **14.6**). It is generally agreed that an appropriate measure is given by some monotonic decreasing function (such as $\sqrt{1 - 2\rho}, \cos^{-1}\rho$ or $-\log\rho$) of the *affinity* $\rho = \sum_{k=1}^{s} \sqrt{p_{ik}p_{jk}}$ between the two populations. Affinity here measures the similarity between the two populations; for some properties of ρ see Matusita (1971) and Ahmad (1982). If we treat $\boldsymbol{x}$ as a degenerate population with probability 1 for its exhibited state m and probability zero for all other states, then its affinity with π_i is $\sqrt{p_{im}}$. The above rule is thus the same as classifying $\boldsymbol{x}$ to the population with which it has greatest affinity, i.e. to the population from which it has least distance.

9.13 If now $\boldsymbol{x}$ has a mixture of categorical and continuous elements, we can use the combination of the above two approaches as embodied in the conditional Gaussian distribution (Part 1, **2.25**). Suppose that c of the variables x_i are continuous and d are categorical $(c + d = p)$. The subvector of categorical variables is treated as a multinomial variable z having s categories, and the remaining continuous variables, denoted by $\boldsymbol{y}$ say, are assumed to have a multivariate normal distribution whose mean and dispersion matrix depend on the state occupied by z. In particular, we assume the probability of z having state m to be p_{im} for population π_i, and the conditional mean and dispersion matrix of $\boldsymbol{y}$ to be $\boldsymbol{\mu}_{im}$, $\boldsymbol{\Sigma}_{im}$ in this state and population. The joint density function of $(z = m, y)$ is thus

$$f_i(m, y) = p_{im}|\boldsymbol{\Sigma}_{im}|^{-1/2}(2\pi)^{-c/2}\exp[-\frac{1}{2}(y - \boldsymbol{\mu}_{im})'\boldsymbol{\Sigma}_{im}^{-1}(y - \boldsymbol{\mu}_{im})] \tag{9.13}$$

in population π_i.

Consider first the case of two populations. Application of (9.6) to the above joint density results in a different allocation rule for each value of m. This allocation rule is given by the quadratic discriminant function of **9.11** but with $\boldsymbol{y}$, $\boldsymbol{\mu}_{im}$, $\boldsymbol{\Sigma}_{im}$ replacing $\boldsymbol{x}$, $\boldsymbol{\mu}_i$, $\boldsymbol{\Sigma}_i$ respectively $(i = 1, 2)$, and with the k on the right-hand side of the quadratic discriminant function being replaced by $k' = kp_{2m}/p_{1m}$. The special case in which $\boldsymbol{\Sigma}_{im} = \boldsymbol{\Sigma}$ for all i, m was named the *location model* by Olkin and Tate (1961). In this case, (9.6) reduces to a different *linear* discriminant function (9.10) for each state m, but with $\boldsymbol{y}$, $\boldsymbol{\mu}_{im}$

replacing $\boldsymbol{x}$, $\boldsymbol{\mu}_i$ and p_{2m}/p_{1m} again multiplying the k on the right-hand side. Furthermore, Krzanowski (1986) demonstrated the equivalence of this latter rule with the one minimising Matusita distance (**14.4**) between $\boldsymbol{x}$ and each of π_1, π_2 in the case of equal costs and equal prior probabilities.

This equivalence leads naturally to the g–population extension: allocate an individual with $z = m$ and with continuous variables $\boldsymbol{y}$ to the population π_j for which

$$d_j = [(2\pi)^c|\boldsymbol{\Sigma}|]^{-1/4}(p_{jm})^{1/2}\exp[-\frac{1}{4}(\boldsymbol{y} - \boldsymbol{\mu}_{jm})'\boldsymbol{\Sigma}^{-1}(\boldsymbol{y} - \boldsymbol{\mu}_{jm})] \qquad (9.14)$$

is maximum.

9.14 The rules developed in **9.11–9.13** are *population* allocation rules, as they have been derived from theoretical considerations, and they all involve unknown population parameters. If the prior probabilities are unknown, then, if desired, they can also be treated as parameters instead of being assumed equal. For practical application, therefore, these rules must be *estimated* in some way. It is generally possible to obtain a sample of individuals from each of the populations; estimation of an allocation rule can then be based on these *training samples*. We suppose in the following that observations $\boldsymbol{x}_{i1}, \ldots, \boldsymbol{x}_{in_i}$ are known to have come from population π_i, $i = 1, \ldots, g$. This notation is applied, if necessary, to the associated variables z and $\boldsymbol{y}$ in the obvious way.

Various different approaches have been proposed for the estimation process. Perhaps the most direct, and simplest, procedure is to estimate all unknown parameters from the training samples and then to replace each parameter of a rule by its estimate. This procedure is often called the *estimative* or *plug-in* approach, and is essentially frequentist in spirit. Bayesian statisticians criticise such an approach, because they claim that each parameter is replaced by just a single value, which is supposed correct but which in practice may be wildly inaccurate. They prefer instead to incorporate variability of parameter estimates through their joint posterior distribution, and then to average the densities of $\boldsymbol{x}$ over this posterior distribution to obtain their predictive distributions. Use of these predictive distributions in (9.6) or (9.9) leads to a *predictive* allocation rule. (A simpler approach, sometimes called the *semi*-Bayesian approach, averages the population allocation rule itself over the posterior distributions to obtain a usable rule.) An entirely different, non-Bayesian, approach is to attempt estimation of the ratio of densities (9.6) more directly, and in the parametric framework this can be done through a likelihood ratio. More specifically, in the two-group case it is done by calculating the likelihood ratio test statistic for the null hypothesis that $\boldsymbol{x}$ belongs to π_1 against the alternative that it belongs to π_2, so this approach is generally called the *hypothesis-testing* one. Moran and Murphy (1979) coined the term *testimative* for it. Finally, given the equivalences of population rules based on (9.6) with their counterparts based on distance measures, it is also possible to take a *distance* approach to estimation of rules. We now consider each of these approaches in turn.

9.15 The simplest approach to start with is the estimative one. Given the existence of distributional assumptions, parameter estimates are most naturally obtained by maximum likelihood; the usual corrections for bias are also generally made for dispersion parameters. If prior probabilities q_i are unknown, and sampling has been done from a *mixture* of the populations, then the sample sizes n_i are random variables with a multinomial distribution and the prior probabilities can be estimated as $\hat{q}_i = n_i/n$ for $i = 1, \ldots, g$. However, if sampling has been conducted separately from each population, then no simple estimates of the q_i are available from the data. In this case we either need extra information or we must make the assumption of equal prior probabilities.

First, suppose that all variables are continuous. Standard theory (Part 1, Chapter 3) yields sample means and covariance matrices as estimates of their population counterparts. If there are just two populations, π_1 and π_2, then from **9.11** we obtain directly the rule: allocate $\boldsymbol{x}$ to π_1 if

$$\frac{1}{2}\log(|\boldsymbol{S}_2|/|\boldsymbol{S}_1|) - \frac{1}{2}[\boldsymbol{x}'(\boldsymbol{S}_1^{-1} - \boldsymbol{S}_2^{-1})\boldsymbol{x} - 2\boldsymbol{x}'(\boldsymbol{S}_1^{-1}\bar{\boldsymbol{x}}_1 - \boldsymbol{S}_2^{-1}\bar{\boldsymbol{x}}_2) + \bar{\boldsymbol{x}}_1'\boldsymbol{S}_1^{-1}\bar{\boldsymbol{x}}_1 - \bar{\boldsymbol{x}}_2'\boldsymbol{S}_2^{-1}\bar{\boldsymbol{x}}_2] > \log k,$$

and otherwise to π_2, where $\bar{\boldsymbol{x}}_j = \frac{1}{n_j}\sum_{i=1}^{n_j}\boldsymbol{x}_{ji}$ and $\boldsymbol{S}_j = \frac{1}{n_j-1}\sum_{i=1}^{n_j}(\boldsymbol{x}_{ji} - \bar{\boldsymbol{x}}_j)(\boldsymbol{x}_{ji} - \bar{\boldsymbol{x}}_j)'$. This is the sample quadratic discriminant function first discussed by Smith (1947). If the population dispersion matrices $\boldsymbol{\Sigma}_i$ are assumed to be equal then the natural estimate of the common quantity is given by the pooled covariance matrix $\boldsymbol{W}$ defined in **9.5**; from (9.10) we then obtain the rule: allocate $\boldsymbol{x}$ to π_1 if

$$(\bar{\boldsymbol{x}}_1 - \bar{\boldsymbol{x}}_2)'\boldsymbol{W}^{-1}[\boldsymbol{x} - \frac{1}{2}(\bar{\boldsymbol{x}}_1 + \bar{\boldsymbol{x}}_2)] > \log k, \tag{9.15}$$

and otherwise to π_2. This rule is essentially Fisher's linear discriminant function of (9.2) but with an extra constant. It dates back to Wald (1944) and Anderson (1951); the left-hand side of (9.15) is known as *Anderson's classification statistic W*. Finally, if $g > 2$ and we assume equal costs due to misclassification and equal population dispersion matrices, then from (9.11) we obtain the rule: allocate $\boldsymbol{x}$ to the population that has the highest value of

$$L_i = \log q_i - \frac{p}{2}\log(2\pi) - \frac{1}{2}\log|\boldsymbol{W}| - \frac{1}{2}\hat{\Delta}^2(\boldsymbol{x}, \pi_i), \tag{9.16}$$

where $\hat{\Delta}^2(\boldsymbol{x}, \pi_i) = (\boldsymbol{x} - \bar{\boldsymbol{x}}_i)'\boldsymbol{W}^{-1}(\boldsymbol{x} - \bar{\boldsymbol{x}}_i)$ and $\boldsymbol{W} = \frac{1}{n_1+n_2-g}\sum_{i=1}^{g}\sum_{j=1}^{n_i}(\boldsymbol{x}_{ij} - \bar{\boldsymbol{x}}_i)(\boldsymbol{x}_{ij} - \bar{\boldsymbol{x}}_i)'$ is the covariance matrix pooled within all g populations.

If there is any reason to suppose that the training data contain outliers, or other types of unusual observation, then any of the above estimates can be replaced by the robust versions outlined in Part 1, Chapter 3 (see Broffitt, 1982). Robust linear and quadratic discriminant functions have also been considered by Randles *et al* (1978), Tiku and Balakrishnan (1984, 1989), Balakrishnan and Tiku (1988) and Todorov *et al* (1994).

Example 9.3
Returning to the bivariate data of Example 9.1, we find the three sample means to be $\bar{\boldsymbol{x}}_1' = (146.19, 14.10)$ for *concinna*, $\bar{\boldsymbol{x}}_2' = (124.65, 14.29)$ for

heikertingeri, and $\bar{x}_3' = (138.27, 10.09)$ for *heptapotamica*. The pooled covariance matrix is

$$W = \begin{pmatrix} 23.02 & -0.56 \\ -0.56 & 1.01 \end{pmatrix}$$

Suppose we now wish to derive an assignment rule for a new individual $\boldsymbol{x}$. If we assume equal prior probabilities in the three groups ($q_1 = q_2 = q_3 = \frac{1}{3}$), then we can evaluate (9.16) for $i = 1, 2, 3$ and assign $\boldsymbol{x}$ to the group π_i yielding largest L_i. Alternatively, we can define $D_{ij} = L_i - L_j$ so that the assignment rule becomes: assign $\boldsymbol{x}$ to π_1 if $D_{12} > 0$ and $D_{13} > 0$, to π_2 if $D_{12} < 0$ and $D_{23} > 0$, and otherwise to π_3. The functions $D_{ij} = 0$ thus define the boundaries of the assignment regions in the sample space. For this set of data we have

$$D_{12} = -132.46 + 0.94x_1 + 0.33x_2,$$
$$D_{23} = 18.34 - 0.50x_1 + 3.87x_2,$$

and

$$D_{13} = -114.13 + 0.45x_1 + 4.19x_2.$$

The boundaries of the assignment regions are lines, and these have been added to the scatter diagram of Fig. 9.1. The result is shown in Fig. 9.2. Applying the assignment rule to this set of data (i.e. resubstituting, cf. **9.38**), we find only one observation incorrectly assigned (from π_1 to π_2).

9.16 The gap between equal dispersion matrices in all groups and arbitrarily different matrices in each group is a wide one, and use of the quadratic discriminant function and its g-group counterpart involves estimation of

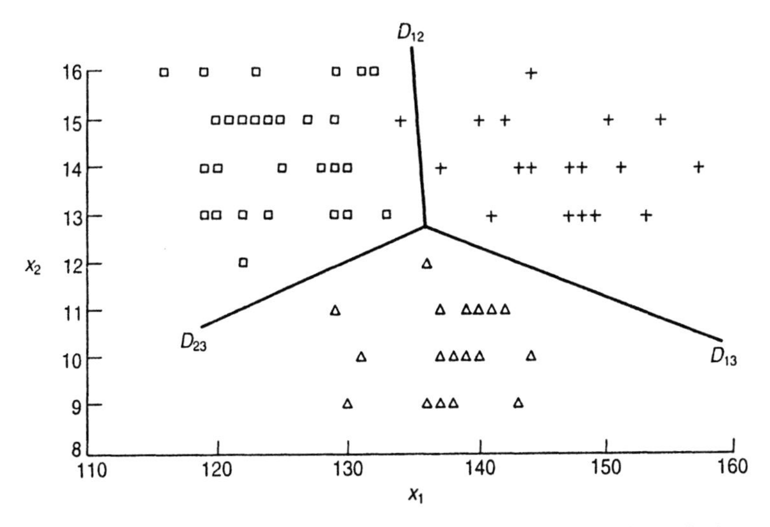

Figure 9.2 Scatter diagram as in Fig. 9.1, but with boundaries of allocation regions marked in. Reproduced with permission from Seber (1984).

many parameters. This is an unsatisfactory feature when training samples are small, because the resulting discriminant functions tend to be unstable. Also, it can be anticipated that *some* similarity will exist among the dispersion matrices of populations in discriminant analysis (Hawkins and Raath, 1982), and any similarity that does exist should be exploited in the analysis. A number of models for the $\boldsymbol{\Sigma}_i$, intermediate between complete equality and arbitrary difference, have thus been proposed and can be applied in the discriminant context. These models include: proportional dispersion matrices (Switzer, 1980; Dargahi-Noubari, 1981; Hawkins and Raath, 1982; Owen, 1984); common principal components (Flury, 1988; Flury and Schmid, 1992; see also **4.5** and **6.28** in Part 1); equal correlation matrices (Manly and Rayner, 1987) and the SIMCA model (Frank and Friedman, 1989). An alternative compromise between linear and quadratic discrimination is struck in *regularised discriminant analysis* as developed by Friedman (1989), who proposed estimating $\boldsymbol{\Sigma}_i$ by

$$(1-\gamma)\hat{\boldsymbol{\Sigma}}_i(\lambda) + \gamma c_i \boldsymbol{I}_p$$

where $\boldsymbol{I}_p$ is the $p \times p$ identity matrix, $c_i = [\text{tr}\ \hat{\boldsymbol{\Sigma}}_i(\lambda)]/p$ and

$$\hat{\boldsymbol{\Sigma}}_i(\lambda) = [(1-\lambda)(n_i - 1)\boldsymbol{S}_i + \lambda(n-g)\boldsymbol{W}]/[(1-\lambda)(n_i-1) + \lambda(n-g)]$$

Here the regularisation parameters λ and γ control shrinkage, the former of the heteroscedastic estimates towards a common estimate and the latter of the result of this first shrinkage towards a multiple of the identity matrix. These parameters are estimated from the training data by minimisation of the cross-validated estimate of overall error rate (see **9.42** below), thereby providing an automatic balance between linear and quadratic discriminant functions. The method extends shrinkage procedures in ridge regression of covariance matrices, as used in discriminant analysis by Di Pillo (1976, 1977), Campbell (1980), Peck and van Ness (1982) and Kimura *et al* (1987). Note, however, that in this formulation regularised discriminant analysis is scale-dependent and therefore only suitable for variables that are either all on the same scale or that have been previously scaled.

9.17 Next, suppose that all variables are categorical. Then the training sample from population π_i can be expressed as an $s_1 \times s_2 \times \ldots s_p$ contingency table with s cells, and with incidence n_{im} in the mth cell ($m = 1, \ldots, s$; $i = 1, \ldots, g$). Standard maximum likelihood procedures on the multinomial probabilities p_{ij} provide estimates

$$\hat{p}_{ij} = n_{ij}/n_i \tag{9.17}$$

Thus, if the individual $\boldsymbol{x}$ to be classified falls in cell m of the multinomial, then substitution of $\hat{p}_{im}$ into the rules of **9.12** provides allocation procedures. In particular, for the two-group case, we allocate to π_1 if

$$n_2 n_{1m}/n_1 n_{2m} > \log k \tag{9.18}$$

to π_2 if the reverse inequality holds, and make a randomised decision if the inequality is replaced by equality.

Unfortunately, this procedure is somewhat illusory. For moderately large values of p, or for moderately small sample sizes, the contingency tables formed from the training samples will be sparse. Thus, some n_{im} will be either very small or zero. If $\boldsymbol{x}$ falls in such cells, allocation will be heavily influenced by chance and may reduce to a purely random assignment. This problem can be overcome by smoothing the estimated probabilities in some way, and various methods of doing this are now available.

9.18 Probably the most popular method of smoothing is to fit a *log-linear* model to the cell means $\eta_{ij} = n_i p_{ij}$ $(i = 1, \ldots, g;\ j = 1, \ldots, s)$. In such a model, $\log \eta_{ij}$ is expressed as a sum of terms representing the main effects of the x_i and the interactions of various orders between them. If there are s such terms then all possible interactions are accommodated, the model is *saturated*, and maximum likelihood estimation leads to the multinomial estimates (9.17). A reduced version of the model is formulated by eliminating higher-order interactions, and typical choices might be to eliminate all interactions (the 'independence' model) or all interactions higher than the first. Optimal model selection can be attempted by graphical modelling (see Chapter 11), while maximum likelihood estimation of parameters in a reduced model can be done either by iterative proportional fitting (see Bishop *et al*, 1975, Section 3.5; or Fienberg, 1980, Chapter 3) or by Newton-Raphson (see Haberman, 1974). The latter method is adopted in the computer package GLIM (1986).

If all the categorical variables x_i are binary, with possible values coded as zero and one, then an alternative parametrisation of the multinomial probabilities p_{ij} was provided independently by Lazarsfeld (1961) and Bahadur (1961). This parametrisation is given by

$$p_{ij} = \prod_{l=1}^{p} \theta_{il}^{x_l^{(j)}} (1 - \theta_{il})^{1 - x_l^{(j)}} [1 + \sum_{l<u} \rho_{i,lu} \zeta_{il} \zeta_{iu} + \sum_{l<u<v} \rho_{i,luv} \zeta_{il} \zeta_{iu} \zeta_{iv} + \cdots$$
$$+ \rho_{i,12\ldots p} \zeta_{i1} \zeta_{i2} \cdots \zeta_{ip}], \qquad (9.19)$$

where θ_{il} is the expected value of x_l in π_i, $\zeta_{il} = (x_l - \theta_{il})/\sqrt{\theta_{il}(1 - \theta_{il})}$, $\rho_{i,lu\ldots v} = E[\zeta_{il} \zeta_{iu} \cdots \zeta_{iv}]$, and $x_l^{(j)}$ is the value of x_l in the jth multinomial state. The $\rho_{i,lu}$ here are the (population) correlations between pairs of binary variables in π_i, so the $\rho_{i,lu\ldots v}$ can be interpreted as higher-order correlations between sets of binary variables in π_i. Once again, inclusion of all these terms leads to a saturated model in which the probabilities p_{ij} are estimated by their multinomial estimates (9.17), while smoothed estimates result from the omission of higher-order terms from the model. Omission of all ρ_i is equivalent to the assumption of independence among the x_i, while a second-order approximation results if all ρ_i except $\rho_{i,lu}$ are omitted. Since these latter terms are just the usual correlations between pairs of x_l, the second-order smoothed estimates of the p_{ij} are obtained by substituting sample means of the binary variables and sample correlations between all pairs of them into (9.19).

Other possible representations of multinomial probabilities may be obtained from orthogonal series expansions (Martin and Bradley, 1972; Ott and Kronmal, 1976; Hall, 1983), Lancaster models (Lancaster, 1969) or latent class models (McLachlan and Basford, 1988, Section 3.5). However, the two described above are the ones most frequently encountered. Extension of regularised discriminant analysis to discrete variables has been considered by Celeux and Mkhadri (1992).

9.19 Estimation procedures in the case where $\boldsymbol{x}$ consists of mixed continuous and categorical variables can be viewed as a combination of the methods in **9.15, 9.17** and **9.18** above. The d categorical variables define an $s_1 \times \ldots \times s_d$ contingency table for each training sample, while the c continuous variables are assumed to have different multivariate normal distributions in each cell of these tables. If training samples are very large, then maximum likelihood estimates of the parameters in allocation rules (9.13) and (9.14) follow directly; estimates of the multinomial probabilities p_{ij} are given by the corresponding sample proportions (9.17), while sample mean vectors and covariance matrices of the continuous variables within each cell of the contingency tables provide estimates of the continuous parameters $\boldsymbol{\mu}_{ij}$ and $\boldsymbol{\Sigma}_{ij}$. In most practical applications, however, some of the contingency table cells will be empty or will have very few observations. For such cells, the parameter estimates and hence allocation rules will be either non-existent or very prone to error. Smoothing is thus again necessary if future individuals are to be classified with any degree of confidence. Log-linear models of low order again provide a suitable approach for the categorical parameters, while linear models of correspondingly low order can be formulated for the continuous parameters and fitted by multivariate regression (Part 1, Chapter 7). For full details of the estimation procedures under the location model assumptions (i.e. constant within-cell dispersion matrices) see Krzanowski (1975, 1980, 1986, 1993a); extension to heterogeneous dispersion matrices is given by Krzanowski (1994b). Missing values can be handled in these models by the E-M algorithm (Little and Schluchter, 1985), and recent developments in graphical modelling can be used to select the most parsimonious models for the smoothing (Krusinska, 1990).

Example 9.4

Krzanowski (1980) describes the fitting of the location model to a set of data obtained from a study of psychosocial influences in breast cancer. Seven continuous variables (the ten-point ratings: acting out hostility, criticism of others, paranoid hostility, self-criticism, guilt, direction of hostility; plus age at menarche), two binary variables (presence/absence of allergy and thyroid disorder) and two three-state variables (degrees of temper and feelings) were measured on each of 137 women with breast tumours, 78 of these being benign (group π_1) and 59 being malignant (group π_2). Each three-state variable was replaced by two dummy binary variables, thus yielding six binary $(x_1, \ldots, x_6)$ and seven continuous variables $(y_1, , y_7)$ in all. Given the constraints imposed by the dummy variables, there were 36 cells in the contingency table formed from the categorical variables. Since 30 of the 78 group π_1 individuals fell in one particular cell, and 17 of the 59 group π_2 individuals did likewise, the remaining individuals of each group were distributed very sparsely among the cells of the two contingency tables. Indeed, most cells in both tables had either 0, 1, 2 or 3 occupants. This distribution meant that a first-order

Table 9.3 Base line constants and main effects of binary variables for the continuous variables in the data dealing with psychosocial influences on breast cancer. Reproduced from Krzanowski (1980) with permission

Group	Cont. Variable	Base Line	Main Effect of x_1	x_2	x_3	x_4	x_5	x_6
π_1	y_1	0.19	2.10	0.30	0.04	2.09	0.82	1.78
	y_2	7.15	−2.62	0.34	−1.23	0.02	0.03	1.08
	y_3	3.95	−3.19	−0.04	−0.26	−0.11	0.35	0.32
	y_4	4.28	1.04	−0.25	−0.88	−1.32	0.22	1.61
	y_5	0.64	0.65	0.14	−0.68	0.38	0.50	1.04
	y_6	−1.91	6.21	−0.84	−1.07	−4.26	−0.24	0.98
	y_7	17.15	−3.49	0.77	−0.13	1.17	−1.16	−1.02
π_2	y_1	3.75	−0.29	−0.37	1.31	2.57	−1.02	0.02
	y_2	12.00	−3.65	−3.54	0.70	3.05	−1.07	−1.52
	y_3	2.89	−0.66	−0.36	−0.33	0.99	−1.16	−0.72
	y_4	21.00	−7.93	−5.22	−0.36	2.77	−4.75	−3.31
	y_5	2.71	−0.51	0.49	−1.30	0.12	−0.67	−0.40
	y_6	26.57	−12.01	−5.81	−3.75	−0.93	−7.04	−4.89
	y_7	13.24	0.36	0.05	−0.50	−0.46	0.61	0.27

model (i.e. base line constants plus main effects only) was the highest order that could be fitted.

Details of all computations are given by Krzanowski (1980). For example, the continuous variable parameters are shown in Table 9.3. The fitted model here is

$$\boldsymbol{\mu}_i = \boldsymbol{\alpha}_i + \boldsymbol{\beta}_{i1}x_1 + \ldots + \boldsymbol{\beta}_{i6}x_6$$

where $\boldsymbol{\alpha}_i$ is the (7-element) base line constant in group π_i and $\boldsymbol{\beta}_{ij}$ is the (7-element) main effect of binary variable x_j in group π_i ($i = 1, 2$; $j = 1, \ldots, 6$). An estimate of the continuous variable mean vector in any given multinomial cell and group is found by picking out the binary variables with value 1 in that cell, and adding their main effects to the base line constant. The values in the table thus illustrate the variation between these mean vectors, and hence the variation in linear discriminant functions between the different binary variable patterns.

The sparseness of the data makes some of the fitted values less than ideal. For example, continuous variables y_1 to y_6 are all on the scale 1–10, but it can be seen that many of the fitted means, e.g. for y_4 and y_6, lie outside this range. Despite this drawback as regards model fit, however, the ultimate discriminant performance of the model proved eminently satisfactory (see Example 9.6).

9.20 Next we turn to the Bayesian methods. For such methods we need first to specify a prior distribution for all the unknown parameters. Let us denote all the parameters by ϕ for simplicity, and their prior distribution by $p(\phi)$. If we further denote the full set of training data by $\boldsymbol{X}$, then the likelihood of these data may be written $L(X, \phi)$. Hence, the posterior distribution for the unknown parameters becomes

$$p(\phi|X) = \frac{p(\phi)L(X, \phi)}{\int p(\phi)L(X, \phi)d\phi} \tag{9.20}$$

Thus, if x is the individual to be classified and $f_i(x|\phi_{(i)})$ is its probability density in population π_i (where $\phi_{(i)}$ is a subset of the parameters ϕ), then the predictive density of x in π_i is given by

$$f(x|X,\phi, \pi_i) = \int f_i(x|\phi_{(i)})p(\phi|X)d\phi, \tag{9.21}$$

and the posterior probability that x belongs to π_i is

$$\Pr(x \in \pi_i|X, \phi, q) \propto q_i f(x|X, \phi, \pi_i). \tag{9.22}$$

If the q_i are also unknown, then we can incorporate them into the process as unknown parameters and assume that they have a prior probability density of the Dirichlet form proportional to $\prod_{i=1}^{g} q_i^{\alpha_i}$ for some specified α_i. If the training data have been obtained by sampling from a mixture of the populations, then the n_i are multinomial with likelihood proportional to $\prod_{i=1}^{g} q_i^{n_i}$, so the posterior density of the q_i is proportional to $\prod_{i=1}^{g} q_i^{n_i+\alpha_i}$. Hence

$$p(q_1, \ldots, q_{g-1}|n_1, \ldots, n_{g-1}, x) \propto \prod_{i=1}^{g} q_i^{n_i+\alpha_i} f(x|q_1, \ldots, q_{g-1}, X, \phi) \tag{9.23}$$

where $f(x|q_1, \ldots, q_{g-1}, X, \phi) = \sum_{i=1}^{g} q_i f(x|X, \phi, \pi_i)$. We thus remove the q_i from the posterior probability that x belongs to π_i on multiplying (9.22) by $p(q_1, \ldots, q_{g-1}|n_1, \ldots, n_{g-1}, x)$ and integrating with respect to all the q_i, to obtain

$$\Pr(x \in \pi_i|X, \phi) \propto \frac{(n_i + \alpha_i + 1)f(x|X, \phi, \pi_i)}{\sum_j (n_j + \alpha_j + 1)f(x|X, \phi, \pi_j)} \tag{9.24}$$

If the training data have been sampled separately from each population then we set $n_i = 0$ in (9.24) and obtain

$$\Pr(x \in \pi_i|X, \phi) \propto (\alpha_i + 1)f(x|X, \phi, \pi_i). \tag{9.25}$$

A predictive density for the observation to be allocated under any specific parametric model is thus obtained by finding $f(x|X, \phi, \pi_i)$ for that model and substituting into either (9.22), (9.24) or (9.25) according to the circumstances involving the q_i.

9.21 The case of continuous normal variables has been fully surveyed by Geisser (1982). If we assume unequal dispersion matrices in the g populations, then a convenient vague prior distribution for the unknown parameters is $p(\mu_1, \ldots, \mu_g, \Sigma_1^{-1}, \ldots, \Sigma_g^{-1}) \propto \prod_{i=1}^{g} |\Sigma_i|^{(p+1)/2}$. This prior can be derived either by application of Jeffreys' principle to μ_i, Σ_i pairs treated independently (Box and Tiao, 1973, p425), or by obtaining the limiting form of the product of g normal-Wishart conjugate priors. Standard manipulations (see, for example,

Aitchison and Dunsmore, 1975) then yield the predictive densities

$$f(\boldsymbol{x}|\bar{\boldsymbol{x}}_i, \boldsymbol{S}_i, \pi_i) \propto \left(\frac{n_i}{n_i+1}\right)^{p/2} \frac{\Gamma(n_i/2)}{\Gamma((n_i-p)/2)|(n_i-1)\boldsymbol{S}_i|^{\frac{1}{2}}} \left[1+\frac{n_i\delta_i^2(\boldsymbol{x}, \bar{\boldsymbol{x}}_i)}{n_i^2-1}\right]^{-n_i/2} \tag{9.26}$$

where $\delta_i^2(\boldsymbol{x}, \boldsymbol{y}) = (\boldsymbol{x}-\boldsymbol{y})'\boldsymbol{S}_i^{-1}(\boldsymbol{x}-\boldsymbol{y})$ for $i = 1, \ldots, g$.

If all the dispersion matrices are assumed equal to $\boldsymbol{\Sigma}$ then the reference prior becomes $p(\boldsymbol{\mu}_1, \ldots, \boldsymbol{\mu}_g, \boldsymbol{\Sigma}^{-1}) \propto |\boldsymbol{\Sigma}|^{(p+1)/2}$ and we obtain

$$f(\boldsymbol{x}|\bar{\boldsymbol{x}}_i, \boldsymbol{W}, \pi_i) \propto \left(\frac{n_i}{n_i+1}\right)^{p/2} \left[1+\frac{n_i\delta^2(\boldsymbol{x}, \bar{\boldsymbol{x}}_i)}{(n_i+1)(n-g)}\right]^{n-g+1/2} \tag{9.27}$$

where $\delta^2(\boldsymbol{x}, \boldsymbol{y}) = (\boldsymbol{x}-\boldsymbol{y})'\boldsymbol{W}^{-1}(\boldsymbol{x}-\boldsymbol{y})$. Murray (1977a) arrives at the same estimate as (9.27) in the case $g = 2$ using invariance criteria based on information measures, and hence provides a frequentist interpretation for a Bayesian method.

Vlachonikolis (1990) considered Bayesian predictive discrimination for $g = 2$ and mixed binary and continuous data using the location model. He adopted the above vague prior density $p(\{\boldsymbol{\mu}_{ij}\}, \boldsymbol{\Sigma}^{-1}) \propto |\boldsymbol{\Sigma}|^{(c+1)/2}$ for the cell means and common within-cell dispersion matrix of the c continuous variables, and prior densities for the cell probabilities p_{ij} of the Dirichlet form $p(\{p_{ij}\}) \propto \prod_{j=1}^{s} p_{ij}^{\alpha_{ij}-1}$. Here, the α_{ij} are positive constants reflecting prior knowledge about the binary variable locations; when no such prior information exists the α_{ij} could be set to a constant value within each population. Vlachonikolis (1990) then obtained expressions for the predictive density of $\boldsymbol{x}$ in each population, under both 'naive' and smoothed parameter assumptions (**9.13**,**9.19**). Predictive methods do not seem to have been developed specifically for the categorical variable case, although the results obtained by Vlachonikolis (1990) for the categorical part of the location model can be used here.

Finally, Geisser (1967,1977) and Enis and Geisser (1970,1974) derived semi-Bayesian allocation rules under various continuous-variable models in the case $g = 2$. Semi-Bayesian rules are ones in which any allocation rule formulated in terms of unknown parameters is integrated over the joint posterior density of those parameters. For example, the homoscedastic normal model in the case $g = 2$ yields the allocation rule (9.10) depending on the quantity $U = (\boldsymbol{\mu}_1 - \boldsymbol{\mu}_2)'\boldsymbol{\Sigma}^{-1}[\boldsymbol{x} - \frac{1}{2}(\boldsymbol{\mu}_1 + \boldsymbol{\mu}_2)]$. The estimative rule (9.15) uses the estimate $V = (\bar{\boldsymbol{x}}_1 - \bar{\boldsymbol{x}}_2)'\boldsymbol{W}^{-1}[\boldsymbol{x} - \frac{1}{2}(\bar{\boldsymbol{x}}_1 + \bar{\boldsymbol{x}}_2)]$; Geisser (1967) showed that, under the usual reference prior, the semi-Bayesian estimate becomes $V - \frac{1}{2}p(\frac{1}{n_1} - \frac{1}{n_2})$. The other references quoted above derive corresponding results for the heteroscedastic case and for the best linear rule minimising predictive error rates.

The main benefits of Bayesian predictive rules are that they temper the very 'sharp' and extreme results that can often be obtained with estimative rules. Aitchison and Dunsmore (1975, p231) illustrate this feature in an application to differential diagnosis of Conn's syndrome.

9.22 The third general approach to the derivation of an estimated parametric allocation rule is the so-called hypothesis testing approach. In the two-group situation, the allocation rule is essentially the generalised likelihood ratio test statistic of the null hypothesis that $\boldsymbol{x}$ belongs with the training set from π_1 against the alternative hypothesis that it belongs with the training set from π_2. This test statistic is obtained as the maximum of the likelihood of the two training samples with $\boldsymbol{x}$ added to the first sample, divided by the maximum of the likelihood of the same two samples but with $\boldsymbol{x}$ now added to the second sample. It was first applied by Anderson (1958, Chapter 6) to the homogeneous multivariate normal situation, leading to allocation of $\boldsymbol{x}$ to π_1 if

$$\frac{1 + n_2(n_2+1)^{-1}(n_1+n_2-2)^{-1}\delta^2(\boldsymbol{x}, \bar{\boldsymbol{x}}_2)}{1 + n_1(n_1+1)^{-1}(n_1+n_2-2)^{-1}\delta^2(\boldsymbol{x}, \bar{\boldsymbol{x}}_1)} \geqslant k' \qquad (9.28)$$

and otherwise to π_2. Here, $\delta^2(\boldsymbol{x}, \boldsymbol{y}) = (\boldsymbol{x}-\boldsymbol{y})'\boldsymbol{W}^{-1}(\boldsymbol{x}-\boldsymbol{y})$ as before, and k' is a suitable constant. The left-hand side of (9.28) has become known in the literature as the Z-statistic; it has the same form as the predictive rule derived from (9.27), and is exactly equivalent to Anderson's W-statistic when $n_1 = n_2$. This result has been subsequently generalised to the homoscedastic normal model with $g > 2$, to the heteroscedastic normal model (McLachlan, 1992, pp. 66–67) and to the location model for mixed data when $g = 2$ (Krzanowski, 1982).

The same sort of idea as applied above to likelihoods has been extended to distances by Dillon and Goldstein (1978). It was noted in **9.12** that the optimal allocation rule (9.12) for categorical data was equivalent to allocation of $\boldsymbol{x}$ to the population from which it had minimum Matusita distance, and problems arising from simple estimative use of this rule were highlighted in **9.17**. Dillon and Goldstein suggested overcoming some of these problems when $g = 2$ in the following way. The Matusita distance between the two training samples is calculated twice, once on adding $\boldsymbol{x}$ to the sample from π_1 and the second time on adding it to the sample from π_2, and $\boldsymbol{x}$ is allocated to the population with which it yields the greater distance. Such a distance approach can be extended in an obvious way if $g > 2$, and to data other than just categorical; see Krzanowski (1987) for appropriate allocation rules.

Semiparametric allocation rules

9.23 Instead of formulating parametric models for the probability distributions of the observations, then deriving optimal allocation rules in terms of the population parameters and obtaining estimates of these rules from the training data, some authors have preferred to formulate a simple parametric expression directly for the allocation rule and then to estimate the parameters of the rule from the training data. This was in fact the procedure adopted by Fisher (1936) in deriving his linear discriminant function (9.2), and other workers have investigated the use of linear functions for general discrimination. Two other functions used in this semiparametric fashion have been quadratic functions for allocation rules, and logistic functions for probabilities of group membership. The motivations for these approaches have been that a single form of rule is thereby obtained for use in a wide class of applications, and that, in general, fewer parameters need estimation than with the fully parametric methods.

The most common way in which linear discrimination is conducted is via Fisher's rule (9.2) or, eqivalently, Anderson's statistic W of (9.15), whether or not the assumptions underlying its derivation are satisfied. Sometimes adjustments to the rule will improve its performance for specific types of data. For example, Knoke (1982) and Vlachonikolis and Marriott (1982) independently suggested augmenting the variable set when both binary and continuous variables are being used, by including squares and cross-products of variables in the discriminant function. Since the most common deviation of data from the necessary linear discriminant assumption is heterogeneity of dispersion matrices, other authors have sought optimal linear functions in such circumstances (where optimality generally means smallest overall error rate). Clunies-Ross and Riffenburgh (1960), Riffenburgh and Clunies-Ross (1960) and Anderson and Bahadur (1962) have considered theoretical aspects, while Freed and Glover (1981), Greer (1984) and Castagliola and Dubuisson (1989) give algorithms for data-based computation.

The difficulties of finding optimal linear functions in the heterogeneous dispersion matrix case become acute when $g > 2$, and this prompted the investigation of optimal quadratic discriminant functions. An approximate solution has been given by Young *et al* (1987), but open questions still remain. However, perhaps the most useful semiparametric methods are those built on logistic functions for probabilities of group membership, so we consider these in more detail next.

9.24 The earliest ideas in *logistic discrimination* may be found in Cornfield (1962), Cox (1966), Truett *et al* (1967), Day and Kerridge (1967) and Walker and Duncan (1967), but the essential features of the method as it is applied today were developed by J. A. Anderson in a series of papers beginning with Anderson (1972) and surveyed in Anderson (1982). In the general case of g groups, the method assumes that the posterior probabilities of group membership of $\boldsymbol{x}$ are given by

$$\Pr(\pi_i|\boldsymbol{x}) = \exp(\beta_{0i} + \boldsymbol{\beta}_i'\boldsymbol{x}) / \sum_{h=1}^{g-1}[1 + \exp(\beta_{0h} + \boldsymbol{\beta}_h'\boldsymbol{x})] \tag{9.29}$$

for $i = 1, \ldots, g-1$, and

$$\Pr(\pi_g|\boldsymbol{x}) = 1 / \sum_{h=1}^{g-1}[1 + \exp(\beta_{0h} + \boldsymbol{\beta}_h'\boldsymbol{x})], \tag{9.30}$$

where $\boldsymbol{\beta}_i = (\beta_{1i}, \ldots, \beta_{pi})'$ $(i = 1, \ldots, g-1)$ are $g-1$ vectors each having p parameters. If we write $\beta_{0i}' = \beta_{0i} - \log(q_i/q_g)$, then this assumption is equivalent to

$$\log[f_i(\boldsymbol{x})/f_g(\boldsymbol{x})] = \beta_{0i}' + \boldsymbol{\beta}_i'\boldsymbol{x} \; (i = 1, \ldots, g-1). \tag{9.31}$$

Thus, logistic discrimination assumes that the log odds for the ith population against the baseline gth population are linear in elements of $\boldsymbol{x}$ for $i = 1, \ldots, g-1$. When $g = 2$, there is just one such linear log odds so that we can drop suffixes in this case and write

$$\log[f_1(\boldsymbol{x})/f_2(\boldsymbol{x})] = \beta_0' + \boldsymbol{\beta}'\boldsymbol{x}. \tag{9.32}$$

Such a model holds for a variety of parametric distributional forms of data, as listed for example by Anderson (1972). However, the beauty of the approach is that the model can be applied empirically whatever the nature of the data. To apply the method, we estimate the parameters β_{ij} from the training samples, substitute the estimates in (9.29) and (9.30), and allocate $\boldsymbol{x}$ to the population for which it has highest posterior probability.

9.25 Maximum likelihood estimation is, in principle, straightforward, although some care is needed in both its justification and execution. We merely give the basic results here; for full details see Anderson (1982) or McLachlan (1992).

First, it is necessary to specify how the training data were obtained, which could be through one of the following sampling designs.

(1) Mixture sampling, in which the individuals are sampled from the mixture distribution (so that the proportion of observations n_i/n from population π_i is an estimate of the incidence of that population).
(2) Separate sampling, in which the n_i are fixed and individuals are sampled separately from each population.
(3) $\boldsymbol{x}$-conditional sampling, in which the values of $\boldsymbol{x}$ are fixed and individuals with the given values are sampled to yield their population membership.

Strictly speaking, only the first two schemes are relevant for discriminant analysis, as the third one arises mainly in bioassay. However, the third scheme provides the central plank of the maximum likelihood estimation so is included in the consideration. Under this scheme, the likelihood of the training data is

$$L_c = \prod_{s=1}^{g} \prod_{i=1}^{n_i} \exp(z_{si}) / \sum_{t=1}^{g} \exp(z_{ti}) \tag{9.33}$$

where $z_{si} = \beta_{0s} + \boldsymbol{\beta}_s' \boldsymbol{x}_{si}$ for $s = 1, \ldots, g-1$, $z_{gi} = 0$, and $\beta_{0i}' = \beta_{0i} - \log(q_i/q_g)$. Thus, L_c is a function of the β_{0s} and $\boldsymbol{\beta}_i$ alone, so maximising it will yield the ML estimates of these parameters. Nothing further is required for discrimination, but if estimates of the β_{0s}' are specifically required, then we need extra information about the q_s.

Anderson (1982) goes on to show that maximum likelihood estimates of the β_{ij} under the other two schemes are also obtained by maximising L_c. For mixture sampling, moreover, n_j/n gives an estimate of q_j, so that β_{0j}' is estimable without any extra information for $j = 1, \ldots, g-1$. However, for separate sampling,

$$\beta_{0j} = \beta_{0j}' + \log(n_j/n) \; (j = 1, \ldots, g-1) \tag{9.34}$$

Thus, the β_{0j}' are estimable directly but, since for discrimination we require $\beta_{0j}' + \log(q_j/q_g)$ as in (9.31), the q_j must be estimated separately.

9.26 The likelihood L_c above is maximised iteratively. Originally, a Newton-Raphson procedure was suggested, but this has more recently been superseded by a quasi-Newton procedure; see Anderson (1972, 1979, 1982) for full numerical details. Both methods enable asymptotic standard errors to be found for the parameters, but two problems may occur during maximisation.

One problem is that of 'complete separation', which occurs when all points in each population are separated from points of all other populations by a hyperplane. Such configurations lead to non-unique maxima at infinity; the estimates of the parameters are then unreliable, but good discriminant functions still emerge. Anderson (1982) gives ways of recognising such situations and stopping the iterations before time is wasted. The second problem arises when there are zero marginal proportions with discrete data, again leading to maxima at infinity. Anderson (1974) suggested an *ad hoc* method for circumventing this problem, based on the approximate assumption that the offending variable is independent of the others conditionally in each group.

Several extensions to the basic logistic method have also been described. Anderson (1975) discussed quadratic logistic discriminant functions, in particular focusing on specific approximations that make parameter estimation feasible, while Anderson (1979) addressed the problem of updating parameter estimates as extra classified individuals become available. Finally, note that the logistic discriminant function is exactly equivalent to the usual linear discriminant function in the case of two homoscedastic normal populations, but parameters are estimated differently in the two techniques. The linear discriminant function is asymptotically more efficient if the assumptions are met (Efron, 1975), but logistic discrimination may be preferable in small samples when distributions are clearly non-normal or dispersion matrices are clearly not equal (see **9.47** for further discussion).

Example 9.5

Anderson (1982) has applied logistic discrimination to the preoperative prediction of patients particularly at risk of postoperative deep vein thrombosis. One hundred and twenty-four patients undergoing major gynaecological surgery were investigated. After surgery, 20 of the patients developed deep vein thrombosis (π_1) while the remaining 104 did not (π_2). Note that this is thus an example of mixture sampling. Preliminary screening identified ten variables as being the best potential predictors of deep vein thrombosis, and then a stepwise logistic procedure described by Anderson (1982) was used to select a predictor set from these ten variables. The five variables selected by this procedure were: euglobulin lysis time (x_1), FR antigen (x_2), age (x_3), presence/absence of varicose veins (x_4) and percentage overweight for height (x_5).

From (9.29), the posterior probabilities $\Pr(\pi_i|\boldsymbol{x})$ depend solely on the index $I(\boldsymbol{x}) = \beta_0 + \boldsymbol{\beta}'\boldsymbol{x}$. Applying maximum likelihood as described above to estimate the logistic parameters, the index $I(\boldsymbol{x})$ was estimated as $-11.3 + 0.009x_1 + 0.220x_2 + 0.085x_3 + 0.043x_4 + 2.190x_5$. The values of this index were calculated for each of the 124 patients in the study, and these values are plotted in Fig. 9.3. Anderson comments that patients with and without deep vein thrombosis are well separated, albeit with something of a 'grey' zone where group membership is equivocal. However, it should be noted that if this logistic function is used to classify future patients to one of the two groups, then much poorer success rates should be expected than those suggested here. The reason is that Fig. 9.3 has been obtained by first selecting those variables that best separate the two groups and then estimating the parameters in such a way as to maximise

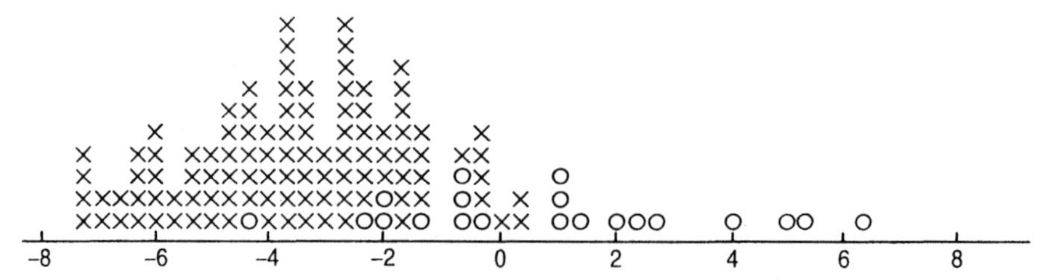

Figure 9.3 Prognostic index for patients at risk from deep vein thrombosis (DVT). Patients with DVT denoted ○, patients without DVT denoted ×. Reproduced with permission from Anderson (1982).

this separation. Future individuals should not be expected to fall as neatly into the derived categories; this point is discussed at greater length below.

Non-parametric allocation rules

9.27 Instead of formulating parametric models for the probability density functions $f_i(\boldsymbol{x})$ of the optimal allocation rules (9.6), (9.8) and (9.9), a more direct and flexible attack would be to estimate these functions *non-parametrically* from the training data and then to apply the estimated functions to the individuals for classification. Such an approach does not impose any presupposed functional forms on the probabilities, but lets the data 'speak for themselves'.

The first ideas in this area were provided by Fix and Hodges (1951, 1952; see also the commentary by Silverman and Jones, 1989), but about 20 years went by before the subject was picked up for intensive study. The 1970s and 1980s saw much development; for full reviews, bibliographies and discussion of non-parametric density estimation in general see Wegman (1972a, 1972b), Fryer (1977), Wertz and Schneider (1979), Silverman (1986) and Devroye (1987), while its specific application to discriminant analysis is covered by Hand (1982), Coomans and Broeckaert (1986) and McLachlan (1992, Chapter 9).

The most widely used method today is probably the *kernel* method, in which the density $f_i(\boldsymbol{x})$ for population π_i is estimated by

$$\hat{f}_i(\boldsymbol{x}) = n_i^{-1} h_i^{-p} \sum_{j=1}^{n_i} K_p\left(\frac{\boldsymbol{x} - \boldsymbol{x}_{ij}}{h_i}\right) \tag{9.35}$$

where p is, as usual, the number of elements of $\boldsymbol{x}$, K_p is a (non-negative and symmetric) kernel function that integrates to one, and h_i is a smoothing parameter (sometimes also called the *bandwidth*). This estimate can be viewed either as a smoothed (Härdle, 1991) version of the histogram estimate for probability densities, or as a generalisation of the 'naive' kernel estimate of Fix and Hodges (1951). Much work has been done on kernel estimates, the main questions revolving around appropriate choice and standardisation of kernel functions, methods of estimating the bandwidth, and development of fast algorithms to do the computations. Space considerations preclude discussion of these aspects here, and the reader is referred to the above-mentioned references for further details. In particular, Silverman (1986) has some good discussions regarding practical implementation of the technique, both in general and also specifically in the context of discriminant analysis. The main point to note is that kernel-based discrimination procedures are now widely available through the ALLOC80 computer program (Hermans *et al*, 1982) and within computer packages such as S (Becker *et al*, 1988) and SAS (1990).

9.28 Another simple non-parametric idea introduced by Fix and Hodges (see Dasarathy, 1991, and Silverman and Jones, 1989) is that of *nearest neighbour* discrimination. The nearest neighbour allocation rule classifies $\boldsymbol{x}$ to the population of its nearest training-set point in data space, while the k-nearest neighbour rule allocates it to the population that is in the majority among the k nearest neighbours. Hills (1967) discussed the use of these rules for the special case of binary data. In general, the k-nearest neighbour rule can be derived from kernel density arguments by choosing the kernel to be constant over the nearest k points and zero elsewhere, or defined formally from a non-parametric regression standpoint (Stone, 1977). The main problem, as can readily be surmised, with nearest neighbour rules is the choice of metric. The obvious choice is the Euclidean metric, possibly after scaling or other standardisation of the feature variables, but see also Todeschini (1989) and Myles and Hand (1990). Hellman (1970), Hall (1981), Luk and MacLeod (1986) and Wojciechowski (1987) all describe variants of the rule.

9.29 Nearest neighbour rules implicitly rely on distances. A more explicit use of distance in non-parametric discrimination has been made recently by Cuadras (1989, 1991, 1992). This approach uses Rao's (1982) work on diversity indices, to build discriminant functions from distances between individuals as follows. Suppose that the symmetric function $\delta(\boldsymbol{x}_1, \boldsymbol{x}_2)$ measures the distance between two individuals $\boldsymbol{x}_1$ and $\boldsymbol{x}_2$ (cf. Chapter 10). Furthermore, suppose that the individuals could have come from any of the g populations $\pi_1, \ldots, \pi_g$ and that the probability measure in π_i is denoted by $dP_i(\boldsymbol{x})$ for $i = 1, \ldots, g$. Then the *diversity coefficient* H_i of π_i is defined as

$$H_i = \int \delta(\boldsymbol{x}_1, \boldsymbol{x}_2) dP_i(\boldsymbol{x}_1) dP_i(\boldsymbol{x}_2) \tag{9.36}$$

i.e. as the average difference between two randomly chosen individuals from π_i. If one individual comes from π_i and the other from π_j then this average difference is

$$H_{ij} = \int \delta(\boldsymbol{x}_1, \boldsymbol{x}_2) dP_i(\boldsymbol{x}_1) dP_j(\boldsymbol{x}_2) \tag{9.37}$$

from which the *dissimilarity coefficient* between π_i and π_j is defined by the Jensen difference

$$\Delta_{ij} = H_{ij} - \frac{1}{2}(H_i + H_j) \tag{9.38}$$

Properties of this coefficient are given by Rao (1982). Now suppose that samples of size n_i are available from each of the populations π_i $(i = 1, \ldots, g)$, that $n = \sum n_i$, and that the symmetric function d_{rs}^2 gives the squared distance between any two sample individuals r, s. Then, by analogy with the above theoretical development, we can define the squared distance between the samples from π_i and π_j by

$$D_{ij}^2 = \frac{1}{n_i n_j} \sum_{r \in \pi_i} \sum_{s \in \pi_j} d_{rs}^2 - \frac{1}{2n_i^2} \sum_{r \in \pi_i} \sum_{s \in \pi_i} d_{rs}^2 - \frac{1}{2n_j^2} \sum_{r \in \pi_j} \sum_{s \in \pi_j} d_{rs}^2 \tag{9.39}$$

The degenerate case of a squared distance $D_{(i)j}^2$ between a single individual $\boldsymbol{x}_i$ and a sample from a population π_j can be obtained from (9.38) by setting $n_i = 1$, thereby yielding

$$D_{(i)j}^2 = \frac{1}{n_j} \sum_{r \in \pi_j} d_{ir}^2 - \frac{1}{2n_j^2} \sum_{r \in \pi_j} \sum_{s \in \pi_j} d_{rs}^2 \tag{9.40}$$

Cuadras (1989, 1991, 1992) has used these distances in deriving allocation rules for a variety of circumstances, and has demonstrated the equivalence of these rules with already existing ones for particular choices of d_{rs}^2. Gower (1989) and Krzanowski (1994a) use similar ideas to extend canonical variate analysis to more general data types. The benefits of this approach are that any of the dissimilarity measures available in cluster analysis (Chapter 10) can be used for the distances d_{rs}^2, including very general ones that can handle any combination of data types, and that missing values can be accommodated without any problems. Distance also plays a role, albeit more implicitly, in the 'ideal point' method of Takane *et al* (1987).

9.30 A final class of non-parametric discrimination methods has come to prominence in the 1980s and 1990s, with the explosion of computer-intensive developments in these years. There are two distinct lines of attack here. The first returns to Fisher's early regression approach, but replaces his linear parametric formulation with a nonlinear flexible one and fits the function to the training data using a variety of smoothing techniques. The main current techniques in this class include projection pursuit regression (Friedman and Stuetzle, 1981; see also Part 1, **4.15**), generalised additive models (Hastie and Tibshirani, 1990) and multivariate adaptive regression splines (Friedman, 1991). These techniques are essentially 'black-box' in nature, i.e. they are numerical, heavily computational, and they do not provide any closed algebraic expressions for allocation rules. We therefore do not give further details here, but leave the reader to follow up the above references.

The second line of attack returns to the space-partitioning idea of Welch (1939), but partitions the feature space by constructing a binary decision tree from the training data in a forward/backward stepwise manner. Although such an approach has a long history, dating back to work in the social sciences by Morgan and Sonquist (1963), Sonquist (1970) and Morgan and Messenger (1973), it was given considerable impetus computationally by Breiman *et al* (1984) under the name of Classification And Regression Trees (CART) and has played a major role in the modern development of machine learning (Quinlan, 1986, 1990, 1993; Michie *et al*, 1994, Chapter 5). This technique can perhaps be seen as a logical development of the non-parametric method suggested by Kendall (1966) and described in Kendall *et al* (1983) (but not mentioned in any of the more recent literature). Kendall's method in the case of two groups is to examine the data on each variable in turn, and to try and separate the training data into their constituent groups by successive partitioning on the variables. Thus, we first identify that variable $x_{(1)}$, and two associated constants $a_{(1)}, b_{(1)}$, such that all the training data with $x_{(1)} < a_{(1)}$ lie in one group, all the training data with $x_{(1)} > b_{(1)}$ lie in the other group, and the smallest possible number of individuals lie in the 'unidentified' region $a_{(1)} \leq x_{(1)} \leq b_{(1)}$. The individuals in this latter region are then subjected to a similar search process using the remaining variables. This procedure is continued, on a decreasing number of individuals and over a decreasing number of variables, until no further classification is possible.

CART essentially automates this procedure, but employs simple binary splitting of each variable (i.e. no 'unidentified' region) in building up the decision tree. Optimal splits are determined by minimising as an objective function, an 'index of impurity of classification'; Breiman *et al* (1984) adopt the Gini index, which involves cross-validated probabilities of misclassification and thus requires a substantial computing effort. Many modifications and refinements to this procedure are possible. For example, Loh and Vanichsetakul (1988) use normal-based linear discriminant analysis in the splitting rule, while Oliver and Hand (1995) discuss such operations as pruning, averaging and fanning of trees. Further reviews and surveys are given by Dietterich (1990), Safavian and Landgrebe (1991), Chou (1991) and Gelfand and Delp (1991).

Distributional aspects of allocation rules

9.31 All classification rules are formulated in terms of a putative 'future individual' $\boldsymbol{x}$ which is assumed to come from one of g populations $\pi_1, \ldots, \pi_g$. Each of these populations can be characterised by a probability distribution for $\boldsymbol{x}$, and this induces a probability distribution for the function on which allocation of $\boldsymbol{x}$ is based. For example, in the two-group case with normal data and homogeneous dispersion matrices, the optimal allocation rule (9.10) is based on the function

$$U(\boldsymbol{x}) = (\boldsymbol{\mu}_1 - \boldsymbol{\mu}_2)'\boldsymbol{\Sigma}^{-1}[\boldsymbol{x} - \frac{1}{2}(\boldsymbol{\mu}_1 + \boldsymbol{\mu}_2)] \tag{9.41}$$

We could, in principle, apply this function whatever the source or form of $\boldsymbol{x}$, and assuming a particular distribution for $\boldsymbol{x}$ will then generate a distribution for $U(\boldsymbol{x})$. However, since this function has been derived by assuming that $\boldsymbol{x}$ has a $N(\boldsymbol{\mu}_i, \boldsymbol{\Sigma})$ distribution in π_i for $i = 1, 2$, it is natural to continue with the same assumption. It follows from standard normal distribution theory (Part 1, Chapter 2) that $U(\boldsymbol{x})$ then has a $N(\frac{1}{2}\Delta^2, \Delta^2)$ distribution in π_1 and a $N(-\frac{1}{2}\Delta^2, \Delta^2)$ distribution in π_2, where $\Delta^2 = (\boldsymbol{\mu}_1 - \boldsymbol{\mu}_2)'\boldsymbol{\Sigma}^{-1}(\boldsymbol{\mu}_1 - \boldsymbol{\mu}_2)$. These two distributions thus specify the 'behaviour' of the allocation rule.

In practice, however, this rule is never used because it contains unknown parameters. Instead, a rule derived from training data has to be employed, such as the estimative rule (9.15) based on the function

$$V(x) = (\bar{x}_1 - \bar{x}_2)' W^{-1}[x - \frac{1}{2}(\bar{x}_1 + \bar{x}_2)] \tag{9.42}$$

There are now two standpoints that could be adopted. One is that, having calculated the sample mean vectors and the pooled dispersion matrix from the available training data, these quantities are fixed for all future allocations and x is the only random quantity in (9.42). Thus, the 'behaviour' of the allocation rule is obtained by calculating the distribution of $V(x)$, treating $\bar{x}_1, \bar{x}_2$ and W as known constants. If we make the same distributional assumptions about x as above, then these *group-conditional* distributions of $V(x)$ are normal, with mean $(\bar{x}_1 - \bar{x}_2)' W^{-1}[\mu_i - \frac{1}{2}(\bar{x}_1 + \bar{x}_2)]$ in population π_i $(i = 1, 2)$ and variance $(\bar{x}_1 - \bar{x}_2)' W^{-1} \Sigma W^{-1}(\bar{x}_1 - \bar{x}_2)$ in each population.

The group-conditional distributions thus describe the behaviour of a *particular* allocation rule. The second standpoint is to treat the training samples, and hence the quantities $\bar{x}_1, \bar{x}_2$ and W, also as random. The distribution of $V(x)$ in this case is the *unconditional* distribution, which describes the behaviour of the allocation rule over all possible future applications (including aquisition of differing training sets of given size). Unfortunately, derivation of unconditional distributions of allocation rules is a very difficult problem, even under the most simplified assumptions about the data. Few exact expressions are available, despite much work in the area. Some key references are John (1959, 1960, 1961), Bowker (1961) and Sitgreaves (1961), with reviews provided by Schaafsma and van Vark (1977) and Raudys and Pikelis (1980). Better progress has been made with asymptotic approximations, but here the derived expressions tend to be rather lengthy. Key references are Okamoto (1963) and Siotani and Wang (1975, 1977), and there is a good review by Siotani (1982). The easiest quantities to obtain are generally the moments of the distributions; calculation of means enables corrections for bias to be obtained, as by Moran and Murphy (1979) for odds ratios. Corrections for bias of logistic discriminant coefficients have been provided by Anderson and Richardson (1979), but by using standard maximum likelihood arguments.

Error rates and their estimation

9.32 Given the difficulties in obtaining distributions of allocation rules, attention is usually turned to simpler summary measures for assessing the worth of a particular rule. By far the most widely used such measures are the probabilities of misclassification, or *error rates*, that the rule gives rise to (but for other possibilities see Breiman *et al*, 1984, Habbema *et al*, 1981, or Hand, 1994). Following the notation of **9.7** and **9.8**, write $p(j|i)$ for the probability that an individual from π_i is allocated to π_j by a given rule. Then the *group-specific* error rates for the rule are $\sum_{j \neq i} p(j|i)$ $(i = 1, \ldots, g)$, while the *overall* error rate can then be defined as $\sum_{i=1}^{g} q_i \sum_{j \neq i} p(j|i)$. Most applications concern the case of just two populations, so the group specific error rates are $p(1|2), p(2|1)$ while the overall error rate is $q_1 p(2|1) + q_2 p(1|2)$.

9.33 Exact analytical evaluation of error rates obviously requires distributional assumptions to be made about all quantities in the rule, so can generally only be done in the context of parametric allocation rules. Unfortunately, just as there are different types of distribution of an allocation rule so also there are different types of error rate that can be calculated, and which should be clearly distinguished. The distinction between them was first made by Hills (1966); we can best establish this distinction by reference to discrimination between two multivariate normal populations. In this case, the optimal allocation rule is given by (9.10). This rule involves unknown parameters, and $\boldsymbol{x}$ is the only random quantity in it. Error rates calculated for this rule are known as *optimal* rates: they represent, in a sense, the best that can be done for such populations. Given training data, we would usually employ the sample rule (9.15). Arguing as above, we could treat the quantities $\bar{\boldsymbol{x}}_1, \bar{\boldsymbol{x}}_2$ and $\boldsymbol{W}$ either as known constants or as random variables. Error rates calculated under the former assumption are known as *actual* rates: they represent the performance of the rule calculated from the given training sets, assuming that the same rule will be used for all future allocations. Treating $\bar{\boldsymbol{x}}_1, \bar{\boldsymbol{x}}_2$ and $\boldsymbol{W}$ as random variables then induces a probability distribution for the actual error rates. Obtaining this distribution is just as troublesome as obtaining the distribution of the allocation rule itself (see the same references as in **9.31**), but moments are again generally tractable. In particular, the mean of the distribution is known as the *expected* error rate: it represents the average performance of the rule over all possible training set realisations.

9.34 We illustrate these ideas for the case of two-group homoscedastic normal discrimination. Here, the optimal allocation rule assigns $\boldsymbol{x}$ to π_i if $U(\boldsymbol{x}) > \log k$, where $U(\boldsymbol{x})$ is given by (9.41). Furthermore, we have already seen that if $\boldsymbol{x}$ has a $N(\boldsymbol{\mu}_i, \boldsymbol{\Sigma})$ distribution in π_i for $i = 1, 2$ then $U(\boldsymbol{x})$ has a $N(\frac{1}{2}\Delta^2, \Delta^2)$ distribution in π_1 and a $N(-\frac{1}{2}\Delta^2, \Delta^2)$ distribution in π_2, where $\Delta^2 = (\boldsymbol{\mu}_1 - \boldsymbol{\mu}_2)'\boldsymbol{\Sigma}^{-1}(\boldsymbol{\mu}_1 - \boldsymbol{\mu}_2)$ is the squared Mahalanobis distance between the two populations. It follows easily from standard theory that the optimal error rates are given by $p_o(2|1) = \Phi[(c - \frac{1}{2}\Delta^2)/\Delta]$ and $p_o(1|2) = \Phi[-(c + \frac{1}{2}\Delta^2)/\Delta]$, where $c = \log k$ and $\Phi(.)$ denotes the standard normal distribution function. For a zero *cut-off* point ($k = 1$), the two group-specific rates are both equal to $\Phi(-\frac{1}{2}\Delta)$.

In the presence of training samples, the estimative allocation rule assigns $\boldsymbol{x}$ to π_i if $V(\boldsymbol{x}) > \log k$, where $V(\boldsymbol{x})$ is given by (9.42). Furthermore, we have already seen in **9.31** that if the training sample means and the pooled dispersion matrix are treated as fixed, then the group-conditional distributions of $V(\boldsymbol{x})$ are normal, with mean $D(\boldsymbol{\mu}_i) = (\bar{\boldsymbol{x}}_1 - \bar{\boldsymbol{x}}_2)'\boldsymbol{W}^{-1}[\boldsymbol{\mu}_i - \frac{1}{2}(\bar{\boldsymbol{x}}_1 + \bar{\boldsymbol{x}}_2)]$ in population π_i $(i = 1, 2)$ and variance $\sigma^2 = (\bar{\boldsymbol{x}}_1 - \bar{\boldsymbol{x}}_2)'\boldsymbol{W}^{-1}\boldsymbol{\Sigma}\boldsymbol{W}^{-1}(\bar{\boldsymbol{x}}_1 - \bar{\boldsymbol{x}}_2)$ in each population. It thus follows that the actual error rates of the estimative rule are given by $p_a(2|1) = \Phi([c - D(\boldsymbol{\mu}_1)]/\sigma)$ and $p_a(1|2) = \Phi([-c + D(\boldsymbol{\mu}_2)]/\sigma)$, where $c = \log k$ as before.

Exact expressions for the expected error rates are available only if the problem is univariate (Sayre, 1980), or if either $\boldsymbol{\Sigma}$ (Moran, 1975) or the $\boldsymbol{\mu}_i$ (Streit, 1979) are assumed known. Otherwise, recourse must be made again to asymptotic expansions (Okamoto, 1963; Wyman *et al*, 1990).

For distributional results relating to the Z-statistic, the studentised Z-statistic and the quadratic discriminant function, see McLachlan (1992, Chapter 4). Vlachonikolis (1985) gives corresponding results for the location linear discriminant function for mixed binary and continuous variables.

9.35 All the error rates introduced above involve the unknown population parameters, however, and so they need to be *estimated* for practical use in a given allocation problem. Once again, the simplest procedure is the estimative one, in which any unknown parameters in an expression are replaced by their estimates from the training data. Thus, in the case of two-group normal homoscedastic populations, putting $\bar{\boldsymbol{x}}_1, \bar{\boldsymbol{x}}_2$ and $\boldsymbol{W}$ in place of $\boldsymbol{\mu}_1$, $\boldsymbol{\mu}_2$ and $\boldsymbol{\Sigma}$ in either the optimal or the actual error rates leads to the estimated error rates $\hat{p}(2|1) = \Phi[(c - \frac{1}{2}D^2)/D]$ and $\hat{p}(1|2) = \Phi[-(c + \frac{1}{2}D^2)/D]$, where $D^2 = (\bar{\boldsymbol{x}}_1 - \bar{\boldsymbol{x}}_2)'\boldsymbol{W}^{-1}(\bar{\boldsymbol{x}}_1 - \bar{\boldsymbol{x}}_2)$ is the squared Mahalanobis distance between the two training samples. If $c = 0$,

$$\hat{p}(1|2) = \hat{p}(2|1) = \Phi\left(-\frac{1}{2}D\right) \tag{9.43}$$

Note that the derivation of these estimators shows them to be estimators of either the optimal or the actual error rates, and not of the expected (i.e. unconditional) error rates.

9.36 The particular estimators (9.43) have been the subject of much study and refinement. The problem is that D^2 as an estimator of Δ^2 is biased upwards, so that the error rates in (9.43) are biased downwards. In practice they can give a very badly overoptimistic impression of the true performance of a given rule, in the sense that the error rates of future classifications are much worse than predicted. An unbiased estimate of Δ^2 is given (Rao, 1973) by

$$D_u^2 = \left(\frac{n_1 + n_2 - p - 3}{n_1 + n_2 - 2}\right)D^2 - p\left(\frac{1}{n_1} + \frac{1}{n_2}\right), \tag{9.44}$$

where p is the number of variables and n_1, n_2 are the training set sample sizes. However, correcting for bias in this way frequently yields negative values of D^2 and hence uncalculable error rates. A partial correction for bias is obtained by using $D_*^2 = (\frac{n_1+n_2-p-3}{n_1+n_2-2})D^2$, but this does not entirely get over the problem. Consequently, many authors have attempted to find better estimators of the error rates for this situation. Alternative single functions of Φ have been proposed by Lachenbruch (1967) and Snapinn and Knoke (1985), while asymptotically unbiased expansions have been given by Okamoto (1963) and McLachlan (1974). The full expressions are given, and compared, by Ganeshanandam and Krzanowski (1990). Corresponding estimates of error rates exist for the other parametric allocation rules, such as the Z-statistic and the quadratic discriminant function; see McLachlan (1992) for details.

9.37 The problem with all the above estimators is that they rely heavily on the distributional assumptions made about the data. We have already seen that distributional assumptions have been instrumental in the formulation of a particular allocation rule. For example, (9.10) is the optimal rule when we are dealing with homoscedastic normal populations, and (9.15) is a reasonable estimate to use when population parameters are unknown. On the other hand, we are perfectly at liberty to use (9.15) on *any* sort of data, simply because of its convenience of use. Thus, there is nothing to stop us using it on multivariate binary data if we wish. (Indeed, Fisher's original derivation suggests that (9.15) should provide a reasonably good *discriminant* function for a wide variety of situations, even if it only gives the best *allocation* rule in a narrow set of circumstances.) However, if we do use it as an allocation rule on binary data, then it would be patently absurd to estimate its error rates by the methods of **9.36**, which assume the data are normal. In general, many different rules can be applied on many different types of data. If we want to judge the performances of these rules, and perhaps compare them in a particular instance, then we need reliable general-purpose methods of error rate estimation that do not depend on distributional assumptions. This consideration has motivated the study of *data-based* estimation of error rates. In all the following we discuss error rates for the case of two-group discrimination, but the extension to $g > 2$ groups should be obvious.

9.38 The earliest (Smith, 1947) such data-based estimator is perhaps the most obvious one: any particular allocation rule derived from training data is re-applied to each individual in the training sets, and the proportion of misclassified individuals from each group is taken as that group's estimated conditional error rate. This method is generally known as the *resubstitution* method; Hills (1966) termed it the *apparent* error rate in his classification scheme. Unfortunately, this method is also highly overoptimistic as an estimator of the error rate, particularly for small samples. This is because the same data are being used both to form the allocation rule and assess its performance; the allocation rule will be formed in such a way as to maximise separation in the training sets, so these sets will be allocated much more correctly than would any future (randomly occurring) individuals from the same populations. The more estimated parameters there are in the allocation rule, the worse is likely to be the distortion in the apparent error rates (which makes resubstitution a particularly misleading method to use when comparing disparate rules on small samples).

To overcome this problem, various authors (e.g. Highleyman, 1962) proposed the *holdout* method, in which the training data are randomly split into two parts; the allocation rule is formed from one part, and its error rates are estimated by applying it to the second part and counting up the misclassifications. However, such a method is inefficient in its use of data. Also, the investigator has a problem in deciding how to split the training sets in order to balance the trade-off between obtaining a good allocation rule and a good estimate of its performance. This method really requires very large samples. More reliable small-sample methods did not become available until the computer revolution brought in its wake various ingenious *resampling* schemes.

9.39 Computer-intensive methods of error-rate estimation centre on three techniques: the jackknife, the bootstrap and cross-validation. Although seemingly different, there are, in fact, close theoretical connections between the three techniques (Efron, 1982). Good introductory accounts of them have been provided by Efron and Gong (1983) and Efron and Tibshirani (1986); see also **6.17** and **6.18** (Part 1) for their use in classical multivariate inference. There are two distinct ways of applying these techniques in the context of error rate estimation: as a method of correcting the bias in the apparent error rate, and as an estimator of error rate in its own right. We now briefly consider each of these ways in turn.

9.40 The original idea behind the *jackknife* (Quenouille, 1956) was as a means of reducing the bias of an estimator from $O(n^{-1})$ to $O(n^{-2})$ (or better). Since the apparent error rate is itself a biased estimator, the jackknife can be applied to it in its original formulation. However, we must distinguish carefully exactly what is being estimated. The basic jackknife procedure is as follows. Compute the group-conditional apparent error rates of the given allocation rule from the n training set observations, denoting the rate for π_i by ea_i $(i = 1, 2)$. Leave out each individual in turn from the training data. For each individual omitted, form a new allocation rule and find its group-conditional apparent error rates from the remaining $n - 1$ individuals. Denote by $ea_i^{(j)}$ the apparent error rate for π_i when the jth of the n individuals has been omitted from the data, and let $ea_i^{(.)} = \sum_{j=1}^{n} ea_i^{(j)}/n$ for $i = 1, 2$. Then the jackknife estimate of the *expected* (i.e. unconditional) error rate in π_i is given by

$$ea_{i,Ju} = ea_i + (n - 1)(ea_i - ea_i^{(.)}) \tag{9.45}$$

for $i = 1, 2$. To obtain the jackknife estimate of the *actual* (i.e. conditional) error rate in π_i, a slight adjustment is needed (Efron and Gong, 1983). Having obtained the allocation rule on omitting the jth individual from the training data, we find the group-conditional error rates when *all* individuals in the data are allocated in turn by this rule (i.e. including allocation of the jth individual as well). Then let $ea_i^{(..)}$ denote the average of the error rates in π_i over all possible omissions; the jackknife estimates of the actual error rates are

$$ea_{i,Jc} = ea_i + (n - 1)(ea_i^{(..)} - ea_i^{(.)}) \tag{9.46}$$

for $i = 1, 2$.

9.41 Bias correction of the apparent error rate can also be obtained easily through bootstrap methods, as follows. A *bootstrap* sample is first obtained, by sampling n individuals at random *with replacement* from the training data, paying attention to the original data aquisition plan (i.e. sampling either separately for each training sample or from the mixture as appropriate). Denote the apparent error rates for the bootstrap sample by $ea_i^{(b)}$ $(i = 1, 2)$. (Thus, $ea_i^{(b)}$ is the proportion of the bootstrap sample from π_i misclassified by the allocation rule computed from the bootstrap sample.) An estimate of the actual error rate for π_i is given by the proportion of the original training set from π_i misclassified by the allocation rule computed from the bootstrap sample; denote this by $a_i^{(b)}$ and compute the difference $d_i^{(b)} = ea_i^{(b)} - a_i^{(b)}$. Repeat this procedure a large number, B, of times. Then the average $\bar{d}_i = \sum_{b=1}^{B} d_i^{(b)}/B$ is the bootstrap estimate of bias in the apparent error rate from π_i, so that the bootstrap bias-corrected estimate of the actual error rate in π_i is $ea_i - \bar{d}_i$.

The only question with this procedure concerns the number B of bootstrap samples that should be taken in order to ensure accuracy of estimation. This question has been studied for general bootstrap estimation by various authors; Efron and Tibshirani (1986) summarise by saying that 50 to 100 replications may be sufficient for bias and standard error estimation, but a larger number, say 350, is needed to give accurate estimates of percentiles or P-values. See also Hall (1986) for a more detailed discussion.

9.42 Turning next to direct estimation of error rates, the first proposal utilising resampling methods was made by Lachenbruch (1967) and Lachenbruch and Mickey (1968). This is the *cross-validation* method, sometimes also called the *leave-one-out* method. As the name suggests it is closely related to the jackknife, and indeed some confusion exists in the literature between the two techniques (see the discussion in McLachlan, 1992, Chapter 10). The phrase 'jackknife estimates', when it appears in the output of a computer package, nearly always means 'leave-one-out' estimates!

The method is very straightforward in principle. Each individual is omitted in turn from the training data, the allocation rule is recomputed from the remaining $n - 1$ individuals, and the omitted unit is classified by this rule. The proportion of units misclassified in this way from each group gives a direct estimate of the actual error rate in that group. The only problems with the procedure (shared by the jackknife methods above) are computational ones. If the allocation rule involves computation of either the inverse or the determinant of a matrix, then recalculation of these quantities each time an individual is omitted from the data set will lead to very heavy computing demands. However, various algebraic identities exist which relieve the computational burden by ensuring that an inverse or determinant only needs to be calculated once for the full data set, and then updated by simple addition/multiplication for each unit omission (see, for example, Bartlett, 1951; Campbell, 1985). Thus, implementation of the process is feasible for most practical situations; for details in the contexts of linear, quadratic, location and kernel discriminant functions see Lachenbruch and Mickey (1968), Fukunaga and Kessel (1971), Krzanowski (1980, 1994b) and Fukunaga and Hummels (1989) respectively. Campbell (1985) applies the procedure to Mahalanobis distance calculations.

The leave-one-out error rate circumvents the bias problems associated with the apparent error rate (because the data classified at each step of the process have not been used in the calculation of the allocation rule), but it incurs the penalty of increased variance (Glick, 1978). Hand (1987) therefore suggests using shrunken estimators, and discusses several possibilities which, although biased, yield smaller mean square error than the leave-one-out estimator. Another variant is the so-called $\bar{U}$ estimator proposed by Lachenbruch and Mickey (1968), which blends the empiricism of the leave-one-out method with the use of the normal distribution. The central limit theorem is applied to the set of linear discriminant functions formed when each individual is left out of the training data in turn, to deduce normality of their means in the different populations. The actual error rate for π_i is then estimated by $\Phi(\bar{w}_i/s_{wi})$, where

$\bar{w}_i$ and s_{wi} are the mean and standard deviation of the n_i W-statistic values (9.15) calculated for each of the omitted observations from π_i.

9.43 The final possibility with resampling-based estimators is to use the bootstrap for direct estimation of actual error rates. Efron (1983) gives a number of variants that could be used, the most successful of which appears to be the so-called *632 estimator*. In this method, a bootstrap sample is drawn from each training set; Efron (1983) describes how this is done for mixture sampling, while Chatterjee and Chatterjee (1983) discuss the separate sampling case. An allocation rule is formed from the bootstrap sample, and the proportion ξ_i of individuals from π_i in the training data that are *not* in the bootstrap sample and are misclassified by this rule is computed. The process is repeated B times, and the average $\bar{\xi}_i$ of the ξ_i values is obtained. Then the 632 estimator of actual error rate in π_i is given by $0.368ea_i + 0.632\bar{\xi}_i$ (the value $0.632 = 1 - \frac{1}{e}$ being the approximate probability that an item appears in the bootstrap sample). Davison and Hall (1992) discuss bias and variance considerations for this and other resampling schemes.

Example 9.6
Return to the data and discriminant function for the two Iris species of Example 9.2. It is evident from the calculations there that the two species are very well separated, which is confirmed by the Mahalanobis distance between the two samples: $D^2 = 103.234$. Thus, if we assume multinormality of the data, then from (9.43) we can estimate the error rates in each of the two populations by $\Phi(-5.08)$. This is a very small number. However, if we apply either resubstitution or the leave-one-out method we obtain no misclassifications at all in either group, an even more unlikely estimate of error rate!

Next, return to the psychosocial influences in breast cancer data of Example 9.4. A first-order location model was fitted to these data in this Example, and some drawbacks in the model fit were pointed out. Nevertheless, applying the leave-one-out method to this fitted model resulted in an estimated overall error rate of 0.332, while coding all binary variables 0/1 and applying the linear discriminant function gave a much worse estimated error rate of 0.423. Indeed, in spite of these drawbacks, the location model produced the best discriminant function of all the ones tried on these data.

9.44 The above account of error rate estimation has been necessarily brief, and does not do full justice to the amount of work done on the topic. Some idea of this volume can be gleaned from the bibliographies provided by Toussaint (1974), Hand (1986) and McLachlan (1986); see also Knoke (1986) and Fukunaga and Hayes (1989).

Given the large number of suggested estimators, much work has also been done on assessing their performance and on comparing the performances of different estimators. A broad comparison, including normal-based as well as computer intensive estimators, has been given by Ganeshanandam and Krzanowski (1990), while the studies of Efron (1983), Gong (1982, 1986), Chernick *et al* (1985, 1986a, 1986b), Jain *et al* (1987) and Fitzmaurice *et al* (1991) focus specifically on the computer intensive methods. Other estimators have been considered by Fitzmaurice and Hand (1987) and Snapinn and Knoke (1985, 1988, 1989).

Comparison of allocation rules

9.45 It is evident from the preceding paragraphs that very many allocation rules have been devised or suggested, and in any particular situation the practictioner will have a number of possible rules to choose between. To help in this choice, there are now various comparative studies in the literature. These studies typically compare several allocation rules in particular contexts, judging their performances in terms of the error rates to which they give rise (but see also Habbema *et al*, 1981, for other possible measures of performance). They usually fall into one of two categories: simulation studies under known population structures, and empirical studies on real data sets. Performance in the former case is assessed by means of actual error rates, and in the latter case by means of estimated error rates. Unfortunately, in many of the early simulation studies the population conditions inadvertently favoured one particular allocation rule and thus gave unreliable results. This feature was present especially in some of the comparisons involving allocation rules based on kernel density estimates, as discussed by Hand (1982) and Murphy and Moran (1986). In the following paragraphs, we mention some of the more important studies and omit any that might contain such 'method' bias.

9.46 Probably the most studied allocation rules are the linear and quadratic functions in their estimative forms (see **9.15**). Direct comparison of the two, under normality but heteroscedasticity, has been conducted theoretically by O'Neill (1986), Critchley *et al* (1987), and Wakaki (1990). Empirical studies have been reported by Gilbert (1969), Marks and Dunn (1974), van Ness and Simpson (1976), Aitchison *et al* (1977), Wahl and Kronmal (1977), van Ness (1979) and Bayne *et al* (1983). The general consensus appears to be that the linear function is preferable if sample sizes are small relative to the number of variables, but if sample sizes are moderate then the optimum choice depends on the degree of heteroscedasticity and on the separation of the populations. Devroye (1988) discusses basing the choice on the estimated error rate.

Robustness of linear and quadratic functions to departures from normality has been studied by many authors. Both types of function have been considered in the case of continuous non-normal variables, the linear function by Lachenbruch *et al* (1973), Ahmed and Lachenbruch (1975, 1977), Subrahmaniam and Chinganda (1978), Chinganda and Subrahmaniam (1979), Brofitt *et al* (1981) and Balakrishnan and Kocherlakota (1985), and the quadratic function by Clarke *et al* (1979), Brofitt *et al* (1980) and Bayne *et al* (1983). Rawlings and Faden (1986) and Joachimsthaler and Stam (1988) also include these allocation rules in a wider comparison with others. The use of the linear function on either categorical or mixed variables has been surveyed by Krzanowski (1977), and further comparisons have been provided by Titterington *et al* (1981), Knoke (1982), Vlachonikolis and Marriott (1982) and Schmitz *et al* (1985).

Comparisons of predictive and estimative versions of the linear discriminant function may be found in Aitchison (1975), Hermans and Habbema (1975), Aitchison *et al* (1977), Han (1979) and Moran and Murphy (1979), while the corresponding comparisons for the location model on mixed data have been provided by Vlachonikolis (1990). Also, assessments of the effectiveness of regularised discriminant analysis have been considered by Friedman (1989) and Rayens and Greene (1991).

9.47 On a more general front, there have been comparisons of logistic discrimination with other procedures and also of kernel discrimination with other procedures. The relative efficiency of logistic discrimination has been studied under normality by Efron (1975), McLachlan and Byth (1979) and Bull and Donner (1987), and under non-normality by Halperin *et al* (1971), Press and Wilson (1978), Crawley (1979), O'Hara *et al* (1982), Hosmer *et al* (1983a, 1983b), Halperin *et al* (1985) and Brooks *et al* (1988). For assessment of kernel density methods on continuous variables, see the previous references to Hand (1982) and Murphy and Moran (1986). Kernel discrimination has been compared with other methods by Aitken (1978), Titterington *et al* (1981) and Hand (1983) on categorical data, and by Vlachonikolis and Marriott (1982) and Schmitz *et al* (1983, 1985) on mixed categorical/continuous data. Finally, comparisons of some of the tree-structured classification methods such as CART (**9.30**) have been conducted by Loh and Vanichsetakul (1988).

Example 9.7
Hand (1983) nicely illustrates the dangers of using resubstitution error rates for the comparison of different allocation rules, by comparing resubstitution and independent test set error rates obtained with the linear discriminant function and two variants of kernel discrimination on various sets of multivariate binary data. To make the point, we simply take results for one kernel method (Aitchison and Aitken, 1976) and one set of data. These data were collected during a study of the effectiveness of a particular form of treatment for nocturnal enuresis; the treatment involves the use of a buzzer that is triggered and wakes the child when the bed becomes wet, and the two classes comprised those children for whom the treatment was a success or failure respectively. The training set comprised 113 subjects, and the test set consisted of a further 62 who were treated while the training set was being analysed.

Resubstitution of the training data into the linear and kernel classifiers respectively yielded 38 and 20 misclassifications, suggesting that the kernel method gave the much superior allocation rule. However, when the two allocation rules were applied to the test set, the numbers of misclassifications were 19 for the linear classifier and 20 for the kernel method, a very different picture!

Missing values

9.48 One frequently encountered practical problem, which is likely to disturb application of any of the above theory, is the absence for some reason of particular data values. There is no problem if such missing values occur only among units to be classified, because in any particular classification the variables are simply redefined to be those for which values are present and the classification procedure is conducted on this reduced variable set. However, missing values among the training data do cause a problem, as it is then not clear how any particular allocation rule is best estimated. A few attempts have been made (e.g. Srivastava and Zaatar, 1972) to obtain maximum likelihood estimates of discriminant functions directly from the available data, by structuring the patterns of missing values appropriately. However, by far the most popular approach is to *impute* values wherever they are missing and then to obtain the allocation rule in the standard way from the 'completed' training data.

Various possible imputation methods have been proposed and studied. Indeed, there is now a very large literature on the topic, as exemplified by the references in Little and Rubin (1987) and recent papers such as Meng and Rubin (1993). We now outline briefly some methods that have found application in discriminant analysis. The oldest, simplest (and crudest) is to replace each missing value by the mean of all values that are present on the corresponding variate in the given group. Once computing facilities became more widespread, the deficiencies of this approach were recognised and more sophisticated methods were proposed. Dear (1959) suggests a method based on principal components: means are first imputed in each group, the augmented data matrix in each group is standardised and the first principal component is extracted from each resulting matrix, new imputed values are obtained by multiplying the appropriate principal component scores (for the unit) and coefficients (for the variable), and the resulting augmented matrices are finally 'de-standardised'.

Buck (1960), by contrast, suggests a method based on multiple regression analysis: means are again imputed first in each group, regression equations are then set up in which each variate containing any missing values is the dependent variable and all other variables are explanatory, parameters in these equations are estimated from the data, each missing value is predicted from the appropriate equation, and these predicted values are the ones imputed in the data. Beale and Little (1975) show that better estimates result on iterating Buck's procedure; rapid convergence is usually obtained, giving predicted values that change minimally between iterations. This iterative scheme is a special case of the E-M algorithm (Dempster *et al*, 1977), and yields maximum likelihood estimates under normality of data. More recently, Little (1988) adapts this procedure for obtaining robust estimates of means and covariance matrices, while Krzanowski (1988b) and Hufnagel (1988) consider other imputation methods.

Whichever imputation method is used, variances and covariances are likely to be systematically underestimated (since the procedure is deterministic). To compensate, degrees of freedom can be adjusted or a random perturbation can be added to each imputed value. Bello (1993a, 1993b) has given further details of all these aspects and has compared the available imputation methods in a number of simulation studies. The first reference gives some general comparisons, while the second focuses on the specific context of discriminant analysis.

Variable selection

9.49 In many situations, the practitioner may wish to use fewer than the full set of available variables for either discrimination or classification. There are various reasons why this might be the case, the two most obvious being the cost of processing the variables or the desire to identify the most relevant ones for separating the groups (but see Schaafsma, 1982, for further discussion). There is also a well-documented problem associated with dimensionality in discriminant analysis (see, for example Jain and Chandrasekaran, 1982): the unconditional error rate of an allocation rule will decrease with the addition of extra variables only up to a certain point, after which addition of further variables will cause it to increase again. This 'optimum performance' threshold will depend on the training sample sizes, the allocation rule used, and the particular features of the populations being discriminated. Thus, selection of variables is also an area that has been much studied in discriminant analysis.

9.50 An important step in the process is the ability to determine whether a new set of q variables offers any additional discriminatory information to the p variables that are already present. Let $\boldsymbol{x}$ denote the vector of all $p + q$ variables and $\boldsymbol{x}_1$ the subvector of p original variables. If we assume multivariate normality of these vectors, then a test of hypothesis can be formulated and carried out to answer the question of interest. In the general g-group situation, this test can be derived using MANOVA principles. The null hypothesis H_0 here is that the conditional means of $\boldsymbol{x}$ given $\boldsymbol{x}_1$ are the same in all g populations. If we denote by $\boldsymbol{B}$ and $\boldsymbol{W}$ the usual between-group and within-group sums of squares and products matrices calculated for the training data from $\boldsymbol{x}$, and by $\boldsymbol{B}_1$, $\boldsymbol{W}_1$ the same quantities calculated from $\boldsymbol{x}_1$, then the likelihood ratio test statistic $\tilde{\Lambda}$ for H_0 is given by

$$1 + \tilde{\Lambda} = (1 + \Lambda)/(1 + \Lambda_1) \tag{9.47}$$

where

$$\Lambda = |\boldsymbol{W}|/|\boldsymbol{B} + \boldsymbol{W}| \tag{9.48}$$

and

$$\Lambda_1 = |\boldsymbol{W}_1|/|\boldsymbol{B}_1 + \boldsymbol{W}_1| \tag{9.49}$$

are the usual one-way MANOVA likelihood ratio test statistics using $\boldsymbol{x}$ and $\boldsymbol{x}_1$ respectively. The general distribution theory of $\tilde{\Lambda}$ is that associated with Wilks' lambda (see Equation A.31, Appendix to Part 1), but simplification occurs if testing whether a single extra variable provides additional discriminatory information. Here $q = 1$, and if n is the total number of units in the training data then $[(n - g - p + 1)(1 - \tilde{\Lambda})]/[(g - 1)\tilde{\Lambda}]$ has an F distribution on $g - 1$ and $n - g - p + 1$ degrees of freedom under H_0.

In the special case $g = 2$, it is more convenient to re-express the above test in terms of Mahalanobis distances. Let Δ and Δ_1 denote the Mahalanobis distances between the two populations based on the full feature vector $\boldsymbol{x}$ and the subvector $\boldsymbol{x}_1$ respectively. Then $\tilde{\Delta} = (\Delta^2 - \Delta_1^2)^{\frac{1}{2}}$ is the extra distance between the two populations when the q new variables are added to the p original ones, and the null hypothesis H_0 above is equivalent to the hypothesis that $\tilde{\Delta} = 0$. Denoting the sample analogues of Δ and Δ_1 by D and D_1 respectively, Rao (1973) shows that this hypothesis can be tested by means of the statistic

$$\tilde{F} = [(n_1 + n_2 - p - 1)(n_1 n_2)(D^2 - D_1^2)]/[q(n_1 n_2 D_1^2 + \{n_1 + n_2\}\{n_1 + n_2 - 2\})], \tag{9.50}$$

which has an F distribution on q and $n_1 + n_2 - 1$ degrees of freedom under H_0.

9.51 A formal mechanism for selecting the 'best' subset of variables in any given problem requires specification of a criterion by which the 'goodness' of any particular subset can be judged, a computing strategy for carrying out the calculations and, possibly, a 'stopping rule' for terminating the process. Three general types of criteria have been found useful when seeking variable subsets in discriminant analysis: descriptive measures of group separation, statistics such as (9.47) and (9.50) above for testing hypotheses about extra discriminatory information, and error rates for judging the worth of an allocation rule. While the particular context may indicate the most appropriate choice of criterion, broadly speaking the first two categories are designed with discrimination in mind while the third one has classification as its primary objective (cf. the distinction made in **9.1**).

As regards computing strategy, the best solution is clearly obtained by computing the value of the chosen criterion for all possible subsets of variables and selecting the subset that yields the optimum criterion value. Various algorithms have been devised for effecting the computations as efficiently as possible (e.g. Furnival, 1971; Garside, 1971), and branch-and-bound methods can remove the need for examining large numbers of subsets (e.g. Ridout, 1988; Krusinska, 1989b). Nevertheless, the volume of computation can still prove prohibitive if there are many candidate variables, and recourse has often to be made to sequential strategies such as forward selection, backward elimination or stepwise selection. A forward selection would start by choosing the best single variable (according to the desired criterion), then adding to it the one out of the remaining $p - 1$ variables that gives the best improvement in performance, then adding to these two the one out of the remaining $p - 2$ that again gave the best improvement, and so on. Backward elimination works in the opposite direction, by starting with all p variables, removing the one that gives the least deterioration in performance, then removing the one out of the remaining $p - 1$ that again causes least deterioration, and so on. Stepwise selection blends both approaches, working essentially in the same way as forward selection, but checking at each stage whether any of the existing variables can be removed without deterioration in performance. All of these strategies require distributional results to be available, as both 'deterioration' or 'improvement' should really be prefixed by 'significant'; the natural stopping rule is then to terminate the process when no addition of variables results in a significant increase of performance, while elimination of any variables produces significant deterioration in performance.

9.52 Use of the test statistic $\tilde{\Lambda}$ as a criterion in an all-subsets approach has been described by McCabe (1975), and in a stepwise approach by McLachlan

(1992, p398). McCabe's idea was to use the statistic in a descriptive fashion, suggesting that subsets of the same size be ordered with respect to their values of $\tilde{\Lambda}$ and that these orderings then be used as a guide to identification of important variables. Further distributional results for formalising the procedure were then supplied by Hawkins (1976) and McKay (1976, 1977); McLachlan provides the necessary distributional results for operation of the stepwise method.

Other criteria that have been used in place of the statistic $\tilde{\Lambda}$ include Hotelling's (1951) trace statistic (reducing to the usual T^2 statistic in the special case of two groups), normal-model error rates (McLachlan, 1976, 1980; Young and Odell, 1986), and Akaike's Information Criterion (Fujikoshi, 1985). The previously mentioned (**9.5**) connection between two-group linear discriminant analysis and multiple regression means that all criteria and methods used for variable selection in the latter context (such as R^2 or partial-F; see Miller, 1990, for further details) have also been adapted to the former context, and have been made widely available through standard computer packages. Costanza and Afifi (1979) have conducted some Monte Carlo comparisons of F-based stopping rules in stepwise analysis, while Rencher (1993) has studied three of the most used criteria (Λ, T^2 and R^2) in much more generality by giving expressions for the contributions of individual variables to each criterion and providing some empirical comparisons on a number of data sets.

The above criteria have all stemmed in some way from normality assumptions for the data, so may not have the widest applicability. Data-based estimators of error rate provide more scope; these have been used by McCabe (1975), van Ness and Simpson (1976), Habbema and Hermans (1977), van Ness (1979), who all used resubstitution error rates (with the bias that this may entail), and Ganeshanandam and Krzanowski (1989) who used leave-one-out error rates. The program ALLOC80 (Hermans *et al*, 1982) is built on kernel density estimates and can thus cater for any type of data, while selection methods for mixed categorical/continuous data based on the location model have been proposed by Krzanowski (1983a, 1995), Daudin (1986) and Krusinska (1988, 1989a, 1989b, 1990).

For further discussion of many of the above topics, see McKay and Campbell (1982a, 1982b).

9.53 Having selected a subset of variables in a practical application, the analyst will want to assess the performance of the chosen variables, and here there are some hidden problems. Murray (1977b) pointed out the extreme danger of simply calculating apparent error rates for the selected variables. Not only has the allocation rule been chosen to be the most favourable one for the training data (giving the optimistic bias in the resubstitution method highlighted in **9.38**), but also the variables selected will be the ones most favourable for the training data. Thus, there is a double helping of over-optimism inherent in the procedure, and the performance of the selected variables on future data will be very markedly poorer. Miller (1990) has discussed extensively the problems caused by such *selection bias*, but in the context of multiple regression.

Following Murray's warning, many subsequent authors have used some form of cross-validation to overcome assessment bias. However, almost invariably this cross-validation has been done on the data *after* the variables have been selected. While this procedure overcomes the error rate estimation bias, it still does not correct for the selection bias. Ganeshanandam and Krzanowski (1989) pointed out that for accurate assessment of the whole selection procedure, the variable selection must be conducted afresh for each unit omission in the leave-one-out process. Different variables may thereby be selected for each unit omission, but simulation studies have indicated that the resulting estimates of error have the right magnitudes. Computing is, of course, very heavy with such a nested cross-validation (particularly so if the selection criterion is itself a cross-validated error rate), but modern computing power makes this progressively less of a problem. Stone (1974) is an early reference that touches upon these aspects.

Example 9.8
Krzanowski (1995) illustrates the extent of selection bias in an application involving mixed categorical and continuous variables. The data set was previously described in Krzanowski (1975); it concerned 186 subjects who had been treated for breast cancer, and was divided into two groups according to whether the treatment was successful (99 subjects) or not (87 subjects). Nine variables were measured on each individual: variables 1–6 were continuous (age at mastectomy, log time to ablation, 17-hydroxycorticosteroids, androsterone, dehydroepiandrosterone and aetiocholanolone) while variables 7–9 were binary (type of mastectomy, type of ablation and presence/absence of lesion on breast).

Using the location model (**9.19**) on the full set of data produced 61 misclassifications with the leave-one-out procedure. Variables were then selected by a backward elimination procedure, the chosen set at each stage being the one that led to fewest leave-one-out misclassifications. Assessing performance by calculating leave-one-out error rates of the selected variables yielded 60, 61, 61, 60 and 64 misclassifications for the best subsets of size 8, 7, 6, 5 and 4 variables respectively. Following the procedure recommended above, however, the nested leave-one-out error rates for subsets of these sizes were 64, 70, 71, 74 and 79 misclassifications respectively. It can therefore be concluded that selection bias is not negligible; that it increases substantially with the degree of selection; that estimated error rates calculated from only the selected variables are over-optimistic; and that a poor subset is likely to be chosen if these error rates are used in determining a stopping rule for the selection procedure.

In the light of these results, studies that compare different discrimination methods and make use of variable selection but do not allow for it in the error rate estimation must be treated with caution.

Some special data structures

9.54 All the methodology discussed to date in this chapter is applicable very generally, and without any specific assumptions about the structure of the data to which it is applied. By far the most popular approach to discriminant anlysis is via either the linear or quadratic discriminant functions of **9.11**, with population means and dispersion matrices being estimated from the data by one of the methods described in **9.15**, **9.21** or **9.22**. However, in certain cases there may be some particular structure in the data that can be directly modelled and which induces a special form of either the population means or dispersion matrices. Use of this special form for the particular structure in question should provide more relevant discriminant functions, and a more efficient analysis, than by just adopting the general all-purpose methodology. Data structures, which induce special covariance forms, are those involving *repeated measures*, *time dependence* or *spatial dependence*, while *growth curves* impose further structure on population means. Some of these topics will be treated again in general terms in Chapter 13, but here we briefly outline the important points in a discrimination context.

9.55 Repeated measures occur whenever the same set of p variables is observed on each sample member more than once. For example, in lactation studies, a set of cows may have the same p blood constituents measured each week for three months. The cows may have been divided into two treatment groups (e.g. two different types of pasture), and it is of interest to discriminate between pastures on the basis of all the measurements. Another example is biomedical diagnosis based on multiple observations of cell membrane thickness, where the muscle capillary is examined at 20 different sites, say, around its perimeter. Information about variability between sites as well as that about mean values could be useful in identifying group membership. In general in such studies, different individuals will be uncorrelated, but there will be correlations among the different variables at each time or site, and also between the different times or sites for each variable.

Study of this problem was initiated by Choi (1972). Suppose that the set of p variables $\boldsymbol{x} = (x_1, \ldots, x_p)'$ is observed on each individual at k time points. Denoting the observations at the tth time point by $\boldsymbol{x}_t = (x_{t1}, \ldots, x_{tp})'$ for $t = 1, \ldots, k$, we can write the complete feature vector of kp variables for each individual as $\boldsymbol{z} = (\boldsymbol{x}_1', \ldots, \boldsymbol{x}_k')$. Choi assumed that the repeated measurements have the same mean and are equicorrelated in a given group, which can be modelled by taking $\boldsymbol{x}_t$ to have the form

$$\boldsymbol{x}_t = \boldsymbol{\mu}_i + \boldsymbol{v}_i + \boldsymbol{\epsilon}_{it} \tag{9.51}$$

in the ith group where $\boldsymbol{v}_i$ has a $N(\mathbf{0}, \boldsymbol{\Psi}_i)$ distribution independently of the $\boldsymbol{\epsilon}_{it}$, which are in turn i.i.d $N(\mathbf{0}, \boldsymbol{\Omega}_i)$ $(t = 1, \ldots, k;\ i = 1, \ldots, g)$. Thus $\boldsymbol{\Psi}_i$ represents the variation in features over sample individuals, while $\boldsymbol{\Omega}_i$ represents the variation between repeated observations on the features of a given individual, in group π_i. Under this model, the full feature vector $\boldsymbol{z}$ has a multivariate normal distribution in π_i, with mean $(\boldsymbol{\mu}_i', \ldots, \boldsymbol{\mu}_i')$ and dispersion matrix $\boldsymbol{\Psi}_i \otimes \mathbf{1}_k \mathbf{1}_k' + \boldsymbol{\Omega}_i \otimes \boldsymbol{I}_k$ for $i = 1, \ldots, g$. Here, $\mathbf{1}_k$ denotes the k-element vector of ones and $\otimes$ denotes the Kronecker product of matrices.

Having established the general structure of the populations, it just remains to apply the standard decision theoretic formulation to obtain discriminant functions and allocation rules, to follow these through in the usual way to obtain error rates, and to provide estimative (or other) rules for use in practice. Of course, a major problem is to obtain compact forms for the various expressions, given the complicated structure above. However, satisfactory progress on all these fronts has been achieved by Choi (1972), Gupta (1980, 1986), Bertolino (1988), Gupta and Logan (1990) and Logan and Gupta (1993), so that the full range of methodology is now available for such data.

9.56 If the k repeated measurements have been made on just a single variable (i.e. $p = 1$), and if they have been made on a single individual at many equally spaced time points, then the resulting observations form a traditional *time series*. We can model such a series $\boldsymbol{x}' = (x_1, \ldots, x_k)$ by

$$x_t = \mu_{it} + \epsilon_{it}, \quad (t = 1, \ldots, k) \tag{9.52}$$

in π_i, where the μ_{it} are constants while the ϵ_{it} are i.i.d. $N(0, \boldsymbol{\Sigma}_i)$ for $i = 1, \ldots, g$. Standard time series methodology (Box and Jenkins, 1976) requires the series to be first preprocessed, e.g. by differencing, so that the resulting values have been reduced to stationarity, in which case we can further assume that $\mu_{it} = \mu_i$ for all t and that the ϵ_{it} arise from some zero-mean stationary process. Box and Jenkins discuss a variety of autoregressive, moving average or mixed processes, which can be used for this modelling. The simplest of these models is the first-order autoregressive process, in which the correlation between two observations u time periods apart is a^u for some parameter a (satisfying $|a| < 1$) that can be estimated from the data. Shumway (1982) provides a comprehensive discussion of frequentist methods for discrimination between time series using such models, while Broemeling and Son (1987) and Marco *et al* (1988) consider Bayesian counterparts.

Another instance of such data occurs when measuring the growth of an individual over a period of time, in which case the series $\boldsymbol{x}$ constitutes a *growth curve* (see also Chapter 13). Successive observations on the individual are again correlated as above, so the same model and assumptions about the ϵ_{it} will be appropriate. Now, however, the means μ_{it} will not be constant in each group but will describe the particular form of growth in that group, and so must be modelled accordingly. The simplest model is a polynomial of appropriate degree, which is the growth model first proposed by Potthof and Roy (1964). Fitting such a model to the training data and developing discriminant functions between growth curves has been discussed by Lee (1982).

It may be that non-stationarity of the data arises through time-varying means, but in a less structured fashion than is the case in growth curves. An instance of such data arises in digitised spectrographs, where some measurement of energy is made at different wavelength values on (typically) various chemical samples. Here 'time' is the wavelength, and the energy varies in some haphazard but non-stationary fashion over the wavelength range. Typically, also, in such data there are very many wavelengths at which the energy is measured so that the data are very high-dimensional. Singularity of dispersion matrices occurs if no structure is imposed on the data, so some form of modelling is imperative. The antedependence models of Gabriel (1962) form suitable non-stationary analogues of the autoregressive models used above, and these models have been adapted to the discrimination context by Krzanowski (1993b). For further discussion of discriminant analysis when the data are high-dimensional, see **14.14** and **14.15**.

9.57 The above models are appropriate whenever a *one-dimensional* dependence exists among the measurements. The most common context for such dependence, as suggested above, is that of time, but it is clear from the application to spectroscopic data that other one-dimensional continua (such as wavelength) can be plausibly modelled in the same way. A case could also be made for similar treatment of spatial observations that are essentially one-dimensional, for example if the chemical constituents of soil samples are measured at a fixed depth along a line transect (e.g. Webster and McBratney, 1981). Spatial data occurring more generally in such studies, or those encountered in modern applications of *image analysis*, however, typically exhibit either *two-* or *three-dimensional* dependence and thus require more complicated methods for their analysis. Examples of image analysis now abound throughout the sciences—remotely sensed images from satellites (Geography), diagnostic body scans (Medicine), photon emission tomography (Nuclear Medicine), automatic object recognition (Computer Science), stellar maps (Astronomy), among others—so that there is now an extensive body of theory on the subject. Once again, therefore, we cannot do more than give an overview of the methodology and leave the reader to follow up the (large body of) relevant literature. For greatest convenience, we shall present this overview in the language of remotely sensed data.

9.58 A typical image, or *scene*, to be analysed will comprise a rectangular array of $m_1 \times m_2$ point observations, or *pixels*. Usually each image will have very many such pixels, values of 64, 128 or 512 for m_1, m_2 being the most common ones (with, often, $m_1 = m_2$). A p-dimensional vector $\boldsymbol{x}$ of values is received from the remote sensor for each pixel, the values being the intensities of electromagnetic energy emanating at different wavelengths from that point on the ground. There are, of course, many practical complications associated with the data collection, and some preprocessing will be necessary before the data can be used for classification (see, for example, Cressie, 1991, p501). We assume that this has been done and that $\boldsymbol{x}$ represents the data to be analysed. The noise in the system means that this information is generally received in a degraded form. The pixels are labelled in some systematic fashion from 1 to $n = m_1 m_2$, so that $\boldsymbol{x}_i$ denotes the observation for the ith pixel in this ordering; $\boldsymbol{z} = (\boldsymbol{x}'_1, \ldots, \boldsymbol{x}'_n)$ denotes the data for the complete scene, and each pixel can belong to one of g groups (*colours*) $\pi_1, \ldots, \pi_g$. The purpose of *image analysis* is to identify the colour of each pixel in a scene, with the help of training data that consist of a number of scenes in which the pixel colours are known. To formalize this objective, let $\boldsymbol{g}_j$ be the group-indicator vector defining the colour of the jth pixel, so that the ith element of $\boldsymbol{g}_j$ is 1 if the pixel has colour π_i and zero otherwise ($i = 1, \ldots, g$), and define $\boldsymbol{g} = (\boldsymbol{g}'_1, \ldots, \boldsymbol{g}'_n)$. The objective of the analysis is thus to provide an estimate $\hat{\boldsymbol{g}}$ of $\boldsymbol{g}$ given the observed data $\boldsymbol{z}$.

There are in general two *strategies* for obtaining this estimate: either by considering all pixels simultaneously (i.e. finding all elements of $\hat{\boldsymbol{g}}$ in one go), or by obtaining the n estimates $\hat{\boldsymbol{g}}_j$ separately and then piecing all the separate estimates together. The former strategy aims for the best reconstruction of the whole scene, while the latter aims for maximising the number of correctly identified individual pixels (so may end up with more of a patchwork-like reconstruction). Different situations may demand different strategies.

In addition to the strategy adopted, there are two possible ways in which the data information may be used. The simplest is to assume that all the feature vectors are independent, and to proceed with any of the general discrimination methods described earlier in the chapter. An allocation rule derived in this way is said to be *non-contextual*, as it treats each pixel in isolation and pays no regard to any information contained in surrounding pixels. Such rules predominated in the early development of image analysis, but are now rarely used. The more realistic approach now adopted recognizes the dependence existing between neighbouring pixels. Spatially contiguous pixels will tend to belong to the same colour group, so a pixel should be classified not just on the basis of its own feature vector but also by using neighbouring feature vectors as well. Such an allocation rule is called *contextual*. To develop a contextual rule, we must first decide on which pixels (and how many) should play a role. A *first-order* model includes just the north, east, south and west neighbours of each pixel in the classification rule, while a *second-order* model adds in the four diagonal neighbours as well. Having determined the order of model, we then need to model the dependence structure.

9.59 Various possible ways of proceeding now exist. One is to use the standard linear discriminant methodology, but to structure the dispersion matrices to reflect the spatial dependencies of the data. This is essentially a generalisation to two-dimensional dependence of the one-dimensional dependence structure in the temporal methods discussed above, with the extra constraint of equal dispersion structures in each group. If the scene contains n pixels with feature vectors $\boldsymbol{x}_1, \ldots, \boldsymbol{x}_n$ then we assume

$$\boldsymbol{x}_t = \boldsymbol{\mu}_i + \boldsymbol{\epsilon}_t \qquad (9.53)$$

for $t = 1, \ldots, n$ in group π_i $(i = 1, \ldots, g)$, where the $\boldsymbol{\epsilon}_t$ are realisations of a zero-mean spatially correlated stationary random process that has the following assumptions. First, it is Gaussian; second, it has a locally spatial isotropic covariance structure, so that $\text{cov}(\boldsymbol{x}_j, \boldsymbol{x}_k) = \rho(d)\boldsymbol{\Sigma}$ for $j \neq k$, where $\rho(.)$ is the isotropic correlation function and d is the Euclidean distance between pixels j and k in the two-dimensional scene; third, there is local spatial continuity, in that if pixel j belongs to group π_i then every other pixel in the chosen size of neighbourhood (first-order, second-order, etc) is assumed to do so as well for estimation purposes. This approach has been developed by Switzer (1980, 1983) and Mardia (1984), who derive allocation rules, their data-based estimation, and associated error rates. Note, however, that while dependence has been introduced here between the feature vectors, no dependence has been introduced into the prior distributions underlying the derivation of the linear discriminant function.

9.60 An alternative approach is thus to develop a Bayes allocation rule (**9.8** and **9.9**) from first principles, modelling departures from independence in the prior distribution of the colours of pixels within a given scene in some way. The most fruitful way of doing this modelling is by means of *Markov Random Fields*. Suppose that $\Pr(\boldsymbol{g})$ defines a probability distribution for $\boldsymbol{g}$ that assigns colours to the pixels in a scene. Let $\boldsymbol{g}_{(j)}$ denote $\boldsymbol{g}$ with $\boldsymbol{g}_j$ deleted, and let $\boldsymbol{g}^{(j)}$

denote the colours of the pixels in a chosen neighbourhood of pixel j. Then the dependence structure is a Markov Random Field if

$$\Pr(\boldsymbol{g}_j | \boldsymbol{g}_{(j)}) = P_j(\boldsymbol{g}_j; \boldsymbol{g}^{(j)}) \tag{9.54}$$

where the function $P_j(.\,;.)$ is specific to the pixel j. The order of the Markov Field is defined to be the order of the neighbourhood chosen.

The general approach is to model the prior distribution $\Pr(\boldsymbol{g})$ of group membership of all pixels by means of an appropriate Markov Random Field, to combine this prior distribution with the likelihood of the data and hence form the posterior distribution $\Pr(\boldsymbol{g}|\boldsymbol{z})$, and then to obtain the estimate $\hat{\boldsymbol{g}}$ as the mode of this distribution. Besag (1986) suggests a particular spatially symmetric model for the prior distribution and describes an iterative algorithm, the ICM algorithm, for obtaining the estimate in a local fashion (i.e. by allocating each pixel individually). Geman and Geman (1984), by contrast, describe a maximum *a posteriori* (MAP) algorithm, directed at finding the global maximum. Besag's method is computationally simpler, but given their reliance on computation both are difficult to describe in compact fashion, so we do not do so here but leave the reader to follow up the individual references. Various modifications and improvements to these algorithms have been subsequently proposed; see McLachlan (1992, Chapter 13) for details and bibliography. See also Ripley (1986, 1988, Chapter 5), and the special issue of the *Journal of Applied Statistics* on Statistical Methods in Image Analysis (1989, **16**, pp. 125–290) for many additional references.

Neural networks

9.61 One consequence of the rapid development of computing power in the 1980s has been the increasing emphasis placed on computer-intensive methods in all areas of Statistics, but particularly in those areas in which derived techniques depend on numerical optimisation of some objective function. Interest in such numerical optimisation has led to a number of innovative ideas in computing, including the development of stochastic optimisation techniques that are modelled on physical or biological processes. Three such methods were briefly touched on in Part 1, namely simulated annealing, neural networks and genetic algorithms (see **4.17**). Neural networks have now found an important role in classification, so it is appropriate to consider them here in a bit more detail. The motivating idea was to mimic the possible learning behaviour of the brain. A number of different models have been put forward to do this (Amit, 1989), but we will focus exclusively on *feed-forward* neural networks based on the *perceptron* model as this is where the major advances in classification have taken place; for a more general account of neural computation, see Hertz *et al* (1991).

Rosenblatt published the *Perceptron Theorem* in 1962, showing that if two data sets were separated by a hyperplane then the perceptron model would find a plane to separate them, but Minsky and Papert (1969) demonstrated that the algorithm would not converge if no separating hyperplane existed. Matters remained more or less at this stage until Rumelhart *et al* (1986) published their 'back propagation' algorithm, since when the subject has proved to be a lively area for research. The original biological or neurophysiological motivating ideas have now faded to a great extent, and all these models are viewed purely as computational algorithms. This is how we treat them here, for the specific objective of classification or discrimination.

9.62 Classification of an individual with feature vector $\mathbf{x}' = (x_1, \ldots, x_p)$ into one of g groups $\pi_1, \ldots, \pi_g$ can be viewed as the mathematical process of transforming the p *input units* $x_1, \ldots, x_p$ into the g *output units* $z_1, \ldots, z_g$ that define the group allocation of the individual, i.e. $z_i = 1$ and $z_j = 0 \ \forall j \neq i$ if the individual is to be allocated to π_i. The *multi-layer perceptron* carries out the transformation by treating the x_i as values of p units in the *input layer*, the z_j as values of g units in the *output layer*, and interposing a number of *hidden layers* of units between these two layers. Usually, each unit in one layer is connected to every unit in adjacent layers and to no other units (although more complicated networks can allow for *skip-layer* connections). The *architecture* of the network is determined by the number of layers, the number of units in each layer, and the connections between units. A simple three-layer network containing four input units, three units in a single hidden layer and full connectivity between layers (but no skip-layer connections) is illustrated in Fig. 9.4. A weight $w_{i(jk)}$ is associated with the connection between the jth unit in the ith layer and the kth unit in the $(i+1)$th layer. The value of any given unit, x_j say, in the ith layer is 'passed on' to the kth unit in the $(i+1)$th layer by transforming it to $f_i(x_j)$ and multiplying it by the appropriate weight. This gives a number of contributions to the kth unit in the $(i+1)$th layer. These are combined additively and a constant α_{ik} can be added on, to give the value $y_k = \alpha_{ik} + \sum_j w_{i(jk)} f_i(x_j)$ for this unit. This process is continued successively from layer to layer until values have been assigned to all units in the network. (Note that the constants α_{ij} can be accommodated by introducing an extra unit in the $(i+1)$th layer that has value 1 and is connected to just the units in this layer with weights α_{ij}.)

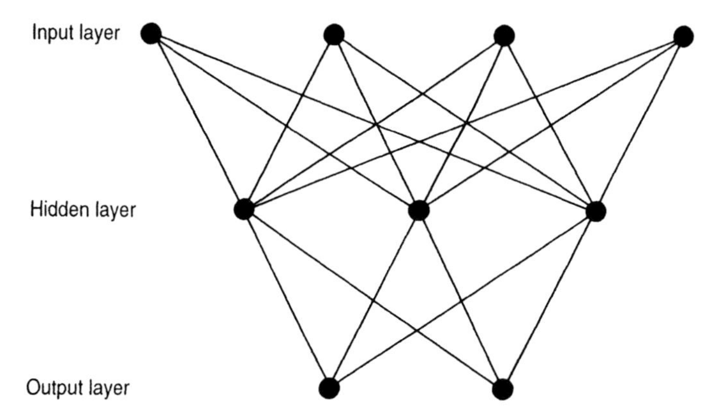

Figure 9.4 A simple multi-layer perceptron.

The functions $f_i(x_j)$ are generally taken to be the identity function in the first layer $[f_1(x_j) \equiv x_j \forall j]$, and threshold functions in the penultimate layer $[f(x_j) = I(x_j > 0)]$ to ensure that the output units have the right form of values. Between hidden layers they are either simple linear or, more usually, logistic functions $[f_i(x_j) = e^{x_j}/(1 + e^{x_j})]$. Determining the best values to use for the weights $w_{i(jk)}$ (and constants α_{ij}) is known as *training the network*. This is generally done by optimising some appropriate objective function which matches the observed and the desired ('target') values of each unit in the network over all n individuals in the training data. The most usual objective function is residual sum of squares, but other possibilities include likelihood (van Ooyen and Nienhuis, 1992), Kullback-Leibler distance (Solla *et al*, 1988) or a robust criterion such as the Huber function (Smith *et al*, 1993).

9.63 We will describe optimisation in the case of the least-squares criterion function. There are n individuals in the training data, the ith of which has feature vector $\boldsymbol{x}_i = (x_{i1}, \ldots, x_{ip})'$. To simplify notation we will redefine quantities in the network slightly. First, we ignore the presence of layers and just focus on units. This means we can remove suffixes relating to layers from all quantities, and write, for example, w_{jk} for the weight connecting units j and k. Secondly, we adopt the terminology used by Smith (1993), writing I_{ij} for the *input value* received by unit j for training set individual i and O_{ij} for the *output value* emanating from the same unit. Thus, $I_{ij} = x_{ij}$ if j is an input unit and $I_{ij} = \sum_k O_{ik} w_{kj}$ otherwise, the summation being over all units in the previous layer connected to j. Similarly, $O_{ij} = I_{ij}$ for an input unit j while $O_{ij} = f(I_{ij})$ otherwise (for the appropriate function f as discussed in **9.62**). If we write T_{ij} for the target output value at unit j for individual i, then the least squares objective function is

$$E = \sum_{i=1}^{n} E_i = \sum_{i=1}^{n} \left[\frac{1}{2} \sum_k (O_{ik} - T_{ik})^2 \right] \tag{9.55}$$

Minimisation of (9.55) can be achieved iteratively using a standard approach such as steepest descent where, in each iteration, all the weights in the network are updated by moving them in the direction of greatest decrease of E.

Identification of this direction requires evaluation of $\frac{\partial E}{\partial w_{kj}}$, where k and j index units in successive layers. Since $E = \sum_{i=1}^{n} E_i$, we thus require to evaluate $\frac{\partial E_i}{\partial w_{kj}}$ for each training set individual i. This expression will differ for output and non-output units. First consider output units. We have, by the chain rule,

$$\frac{\partial E_i}{\partial w_{kj}} = \frac{\partial E_i}{\partial I_{ij}} \frac{\partial I_{ij}}{\partial w_{kj}} = \frac{\partial E_i}{\partial O_{ij}} \frac{\partial O_{ij}}{\partial I_{ij}} \frac{\partial I_{ij}}{\partial w_{kj}} \tag{9.56}$$

and from (9.55)

$$\frac{\partial E_i}{\partial O_{ij}} = (O_{ij} - T_{ij}) \tag{9.57}$$

Also $\frac{\partial O_{ij}}{\partial I_{ij}} = f'(I_{ij})$ and $\frac{\partial I_{ij}}{\partial w_{kj}} = \frac{\partial}{\partial w_{kj}} (\sum_l O_{il} w_{lj}) = O_{ik}$, so that

$$\frac{\partial E_i}{\partial w_{kj}} = (O_{ij} - T_{ij}) f'(I_{ij}) O_{ik} \tag{9.58}$$

This expression can be evaluated for all j, k where j indexes output units and k indexes units in the previous layer.

When j applies to non-output units, it is necessary to re-define $\frac{\partial E_i}{\partial O_{ij}}$ to be

$$\frac{\partial E_i}{\partial O_{ij}} = \sum_l \frac{\partial E_i}{\partial I_{il}} \frac{\partial I_{il}}{\partial O_{ij}} = \sum_l \frac{\partial E_i}{\partial I_{il}} w_{jl} \qquad (9.59)$$

with l indexing units in the succeeding layer. Thus, for non-output units we have

$$\frac{\partial E_i}{\partial w_{kj}} = f'(I_{ij}) O_{ik} \sum_l \frac{\partial E_i}{\partial I_{il}} w_{jl} \qquad (9.60)$$

To state the algorithm in simple terms, first set

$$\partial_{ij} = -\frac{\partial E_i}{\partial I_{ij}} \qquad (9.61)$$

(the minus sign ensuring a downward gradient). Then at each iteration we change w_{kj} to $w_{kj} + \partial_{ij} O_{ik}$, where ∂_{ij} is $-(O_{ij} - T_{ij}) f'(I_{ij})$ for output units j and $f'(I_{ij}) \sum_l \partial_{il} w_{jl}$ for non-output units.

This iterative process is known as the *generalised delta rule*, or the *back-propagation algorithm*. Since the group membership of each training set individual is known, the target outputs from the network are known for these individuals. A forward pass of the algorithm generates a set of derived outputs that are compared with the known targets to yield the error value from (9.55). The reverse pass then propagates the error signal back towards the input whilst computing the gradient vector $-(\frac{\partial E_i}{dw_1}, \ldots, \frac{\partial E_i}{dw_m})$, where m is the total number of weights. This vector yields an updated set of weights through the above formulae, and the next iteration begins.

Second derivatives of E with respect to weights w_{ij} have been given by Bishop (1992), enabling the Hessian matrix to be evaluated for estimation of variances and covariances of weights and for checking on whether the global minimum has been reached. Also, a number of ideas have been put forward to speed up convergence of the process; see Jacobs (1988) for a review.

9.64 The main practical problem is deciding when to stop the process. Clearly, each iteration improves the value of the criterion used and if this criterion is (9.55) then its value is proportional to the resubstitution error rate of the network. One possible strategy is therefore to plot the resubstitution error rate for each iteration, and to stop the process when this rate is sufficiently small. In practice, the complement of the error rate (i.e. the proportion of correctly classified training individuals) is often used instead, and the process stopped when this proportion is sufficiently close to one.

There are several problems with such an approach. First, the proportion of correctly classified training individuals produces a monotonic increasing asymptotic curve (asymptote at one), and such a curve is not ideal for deciding on a stopping rule. (This situation is exactly analogous to the use of R^2 as a criterion for variable selection in multiple regression; see Miller 1990.) Secondly, resubstitution rates are highly biased as measures of performance, and this bias will depend heavily on the architecture of the network (i.e. the number of weights estimated from the data). For these reasons, it is preferable to split the training data arbitrarily into two sets. The above procedure is carried out on the first of these sets (which retains the title 'training set'), while the second set is not used in the training phase but only to evaluate the performance of the network at each iteration. This second set is the 'validation set', and for reasons discussed earlier in the chapter it provides a better estimate of performance of the network. It also tends to peak at a certain number of iterations and thereafter decline (when resubstitution performance is being overfitted), so the process should be stopped when this peak is reached. A typical plot of the two proportions of correct allocation in a practical example is shown in Fig. 9.5, taken from Smith (1993). Here, the cross-validation performance has only been evaluated from iteration 800 onwards, and the graph suggests that training of the network should be stopped after about 25 000 iterations.

The final point with such a procedure is estimation of its performance. Usually this would be done by splitting the data into two portions, training the network on one portion and estimating performance of the trained network on the other. Since the validation set has actually been used to determine the stopping rule, however, the performance of the network as measured by that of the validation set on the graph is still a biased estimate. Logically we should therefore split the training data into *three* sets: a training set, a validation set, and a test set. The network is trained on the first set, the stopping rule is determined from the second set, and the performance of the chosen network evaluated on the third set. (Another difficulty is that when the data are remotely sensed and strongly autocorrelated, as well as falling into compact geographical groups, even randomly selected training and test sets can be misleading.) Little work appears to have been done to date, however, on investigating such questions or on implementation of other approaches to the evaluation of classification performance of a network.

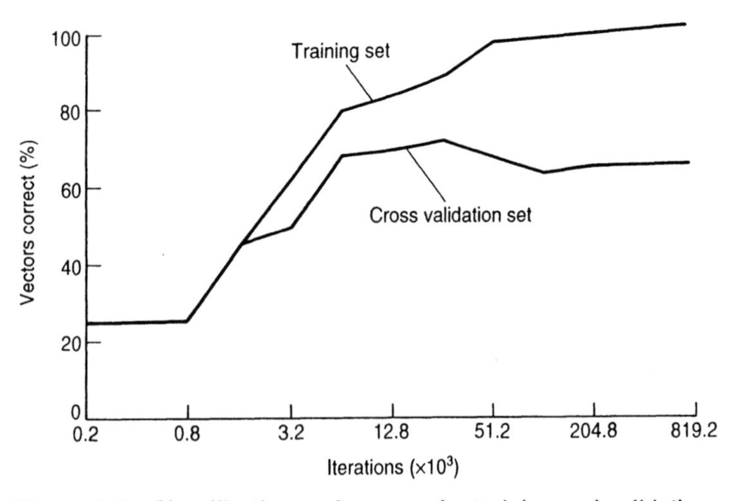

Figure 9.5 Classification performance for training and validation sets as training progresses.

Clearly many other open questions remain in the area of neural networks, and the above account has only provided a general introduction to the topic. For fuller accounts, and for an indication of remaining problems, see Ripley (1993, 1994), Cheng and Titterington (1994) and Michie *et al* (1994). In the end, however, it should be emphasised that all networks are 'black-box' routines, i.e. they only give the end result and no information is available about *how* the classification has been achieved.

Example 9.9
Smith *et al* (1993) have investigated the use of neural networks for classification of underwater cetacean sounds. The data arose from recordings of a family of Atlantic Bottlenose dolphins taken in the inshore waters around Newport, Wales. The initial recordings were made on an analogue compact cassette tape, the raw data were processed by means of the discrete Fourier transform into the frequency domain, and dimensionality was reduced by selecting 32 frequency bands for each of nine time points. Full details of all these steps are given by Smith (1993); for present purposes it is sufficient to note that each training set individual thus comprises $32 \times 9 = 288$ values $(x_1, \ldots, x_{288})$. Given the noise inherent in the data collection process, it was desired to build a classifier that would distinguish between the four states 'no signal', 'whistle', 'click' and 'bark' (the latter three being the most common sounds emitted by dolphins). Thus a neural network needs to convert the 288 input units $(x_1, \ldots, x_{288})$ for any given recording into three output 0/1 units which determine the classification of that recording according to the coding: 0,0,0 for 'no signal'; 1,0,0 for 'whistle'; 0,1,0 for 'click'; and 0,0,1 for 'bark'.

Smith *et al* investigated different network architectures and different objective functions for efficiency of classification. The objective functions compared were least squares (equation 9.55, denoted L_2), least absolute deviations (L_1), and the Huber (1981) criterion. The latter uses a data-based mixture of L_1 and L_2: L_2 if $|O_{ij} - T_{ij}| \leqslant k$ and a scaled version of L_1 if $|O_{ij} - T_{ij}| > k$; values $k = 0.2, 0.3, 0.4, 0.5$ were all investigated and these are denoted $H_{0.2}, \ldots, H_{0.5}$ respectively. The architectures compared, all designed to reduce 288 input units to three output units, were as follows: the simple perceptron, in which there are no hidden units ($288 \rightarrow 3$); two multilayer perceptrons, the first with 10 units in a single hidden layer ($288 \rightarrow 10 \rightarrow 3$) and the second with 20 units in a single hidden layer ($288 \rightarrow 20 \rightarrow 3$); and two time-delay neural networks (in which the connections between units take the temporal structure of the data into account), the first having 35 and nine units in successive hidden layers ($288 \rightarrow 35 \rightarrow 9 \rightarrow 3$) and the second having 50 units in the first hidden layer and nine again in the second ($288 \rightarrow 50 \rightarrow 9 \rightarrow 3$). These five architectures are denoted *P*, *MLP*1, *MLP*2, *TDNN*1 and *TDNN*2 respectively.

The available data consisted of 200 vectors, each containing the 288 input values. These data were randomly divided into two sets, a training and a test set, each of 100 individuals. The network was trained on the first set and its performance was evaluated on the second set, the cross-validation method outlined earlier being incorporated into the training

Table 9.4 Percentage correct classification over the test set for dolphin sounds, for five random weight starts in each architecture/objective function combination. Taken with permission from Smith *et al* (1993)

Network		L_1	L_2	$H_{0.5}$	$H_{0.4}$	$H_{0.3}$	$H_{0.2}$
P	Mean	58.95	59.20	59.00	59.00	59.65	59.55
	Best	59.50	59.50	59.25	59.25	60.50	61.00
	S.D.	0.48	0.27	0.25	0.43	0.55	0.87
MLP1	Mean	70.20	67.90	69.55	69.80	70.60	68.00
	Best	71.25	69.50	70.25	70.50	71.75	69.75
	S.D.	0.74	1.28	0.69	0.54	0.80	1.19
MLP2	Mean	68.20	68.55	67.20	67.10	68.75	66.90
	Best	69.50	69.50	69.25	68.00	69.75	68.00
	S.D.	0.89	0.80	1.80	0.65	1.05	0.91
TDNN1	Mean	71.25	71.20	74.35	73.00	73.40	72.90
	Best	75.75	75.50	76.50	77.50	78.25	74.50
	S.D.	3.08	2.99	1.52	3.02	2.79	1.17
TDNN2	Mean	73.10	70.95	71.95	71.35	71.40	69.30
	Best	77.25	72.00	72.75	73.25	73.75	73.75
	S.D.	2.95	1.67	0.76	1.53	1.39	2.94

algorithm. Table 9.4 summarises the results, each method/architecture combination being run five times with different random weight starts. The results show that classification performance seems to improve in general as the number of hidden layers increases (and the temporal structure in the data is allowed for), although it should be noted that variability also increases at the same time. As regards objective functions, performance of the Huber criterion is generally better by the order of 1–3 per cent over the L_1 and L_2 functions, although L_1 turns out to be the winner in the larger *TDNN* architecture. While the *bending point k* for the Huber function is clearly important, no single value is superior in all cases and 0.3–0.5 seems to be a good all-round range for these data.

9.65 As can be seen from the above overview, very many different methods and techniques are now available for discriminant analysis, and the practical user can be forgiven feelings of confusion or indecision when faced with the need to select a method for use on a particular set of data. So what considerations might affect his or her choice?

Clearly, an overriding one will be the availability of software. Here, it must be confessed that despite all the research done over the past 30 years, relatively few of the methods have found their way into standard statistical packages such as SAS, SPSS, BMDP, MINITAB or GENSTAT, while no special purpose software appears to have been written for discriminant analysis in the way that it was for multidimensional scaling [MDS(X)] or cluster analysis [CLUSTAN]. Consequently, the analyst may be forced into considering only those methods that are most easily accessible. The linear and quadratic discriminant functions can be obtained with most packages, and both resubstitution and leave-one-out error estimates are usually provided with these techniques. Resampling methods such as bootstrapping can generally be implemented fairly easily on standard packages, if desired, which provides extra scope for error rate estimation or sensitivity analysis (but care must be taken when using such methods on eigenvalue/eigenvector operations because of problems caused by arbitrary sign choice and reversals of close eigenvalues). Augmenting the linear discriminant function (**9.23**; Knoke, 1982; Vlachonikolis and Marriott, 1982) will often provide good results for mixtures of binary and continuous variables, and variable selection routines are usually also available alongside their regression counterparts (but note the cautionary remarks in **9.53** regarding assessment of performance of such procedures). Beyond these functions, accessibility becomes patchy. Logistic discrimination is probably the next most available method but, given its computational basis, leave-one-out error estimates may not be provided automatically. Indeed, these estimates are often only obtainable by physically rerunning the program on leaving each individual out of the training set, which is a crude and time-consuming process. Kernel discriminant analysis has been made more easily accessible through the ALLOC80 package (**9.27**), which does allow leave-one-out error rate estimation, but other techniques have generally only been implemented in special-purpose programs written by their originators.

Other important practical considerations in the choice of discrimination method are the relationship between sample sizes and number of variables, and the ease of interpretation of the computed functions. In general, sample sizes should be considerably greater than the number of unknown parameters, and if they are not then the resulting discriminant functions may be unstable. For example, the quadratic function may not give good results when sample sizes are small, and the linear discriminant function may be preferable in such cases even when necessary assumptions (e.g. equal population dispersion matrices) do not hold. Also, in general, the simpler the function, the more easily it is understood and the more interpretable will be its parameters. At the extreme come the 'black-box' techniques, which give no information about the discriminant functions but simply provide the end-product classifications.

In the end, the user may have to try a selection of methods on a given data set and choose among them on the basis of their interpretabilities and their classification performances.

9.66 The theory developed in this chapter has assumed that any classification is a *forced* one (i.e. every individual $\boldsymbol{x}$ *must* be classified into one of the *a priori* groups), and that the correct group identification is available for all individuals in the training data. In practice, either of these assumptions may be open to doubt, so to conclude the chapter we mention briefly what actions can be taken in such cases.

First, occasions may arise when the researcher wishes to reserve judgement on the classification of an individual until further, more conclusive, data have been gathered on it. A simple procedure is to decide what level of misclassification probabilities can be tolerated, use these to define the regions R_i of the sample space for allocation of $\boldsymbol{x}$ (**9.9**), and designate the remainder of the sample space as a 'region of doubt' in which no classification is performed. Alternatively, one can look for a constrained classification rule in which restrictions on misclassification probabilities are explicitly built in (Anderson, 1969; Habbema *et al*, 1974; McLachlan, 1977).

Next, there may be occasions when the individual $\boldsymbol{x}$ does not come from any of the *a priori* groups, so no allocation should be made. Various *typicality indices* have been defined, measuring how typical $\boldsymbol{x}$ is of each group in turn. If this typicality index is low for each group, then no classification should be made for that individual. Possible indices are ones based on Mahalanobis distance (Cacoullos, 1965a,b; Srivastava, 1967; McLachlan, 1992, p183), or ones based on predictive assessment (Aitchison and Dunsmore, 1975, Chapter 1; Aitchison *et al*, 1977; Exercise 9.8).

Turning next to the training data, it may happen that some of the individuals have been initially misallocated and hence have the wrong group label. Various authors have studied the consequences of such misclassifications on discrimination (see, for example, Lachenbruch, 1979). The general conclusion (McLachlan, 1992, p36) is that ignoring these errors has little effect when the misclassifications are non-random (i.e. when the misclassification probability depends on the feature vector $\boldsymbol{x}$), but can be serious when the misclassifications are random (i.e. when the misclassification probability does not depend on $\boldsymbol{x}$). Krishnan and Nandy (1987) and Titterington (1989) describe methods for taking random misclassification of training data into account in the construction of an allocation rule.

Finally, the problem of unclassified observations among the training data can arise, for example, in medical data with post-mortem diagnosis or in remote sensing applications. An empirical procedure for handling such observations in the case of linear discriminant analysis has been proposed by McLachlan (1975): first estimate the linear discriminant function using the observations of known origin only and use this function to classify the other observations, then incorporate these observations into their assigned groups and re-estimate the discriminant function, and continue iterating on these steps if further unclassified observations become available. However, if unclassified observations are sampled randomly from the mixture of populations, then they contain information about the distributions associated with each of the populations as well as about the mixing proportions. A better approach might therefore be to analyse the data by mixture separation methods, and to substitute the resulting estimators into the chosen discriminant function. Some relevant results have been given by Ganesalingam and McLachlan (1978, 1979) and O'Neill (1978) for normal discrimination, by Murray and Titterington (1978) for kernel discrimination, and by Anderson (1979) for logistic discrimination. Mixture separation is one of the possible methods of cluster analysis, however, so details of this approach are deferred to the next chapter.

Exercises

9.1 Suppose that $\boldsymbol{B} = \boldsymbol{A} + \boldsymbol{u}\boldsymbol{u}'$, where $\boldsymbol{A}$ is a non-singular $(p \times p)$ matrix and $\boldsymbol{u}$ is a $(p \times 1)$ vector. Show that $\boldsymbol{B}^{-1} = \boldsymbol{A}^{-1} - k\boldsymbol{A}^{-1}\boldsymbol{u}\boldsymbol{u}'\boldsymbol{A}^{-1}$, where $k = 1/(1 + \boldsymbol{u}'\boldsymbol{A}^{-1}\boldsymbol{u})$. (This result is useful in some of the questions below).

(Bartlett, 1951)

9.2 Let $U(\boldsymbol{x})$ be given by equation (9.41). Show that $U(\boldsymbol{x})$ has a $N(\frac{1}{2}\Delta^2, \Delta^2)$ distribution if $\boldsymbol{x}$ is distributed $N(\boldsymbol{\mu}_1, \boldsymbol{\Sigma})$, and a $N(-\frac{1}{2}\Delta^2, \Delta^2)$ distribution if $\boldsymbol{x}$ is distributed $N(\boldsymbol{\mu}_2, \boldsymbol{\Sigma})$, where $\Delta^2 = (\boldsymbol{\mu}_1 - \boldsymbol{\mu}_2)'\boldsymbol{\Sigma}^{-1}(\boldsymbol{\mu}_1 - \boldsymbol{\mu}_2)$.

Hence show that, if $U(\boldsymbol{x}) > c$ is the function used to discriminate between the two normal populations above (cf. **9.34**), then the optimal error rates are given by $p_o(2|1) = \Phi[(c - \frac{1}{2}\Delta^2)/\Delta]$ and $p_o(1|2) = \Phi[-(c + \frac{1}{2}\Delta^2)/\Delta]$, where $\Phi(.)$ denotes the standard normal distribution function.

9.3 Suppose that the p-element $\boldsymbol{x}$ is distributed $N(\boldsymbol{\mu}, \boldsymbol{\Sigma}_i)$ in population π_i $(i = 1, 2)$, where $\boldsymbol{\Sigma}_i = \sigma_i^2[(1 - \rho_i)\boldsymbol{I} + \rho_i\mathbf{1}\mathbf{1}']$ and $\mathbf{1}$ denotes the p-vector all of whose elements are 1. Show that the optimal (i.e. Bayes) discriminant function is given, apart from an additive constant, by $-\frac{1}{2}(c_{11} - c_{12})Q_1 + \frac{1}{2}(c_{21} - c_{22})Q_2$, where $Q_1 = \boldsymbol{x}'\boldsymbol{x}$, $Q_2 = (\mathbf{1}'\boldsymbol{x})^2$, $c_{1i} = [\sigma_i^2 + (1 - \rho_i)]^{-1}$, and $c_{2i} = \rho_i[\sigma_i^2(1 - \rho_i)\{1 + (p - 1)\rho_i\}]^{-1}$.

(Bartlett and Please, 1963; Han, 1968)

9.4 Suppose now that the p-element $\boldsymbol{x}$ is distributed $N(\boldsymbol{\mu}_i + \boldsymbol{B}'\boldsymbol{\theta}_i, \boldsymbol{\Sigma})$ in population π_i, where $\boldsymbol{B}'$ is a known $(p \times m)$ matrix and $\boldsymbol{\theta}_i$ is a vector of m unknowns $(i = 1, 2)$. Show that the form of the optimal discriminant function for this case is

$$(\boldsymbol{\mu}_1 - \boldsymbol{\mu}_2)'(\boldsymbol{\Sigma}^{-1} - \boldsymbol{\Sigma}^{-1}\boldsymbol{B}'(\boldsymbol{B}\boldsymbol{\Sigma}^{-1}\boldsymbol{B}')^{-1}\boldsymbol{B}\boldsymbol{\Sigma}^{-1})\boldsymbol{x}$$

(Rao, 1966a)

9.5 Consider testing the null hypothesis that $\boldsymbol{x}, \boldsymbol{y}_1, \ldots, \boldsymbol{y}_n$ are i.i.d $N(\boldsymbol{\mu}_1, \boldsymbol{\Sigma})$, independently of $\boldsymbol{z}_1, \ldots, \boldsymbol{z}_m$, which are i.i.d. $N(\boldsymbol{\mu}_2, \boldsymbol{\Sigma})$, against the alternative hypothesis that $\boldsymbol{y}_1, \ldots, \boldsymbol{y}_n$ are i.i.d $N(\boldsymbol{\mu}_1, \boldsymbol{\Sigma})$, independently of $\boldsymbol{x}, \boldsymbol{z}_1, \ldots, \boldsymbol{z}_m$, which are i.i.d. $N(\boldsymbol{\mu}_2, \boldsymbol{\Sigma})$. Show that the likelihood ratio statistic for these composite hypotheses is

$$\frac{1 + \frac{m}{m+1}(\boldsymbol{x} - \bar{\boldsymbol{z}})'\boldsymbol{\Sigma}^{-1}(\boldsymbol{x} - \bar{\boldsymbol{z}})}{1 + \frac{n}{n+1}(\boldsymbol{x} - \bar{\boldsymbol{y}})'\boldsymbol{\Sigma}^{-1}(\boldsymbol{x} - \bar{\boldsymbol{y}})}$$

(Anderson, 1958)

9.6 Let $\boldsymbol{y}_1, \ldots, \boldsymbol{y}_n$ be a p-variate sample from π_1, and $\boldsymbol{z}_1, \ldots, \boldsymbol{z}_m$ a corresponding p-variate sample from π_2. Write $\bar{\boldsymbol{y}}, \bar{\boldsymbol{z}}$ for the two sample means, $\boldsymbol{W}$ for the pooled within-sample covariance matrix (divisor $n + m - 2$), and $D^2 = (\bar{\boldsymbol{y}} - \bar{\boldsymbol{z}})'\boldsymbol{W}^{-1}(\bar{\boldsymbol{y}} - \bar{\boldsymbol{z}})$ for the between-sample Mahalanobis squared distance. Now suppose that $\boldsymbol{y}_j$ is omitted from the data, set $\boldsymbol{u} = \boldsymbol{y}_j - \bar{\boldsymbol{y}}$, and denote the new values of the above statistics by $\bar{\boldsymbol{y}}_{(j)}$, $\boldsymbol{W}_{(j)}$ and $D^2_{(j)}$. Show that:

(1) $\bar{\boldsymbol{y}}_{(j)} = \bar{\boldsymbol{y}} - \boldsymbol{u}/(n - 1)$,
(2) $(n + m - 3)\boldsymbol{W}_{(j)} = (n + m - 2)\boldsymbol{W} - n\boldsymbol{u}\boldsymbol{u}'/(n - 1)$,

(3) $(n+m-3)^p|W_{(j)}| = (n+m-2)^p|W|[1 - n(n-1)^{-1}(n+m-2)^{-1}(y_j - \bar{y})'W^{-1}(y_j - \bar{y})]$.

(Lachenbruch, 1975; Campbell, 1985)

9.7 For the set-up as in Exercise 9.6, show that the linear discriminant function when y_j is omitted may be written as

$$(\bar{y} - \bar{z})'W_{(j)}^{-1}\left[y_j - \frac{1}{2}(\bar{y} + \bar{z}) + \frac{u}{2(n-1)}\right] - \frac{u}{(n-1)}$$

Derive corresponding expressions for the quadratic discriminant function and the hypothesis-testing function (9.28).

(Lachenbruch, 1975; Campbell, 1985)

9.8 For g-group discrimination, an individual with observation vector y is said to be *more typical* of group π_i than an individual with observation vector x if $f(y|\pi_i) > f(x|\pi_i)$. The set $R_i(x)$ of all observation vectors more typical of π_i than x is thus the set of all y satisfying this criterion, and the *atypicality* for π_i of x is the probability $\Pr[R_i(x)|\pi_i]$ of obtaining an observation vector more typical than x.

Show that, for predictive discrimination with continuous p-variate normal data, the predictive distributions (9.26) lead to atypicality

$$B\left(\frac{1}{2}p, \frac{1}{2}[n_i - p]; \frac{w_i[x]}{w_i[x] + [n_i^2 - 1]/n_i}\right)$$

for π_i of x, where $w_i[x] = (x - \bar{x})'S_i^{-1}(x - \bar{x})$ and B denotes the incomplete beta function defined by

$$B(a, b; c) = \int_0^c u^{a-1}(1-u)^{b-1}du/B(a, b) \; (0 \le c \le 1)$$

(Aitchison *et al*, 1977)

9.9 Suppose that x is a vector of p binary variables, each scored 0 or 1. A training set of n individuals thus yields n patterns of binary variable values. Let $p(x_0)$ denote the probability that an individual chosen at random from a certain population exhibits the pattern x_0. Consider estimation of $p(x_0)$ from the training set by $\hat{p}(x_0) = n^{-1}\sum_{j=0}^{l} w_j n_j(x_0)$, where $n_j(x_0)$ is the number of individuals differing from x_0 in their patterns by exactly j values (i.e. the *j-near-neighbours* of x_0) and w_j are weights to be determined.

Find the minimum mean-squared error weights, i.e. the weights minimising $\sum E\{\hat{p}(x_0) - p(x_0)\}^2$ where the summation is taken over all possible patterns x_0.

(Hall, 1981)

9.10 Consider two-group linear discriminant analysis, with training samples of sizes n_1 and n_2 from π_1, π_2 respectively and allocation of the p-variate x to

π_1 or π_2 according to $V(\boldsymbol{x}) \geqslant 0$ or < 0 respectively, where $V(\boldsymbol{x})$ is defined by (9.42). Consider estimating the error rate from π_1. The class of *normally smoothed estimators* is defined by $\alpha^S = \sum_{j=1}^{n_1} \Phi(-V[\boldsymbol{x}_j]/bD)$, where $\Phi(.)$ is the standard normal distribution function, $D^2 = (\bar{\boldsymbol{x}}_1 - \bar{\boldsymbol{x}}_2)' \boldsymbol{W}^{-1} (\bar{\boldsymbol{x}}_1 - \bar{\boldsymbol{x}}_2)$ is the Mahalanobis squared distance between the two training samples, and b is the smoothing parameter to be estimated.

Approximating the distribution of $\boldsymbol{x}_j$ by a normal distribution with mean $\bar{\boldsymbol{x}}_1$ and dispersion matrix $(n_1 - 1)\boldsymbol{W}/n_1$, show that

$$E[\alpha^S | \bar{\boldsymbol{x}}_1, \bar{\boldsymbol{x}}_2, \boldsymbol{W}] = \Phi\{-[n_1/(n_1 b^2 + n_1 - 1)]^{1/2} D/2\}.$$

Another estimate is obtained from **9.35** as $\alpha^{DS} = \Phi(-\frac{1}{2} D_*)$. Show that

$$E[\alpha^{DS} | \bar{\boldsymbol{x}}_1, \bar{\boldsymbol{x}}_2, \boldsymbol{W}] = \Phi\{-[(n_1 + n_2 - p - 3)/(n_1 + n_2 - 2)]^{1/2} D/2\}.$$

Hence deduce $b = \{[(p+2)(n_1 - 1) + n_2 - 1]/[n_1(n_1 + n_2 - p - 3)]\}^{1/2}$ as an estimate of the smoothing parameter.

(Snapinn and Knoke, 1985)

10

Cluster Analysis

Introduction

10.1 Cluster analysis is the name usually given by statisticians to techniques of data analysis that divide the data into groups. These are usually groups of individual multivariate observations, but cluster analysis of variates is also possible. The techniques are sometimes referred to as classification, but that name is also applied to discriminant analysis, particularly to so-called 'black box' methods of discriminant analysis, which classify new observations without an explicit and intelligible rule. In the pattern recognition literature, cluster analysis is referred to as 'unsupervised learning', as opposed to 'supervised learning', which is used for those techniques in which a training set is available.

The distinction between cluster analysis and discriminant analysis is that the former involves no observations known to belong to specified groups, and usually no prior knowledge of the number of groups. This distinction is sometimes blurred; particularly with remotely sensed data, the number of observations definitely classified may be a tiny proportion of the whole data set, and there may be classes not represented among those observations.

Cluster analysis is also closely linked to ordination methods and methods of graphical presentation (Part 1, Chapters 3 and 5). The aim is subdivision of the data into a number of groups, and graphical methods can do much to support a proposed subdivision. In fact, formal statistical inference about the number of groups and the positions of the boundaries is usually impossible and inappropriate, and some illustration to show how closely the data are clustered is almost essential.

There is an important distinction between the idea of finding 'natural groups' in the data, and that of subdividing the data merely for convenience. Kendall (1980) called the latter 'dissection', and we shall follow that convention. Methods for the two problems are closely related, but the number of groups in dissection is chosen for convenience, with no implication that there is any optimal solution.

Natural groups

10.2 One of the problems of cluster analysis is that of deciding what a scientist means when he or she asks whether there is a natural grouping of the data. A tentative suggestion is that natural groups are patches of high density surrounded by space of lower density in the p-dimensional space defined by the p-variate observations. This suggestion is expressed formally by Hartigan (1975a). Suppose the data are a random sample from a population with density $f(\boldsymbol{x})$. Then a high-density cluster of density f^* in the population is defined as a maximal connected set of the form $\{\boldsymbol{x} | f(\boldsymbol{x}) \geqslant f^*\}$.

As a general definition for all the situations in which cluster analysis is used, this is unsatisfactory.

(1) It applies only to continuous variates with a density.
(2) The number of modes, or the number above a certain level, is not invariant under nonlinear transformations.
(3) Many data sets are not random samples from a population; they may be the complete population, or a selected sample.
(4) Identifying modes is a notoriously difficult problem, even in the univariate case (Silverman, 1981).

Nevertheless, it is useful to think about the association of natural clusters and modes. A projection that shows three distinct and well-separated clusters is reasonably convincing evidence that there are (at least) three natural groups in the data, while a grouping that does not show distinct clusters in any projection is probably not a natural subdivision.

Most methods of cluster analysis tend to work well for clusters of particular shapes. This is important, because scaling the whole data set before analysis will not, as a rule, make the groups compact. Methods that favour spherical clusters and fail with long ellipsoids are generally at a disadvantage. Clusters of unusual form—L-shaped, or annular—are difficult to identify. Most cluster analysis methods will subdivide them; only non-parametric mode-hunting is likely to isolate them correctly, and then only in very large samples.

The great number of methods that have been proposed for cluster analysis can be divided into four groups.

(1) Strictly hierarchical methods, dependent on dissimilarity or similarity measures. These provide a complete scheme of division into all possible numbers of groups from 1 to n, in which the g-group partition is obtained by fusing two of the groups in the $(g + 1)$-group partition, or by splitting one group in the $(g - 1)$-group partition. They are nearly always *agglomerative*; starting from n individuals, they build up groups until all the individuals are in a single group. The opposite (*divisive*) procedure, starting with a single group and fragmenting it, is also possible, but is less prevalent in practice. The characteristic end-product, in either case, is a

dendrogram, a tree diagram showing how the groups build up or how the subdivision proceeds, and the partition into a given number of groups is found by cutting the dendrogram at an appropriate level. These methods started to appear mainly in the ecological and taxonomic journals in the 1950s. They remain the most popular group of techniques for cluster analysis.

(2) Optimal partitioning. The data are subdivided into g groups to optimise some criterion, often based on the within-group covariance matrix. MacQueen (1967) introduced the k-means procedure, and many variations have since been suggested. This is not a hierarchical method in the strict sense; the optimal division into $g + 1$ groups cannot, in general, be derived by splitting one group in the solution with g groups.

(3) Fitting distribution mixtures. In practice, this is confined to fitting mixtures of multivariate normal distributions, with different means, and covariance matrices either separately estimated or assumed to be the same. The end-product is not a partitioning of the data, but a set of posterior probabilities of group membership; these can, of course, be used to assign individuals to the groups for which the probabilities are highest. The idea goes back to Pearson (1894), who fitted univariate normal mixtures by the method of moments. The first practical algorithm for maximum likelihood fitting was due to Wolfe (1965).

(4) Non-parametric methods involving local density estimation. Wishart (1969) first developed an algorithm for direct search for modes. Since the early 1980s, many sophisticated and computer-intensive methods have been suggested that avoid unrealistic distributional assumptions and that do not use distance measures determined before the clusters are formed. They have the disadvantage that the data structure is usually not very apparent from the resulting classification.

Agglomerative hierarchical clustering

10.3 The algorithm for agglomerative hierarchical clustering is extremely simple. The original n items are characterised by a dissimilarity matrix, regarded as a distance matrix in some high-dimensional space. (Sometimes a similarity matrix is used, but it is easy to convert from one to the other; see Part 1, Chapter 5.) The two nearest items are joined, and a new $(n - 1) \times (n - 1)$ dissimilarity matrix is formed. This step is repeated until the data are condensed to a single group. The record shows the level of dissimilarity at which each join is made, and can be used to construct a dendrogram; examples are given later.

The procedure requires an appropriate measure of:

(1) the distance between pairs of the original items, represented, as a rule, by p-variate vectors;

(2) the distance between two groups, each consisting of one or more of the original observations.

It is worth emphasising that once the data have been used to construct a dissimilarity matrix, the original observations are forgotten. Apart from the difficulty of computing with two sets of data, it could be quite misleading to calculate the distance between groups using the original coordinates. Unless the dissimilarities have been calculated as Euclidean distances in terms of the coordinates, and there are no missing data, the dissimilarity matrix will not represent a set of Euclidean distances in p-dimensional space, or, perhaps in any Euclidean space, and reference back to the data could lead to contradictions.

The great advantage of this hierarchical distance class of methods is the flexibility in the use of dissimilarities. In terms of the original variables, missing data present a problem, and combinations of continuous, binary and multicategory variables can hardly be analysed in a reasonable way. These difficulties disappear when the original coordinates are replaced by a dissimilarity matrix. Provided the dissimilarities are a fair reflection of the perceived relationships among the objects, the features of the original data can be ignored.

The disadvantage is that dissimilarities must be decided before the grouping is known, and they depend on the scale of the original variables and on the correlations between them. If, for example, the data were continuous and roughly normally distributed, the Mahalanobis distance might be a suitable dissimilarity measure. As the groups are unknown, this distance cannot be calculated, and standardising on the basis of the whole data set may lead to long, thin clusters, and to inefficient clustering.

The choice of an appropriate transformation of the dissimilarity matrix when two groups are combined also has an important effect on the clusters that are eventually recovered. There are two ways in which this happens.

(i) Some methods tend to produce clusters of approximately equal size, and small clusters, even if well separated from the rest, are absorbed into larger groups. Thus, some of the structure in the data may be missed. This tendency is usually an advantage in dissection problems; a classification is more convenient if the classes are roughly equal in size.

(ii) A few points intermediate between well-defined clusters may, in some methods, cause them to merge. This feature is referred to as 'chaining'; it is often thought of as a lack of robustness; a small change in the data gives a different outcome. Again, it is not necessarily a disadvantage; in taxonomic problems the discovery of a 'missing link' may quite reasonably change the perception of the relationships among species.

Divisive methods will not be discussed further in this section. They are far more computer intensive than agglomerative methods, and seem to have no clear advantages. It is sometimes claimed that divisive techniques give a clearer picture of the early stages of the divisive procedure, when the number of groups is small, while agglomerative methods may have accumulated inaccuracy by the time they reach this stage, but that has never been established. An early paper by Edwards and Cavalli-Sforza (1965) discussed the technique. They attempted all possible dichotomies of the data at the first stage, followed by all possible subdivisions of the two groups formed, and so on; the task rapidly became impossible, even for quite small data sets. MacNaughton-Smith *et al* (1964) derived a workable—but sub-optimal—algorithm, and this is the basis for the divisive procedure DIANA described by Kaufman and

Rousseeuw (1990). Guénoche *et al* (1991) describe another divisive algorithm.

The dendrogram

10.4 A hierarchical clustering is usually represented by a *dendrogram*, or tree diagram. This is a structure familiar to taxonomists, having the same form as the phylogenic trees used to represent relationships among groups of organisms, or the identification trees of botanists.

Conventions vary, but typically the dendrogram is presented with the 'root' at the top, and the ordinate scale showing the dissimilarity level at which each pair of groups became joined.

Example 10.1
Figure 1.4 in Part 1, reproduced here as Fig. 10.1, gives an example of a dendrogram based on the data in Table 1.5 (Part 1). This represents a nearest neighbour, or single linkage, cluster analysis (see **10.9** below). The first step is to calculate a distance matrix from the data of Table 1.5. This was done by standardising each variable to unit variance, and then calculating the Euclidean distance (Manly, 1986, pp.45–46) between every pair out of the seven species: Modern dog (Md), Golden jackal (Gj), Chinese wolf (Cw), Indian wolf (Iw), Cuon (C), Dingo (D), Prehistoric dog (Pd); to give the dissimilarity matrix:

	Md	Gj	Cw	Iw	C	D	Pd
Md	—	2.07	5.81	3.66	1.63	1.68	0.72
Gj	2.07	—	7.69	5.49	3.45	3.44	2.58
Cw	5.81	7.69	—	2.31	4.94	4.55	5.52
Iw	3.66	5.49	2.31	—	3.14	2.37	3.49
C	1.63	3.45	4.94	3.14	—	1.80	1.38
D	1.68	3.44	4.55	2.37	1.80	—	1.84
Pd	0.72	2.58	5.52	3.49	1.38	1.84	—

In Fig. 10.1, the ordinate scale shows the distance at which groups join. The first step combines Modern and Prehistoric Thai dogs, at a level $d = 0.72$. The next step combines them with Cuon, at a level 1.38. This is the distance between Cuon and Prehistoric dog; the distance between Cuon and Modern dog is greater. The level at which two items join the same group, say d_{ij}^+, depends on the cluster method used, but in general is different from their dissimilarity d_{ij}—in fact, for single linkage clustering, $d_{ij}^+ \leqslant d_{ij}$.

The distances d_{ij}^+ are always symmetric, and satisfy the metric inequality

$$d_{ij}^+ \leqslant d_{ik}^+ + d_{jk}^+.$$

They also satisfy the *ultrametric inequality*,

$$d_{ij}^+ \leqslant \max(d_{ik}^+, d_{jk}^+).$$

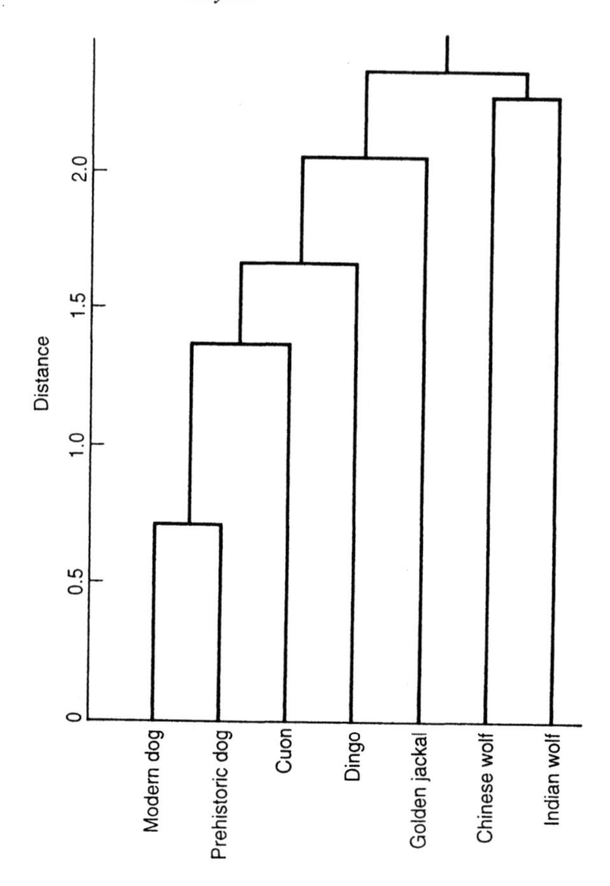

Figure 10.1 Dendrogram from nearest neighbour cluster analysis of canine data. Reproduced with permission from Manly (1986).

This implies that the three distances form an isosceles triangle with one side shorter than the other two. For example, Modern dog and Prehistoric dog join at $d^+ = 0.72$, and each of them joins Cuon at level $d^+ = 1.38$.

Figure 10.1 is constructed from the bottom up, but the natural way to read it is from the top down. The data are divided into two groups by cutting between 2.37 (where the wolves join the rest, to form one single group) and 2.31 (where the wolves join each other). The two groups are then the wolves and the others. A division into three groups would give two groups with a single item, Chinese and Indian wolf, and one group containing the other five items.

Notice that the ordering of the items along the abscissa scale is largely arbitrary. The dendrogram is not a rigid tree; rather, it can be thought of as a hanging mobile, free to rotate about any of its joints. It gives a complete picture of the clustering process, but not of the dissimilarity matrix.

The information in the dendrogram may be presented in other ways. Kaufman and Rousseeuw (1990) describe variants they call banners, icicle plots, and Ward-type plots. These are less familiar, and less attractive, than the dendrogram.

Dissimilarities between individual items; continuous variables

10.5 In one sense, the choice of a similarity or dissimilarity measure between items is not the job of the data analyst. A measure is 'good' if it accurately reflects the judgement of experts in the field, and a final decision must satisfy that criterion. Occasionally, data consist of subjective assessments of similarity or dissimilarity. More often, the data are observations on a number of variables, which may be continuous, binary, categorical or ordered categorical, or a mixture of some or all of these. This data set has to be transformed into dissimilarities in a way that will satisfy the experts, and problems arise that have no clear cut solutions, but must be resolved in accordance with the characteristics of the particular data set.

In this chapter, all measures of similarity and dissimilarity will be assumed symmetric, $d_{ij} = d_{ji}$, $s_{ij} = s_{ji}$.

Consider first the case of continuous variables.

(1) Scaling and transformations. Usually, variables with highly skewed distributions should be transformed to a roughly symmetrical form. Further, unless all the variables are of the same type with similar variability, linear scaling is needed—compare the corresponding problem in principal component analysis (Part 1, Chapter 4). Usually, variables are standardised to have equal variance, but some authors prefer standardisation by a robust measure of scale (Kaufman and Rousseeuw, 1990) or by range (Milligan and Cooper, 1988). The former choice gives extra weight to variables with outliers, while the latter tends to downweight them. If data are examined, checked for outliers, and transformed if necessary, the difference is probably unimportant. Optimal scaling is discussed by van Buuren and Heiser (1989).

(2) Choice of variables. The choice of variables will determine what features control the grouping. If a number of highly correlated variables are included, they will tend to dominate the clustering procedure. This is not necessarily a bad feature. Variables are often correlated because they take different values in different groups. Recent papers include Fowlkes *et al* (1988) and Green *et al* (1990).

 Some care should be taken to avoid unwanted clusters. The patients admitted to hospital with a particular disorder often constitute two natural clusters, male and female. Analysis of medical records might show these clusters, which should not be interpreted as two variants of the disorder—as has actually happened in some analyses.

(3) Weighting. The choice of weights for the different variables should reflect their relative importance in deciding on the similarity or dissimilarity of

individuals. Often weights are taken to be equal. This may be reasonable, but it should be a conscious decision. See De Soete *et al* (1985) and Milligan (1989) for more technical discussions.

(4) Distance. Gower (1985) lists ten different distance or dissimilarity measures for continuous variables. For use as dissimilarity coefficients, it is convenient to consider an *average* of the distances along the different axes; thus the ordinary Euclidean distance is replaced by

$$d_{ij}^2 = \frac{1}{p}\sum_{k=1}^{p}(x_{ik} - x_{jk})^2 \qquad (10.1)$$

The inclusion of the factor $1/p$ means that dissimilarities do not increase indefinitely as extra (standardised) variables are included, and if observations are missing, p can be replaced by the number of differences actually in the sum (see Part 1, Chapter 5).

The Manhattan or city-block distance is given by

$$d_{ij} = \frac{1}{p}\sum_{k=1}^{p}|x_{ik} - x_{jk}| \qquad (10.2)$$

This is an L_1 metric, and is usually thought to be more robust against outliers than L_2 metrics such as the Euclidean distance; see Kaufman and Rousseeuw (1990), but see also the review of that book by Sarle (1991).

The Minkowski metric

$$d_{ij}^r = \frac{1}{p}\sum_{k=1}^{p}|x_{ik} - x_{jk}|^r \qquad (10.3)$$

includes both the Euclidean and city-block distances, and satisfies the metric inequality provided $r \geqslant 1$. It is seldom used except for these two cases.

The Canberra metric is associated with the work of Lance and Williams (1967a), though in fact it was first used by Bray and Curtis (1957):

$$d_{ij} = \frac{1}{p}\sum_{k=1}^{p}\frac{|x_{ik} - x_{jk}|}{|x_{ik}| + |x_{jk}|} \qquad (10.4)$$

This is the form suggested by Gower (the original coefficient was unsuitable for negative observations). The idea behind the measure is that standardisation depends only on the two items involved; if further data are collected, they can be incorporated into the dissimilarity matrix without changing the values of the existing entries. It can be used for continuous or discrete variables, or for a mixture of the two.

(5) Missing values. Almost all large multivariate data sets involve missing values, and discarding all items with data missing on any variable can lose useful information. There are three types of missing observations:
(i) Data missing at random. This implies that the probability of a missing value of x_i is independent of the value of x_i or of the other variables recorded (or unrecorded) on the same individual; the probability is usually different for different variables. In this situation, imputation is

possible, and is often used in regression or discrimination problems. In cluster analysis, imputation is inappropriate and can be misleading. It leads to underestimation of the distances of points with missing values from their neighbours (see Part 1, Chapter 5).
(ii) Data structurally missing. If x_i is a binary variable representing presence or absence of a certain feature, and x_j is a measurement on that feature, $x_i = 0$ obviously entails that x_j is missing. This type of missing data is common in numerical taxonomy.
(iii) Data unrecorded, but not necessarily at random. In medical statistics, the fact that a test was not carried out may imply either that the patient was not well enough to take it, or that it was considered unnecessary.

Usually, the best solution is to calculate the distances between each pair of individuals separately over those variables that have both values present, using divisor $p - r$ if r variables have thus not been used. This is unlikely to be seriously misleading unless there is a high proportion of missing data.

Dissimilarities between individual items; discrete variables

10.6 For discrete observations, it is usually more convenient to consider measures of similarity, and transform them to dissimilarities by a reflection, a suitable order-reversing transformation.

For unordered categorical data, it is natural to score similarity 1 for observations falling in the same category, and 0 otherwise; the values can be weighted or scaled as required. For ordered categories, the easiest procedure is to assign scores to reflect the distances between different categories, and treat them as continuous variables.

Similarities for binary values have been studied since the work of Yule (1900, 1912) and Pearson (1904). Suppose binary variables are coded 1 or 0. One major question is whether the 'similarity' of two items is the same when both are 1 (a positive match) as when they are both 0 (a negative match). Often, particularly when the two states are about equally probable, the answer is 'yes'. If, on the other hand, 1 represents a rare characteristic such as the ability to secrete an unusual enzyme, a positive match suggests a close relationship, while a negative match gives little or no information about similarity. Finally, if the binary variables represent a coding of a categorical variable with more than two states, clearly only a positive match represents similarity.

Gower (1985) lists 15 different similarity measures for binary variables. These are, for the most part, defined to range from 0 to 1, and are based on a 2×2 table of the form

		x_i	
		1	0
x_j	1	a	b
	0	c	d

Table 10.1 Similarity coefficients for dichotomous variables

S_1	$\frac{a+d}{a+b+c+d}$	simple matching
S_2	$\frac{a}{a+b+c}$	Jaccard
S_3	$\frac{a}{a+b+c+d}$	
S_4	$\frac{ad-bc}{\sqrt{(a+b)(a+c)(b+d)(c+d)}}$	Pearson
S_5	$\frac{ad-bc}{ad+bc}$	Yule

Thus, a represents the number of variables on which $\boldsymbol{x}_i$ and $\boldsymbol{x}_j$ both score 1, and $a+b+c+d$ is the total number of variables in this group.

In Table 10.1, S_1, S_4 and S_5 represent 'invariant' coefficients, in which the scores 0 and 1 are interchangeable. S_1 has the range 0 to 1, and the other two have the range -1 to 1. The Jaccard coefficient (Jaccard, 1908) regards positive matches only as indicating similarity, with negative matches treated as missing, and S_3 treats negative matches as mismatches. S_4 is known as Pearson's ϕ, and was introduced by Pearson (1904) as a measure of association for contingency tables; in general, $\phi^2 = X^2/N$, where X^2 is the Pearson chi-squared statistic and N is the total number of observations. For the 2×2 table, it is simply the product-moment correlation coefficient between the two binary variables. S_5 is one of several similarity measures suggested by Yule (1900, 1912).

Many other coefficients of this type have been proposed, but the main difference is between the invariant type and those that count only positive matches. The distinction between the variables for which each is appropriate is usually clear, and the first step should be to assign each binary variable in the data to one group or the other. The choice of the actual coefficient for each group is less important.

Dissimilarities between individual items; mixed data

10.7 There is no difficulty, in principle, in calculating a dissimilarity for mixed continuous and discrete variables. Euclidean or Manhattan distance can be defined in the usual way, with the contribution from the binary variables being taken as zero either when they take the same value, or only when there is a positive match. The Canberra metric is also suitable for mixed continuous and discrete variables, when the variability increases with magnitude. In most applications, there are a number of binary variables and some continuous variables; this is typical, for example, in medical problems, where the binary variables represent the presence or absence of symptoms, and continuous variables are measurements such as temperature, blood pressure, and chemical composition of blood or urine.

Care must be taken to ensure that reasonable weights are given to the different classes of variables. This is particularly difficult when there are few binary variables. For example, if there is a single binary variable and p

continuous variables, the distribution of distances is apt to be dominated by the binary variable, and it is hard, in any case, to to think of a framework that does not recognise the binary variable as defining two natural groups. Cluster analysis is reasonable only when the number of classes defined by the binary variables is much larger than the number of clusters.

Gower (1971) suggests a general similarity measure suitable for mixed data and taking due account of missing observations. Define

$$s_{ij} = \frac{\sum_{k=1}^{p} w_{ijk} s_{ijk}}{\sum_{k=1}^{p} w_{ijk}} \tag{10.5}$$

where s_{ijk} is a measure of similarity based on the variable x_k. Gower proposes that s_{ijk} should be 1 for binary data with a positive match, and for categorical data when items i and j fall in the same category, and zero otherwise; while for continuous variables $s_{ijk} = 1 - |x_{ik} - x_{jk}|/R_k$, where R_k is the range of x_k. These similarities, however, can be varied; in particular, for some binary variables an invariant similarity may be appropriate.

The similarity s_{ij} is thus a weighted mean of similarities on the individual variables. The weights are defined by

$$w_{ijk} = 0 \text{ when } x_k \text{ is missing on item } i \text{ or } j,$$
$$w_{ijk} = w_k \text{ otherwise (often } w_k = 1).$$

This similarity may be converted to a dissimilarity. Gower (1971) shows that the similarity matrix defined in this way is positive semi-definite, provided that the dissimilarity for continuous variables is defined as shown, and that there are no missing values. If it is positive semi-definite, $d_{ij} = (1 - s_{ij})^{1/2}$ is a Euclidean distance, in the sense that there is an exact conformation of points in $n - 1$ dimensions or fewer, with exactly this matrix of Euclidean distances (see Part 1, **5.6**, Theorem 6: Gower and Legendre, 1986).

Distance between groups

10.8 Once a measure of distance or dissimilarity between two *items* has been chosen, it must be used to define a distance between two *groups*, of, say, n_1 and n_2 items. This is important because it affects the sequence in which the clusters form; some definitions favour the formation of one or two large clusters with outlying single points, while others tend to produce smaller, roughly equal, clusters that then join up.

It seems natural to define the distance between two groups either as some sort of average of between-individual distances, or as a distance between centres of clusters. In fact, a large number of suggestions have been tried in practice, and the problem has given rise to much controversy. We shall start by considering the two extreme cases:

- single linkage, or nearest-neighbour clustering, in which the distance between two groups is defined as the smallest distance between a member of one group and a member of the other;

- complete linkage, or furthest-neighbour clustering, in which the distance between two groups is defined as the largest distance between a member of one group and a member of the other.

These two methods produce quite different patterns of clustering when applied to data sets in which there is no particular structure. Most of the other definitions that have been used are intermediate between them. It is important to emphasise, however, that where the data are strongly clustered, all methods are likely to give similar results; the major differences occur when in fact there is no natural clustering.

Single linkage clustering

10.9 Agglomerative clustering proceeds (**10.3**) by successively joining the items or groups that are closest, and single linkage defines the distance between two groups as the distance between the nearest points in those groups. Each time a point is added to a group, the distances of that group from all the other groups are either reduced or unchanged. This implies a tendency for large groups to grow and join up, while isolated points tend to remain unattached. The method was popularised by the work of Sokal and Sneath (1963) and Jardine and Sibson (1971). It was apparently first suggested by Florek *et al* (1951).

There is one important property of single linkage clustering that has been urged as a reason for preferring it to other methods. Suppose two distances are equal, say $d_{AB} = d_{CD}$ or $d_{AB} = d_{AC}$. If these are the smallest distances between groups, one choice must be made for the next fusion. It is easy to see that it does not matter which choice is made; the following fusion will be the one corresponding to the other equal distance. It is not so obvious, but is true, that this property is unique to single linkage clustering. With any other method, the whole course of the subsequent clustering may depend on which choice is made.

Tied distances are mainly confined to clustering based on binary data. If continuous variables, recorded with reasonable precision, are included, such ties are extremely unlikely. Nevertheless even in this case the property is important; it implies that small perturbations of the distances are unlikely to affect the general pattern of clustering. Single linkage clustering is robust against variations of this sort.

On the other hand, single linkage clustering is notoriously liable to 'chaining'. The addition of one or two points between two well-defined groups can cause them to join at an early stage. In the context of mixtures of continuous distributions, this is a type of non-robustness; the clustering is quite altered by a small addition to the data.

This last point, as has already been mentioned, is a feature of the method that is not necessarily bad. In taxonomy, the discovery of a form intermediate between two groups may profoundly affect our understanding of the relationship between them. It is no coincidence that the proponents of single linkage clustering have been, for the most part, working in the field of numerical taxonomy.

Complete linkage clustering

10.10 In complete linkage clustering, the distance between two groups is taken as the greatest distance between an item in one group and an item in the other. This means that each time an item is added to a group, the group's distance from all the other groups either increases or remains the same. The larger a group grows, the less is its tendency to grow further. This implies that, in unstructured data, items will join up into small groups that will then combine into larger groups; at each stage in the clustering the group size will vary far less than with single linkage clustering.

This feature means that complete linkage is particularly suitable for dissection. If the aim is not to detect natural clusters but to divide unstructured data into convenient groups, single linkage will fail to give any useful results, while complete linkage will give clusters of sizes that are not wildly different. On the other hand, if the data consist of groups of very different sizes, complete linkage will probably fuse the smallest clusters before the largest are complete.

Glasbey (1987) uses complete linkage as a multiple stopping rule for single linkage clustering.

Example 10.2
The two dendrograms in Fig. 10.2 are based on the same set of data. They refer to soils in Glamorgan; the data are from Rayner (1966), and the histograms are reproduced from Gordon (1981). For further discussion of the data, see Krzanowski (1988a, Chapters 2 and 3).

The first dendrogram is based on single linkage clustering, the second on complete linkage clustering. Both show two fairly well defined groups, and an outlier. Within the groups, complete linkage clustering appears to show sub-clusters rather similar in size, while single linkage tends to join points to groups one at a time. It seems reasonable to infer that the two groups represent 'natural clusters' in the data, and that there is one point that belongs to neither cluster. The structure within the clusters, on the other hand, is what might be expected from applying the two clustering methods to unstructured data, and probably corresponds to random variation within the clusters.

'Flexible strategy'

10.11 Lance and Williams (1967b) introduced a general rule for defining the distance between groups based on an update equation. This rule gives the distance $d_{C(AB)}$ of group C from a group formed by combining groups A and B. The new distance is given in terms of the original distance matrix before combining A and B, and involves four parameters, which are either constants or depend on the numbers in the groups. The general rule includes as special cases a number of methods that had been already suggested, such as single linkage and complete linkage.

The rule is

$$d_{C(AB)} = \alpha_A d_{CA} + \alpha_B d_{CB} + \beta d_{AB} + \gamma |d_{CA} - d_{CB}|, \tag{10.6}$$

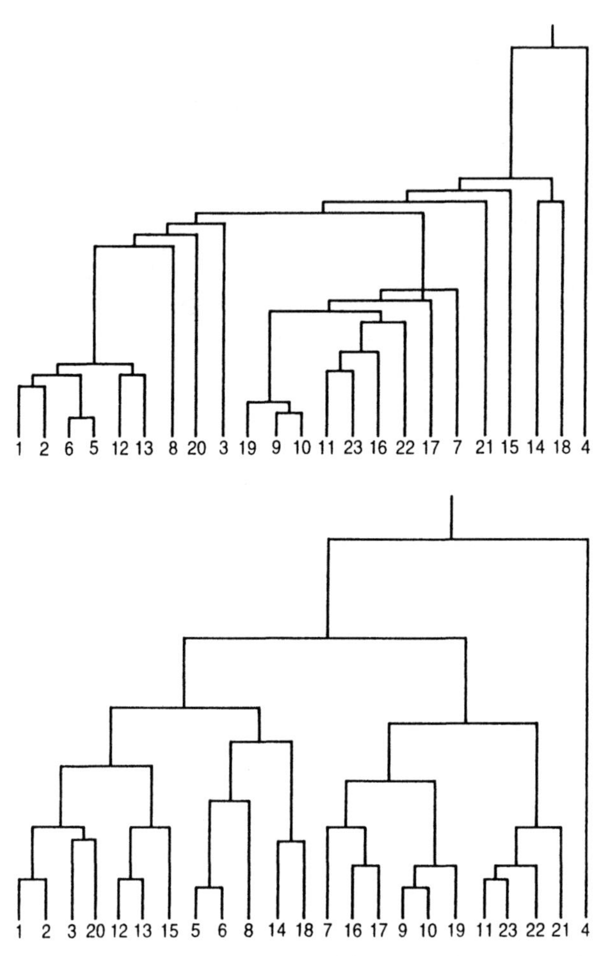

Figure 10.2 Single linkage (above) and complete linkage (below) dendrograms based on the same data set for Glamorgan soils. Reproduced, with permission, from Gordon (1981).

where d_{IJ} is the distance between groups I and J, and the four parameters are $\alpha_A, \alpha_B, \beta$ and γ.

Table 10.2 shows some of the possible values that may reasonably be chosen for the four parameters, with the names given to the corresponding clustering methods. The name 'flexible strategy' is applied to the general set of rules given by the equation, or sometimes to the last rule suggested by Lance and Williams (1967b).

The single linkage (nearest-neighbour) and complete linkage (furthest neighbour) rules have already been discussed.

The centroid rule defines the distance between two groups as the distance between their centroids, and when two groups merge the resulting group is

Table 10.2 Coefficients for 'Flexible Strategy'

Cluster method	α_A	α_B	β	γ
Single linkage	1/2	1/2	0	−1/2
Complete linkage	1/2	1/2	0	1/2
Centroid	$n_A/(n_A+n_B)$	$n_B/(n_A+n_B)$	$-\alpha_A\alpha_B$	0
Median	1/2	1/2	−1/4	0
Group average	$n_A/(n_A+n_B)$	$n_B/(n_A+n_B)$	0	0
Ward's method	$\frac{n_C+n_A}{n_C+n_A+n_B}$	$\frac{n_C+n_B}{n_C+n_A+n_B}$	$-\frac{n_C}{n_C+n_A+n_B}$	0
Lance and Williams	$(1-\beta)/2$	$(1-\beta)/2$	<1	0

centred at the centroid of the complete set of individuals. This is the case, at least, when the items are defined by their coordinates. If the distances are *squared* Euclidean distances defined from the coordinates, the parameters in Table 10.2 define this clustering method. The rule in the table is, however, more general. It gives a reasonable clustering rule regardless of whether the distance matrix is Euclidean. The method was first proposed by Sokal and Michener (1958).

The median rule (Gower, 1967) defines the centre of a combination of two groups as the mid-point of the join of their centres, regardless of their size. In the centroid method, a small group combined with a large one has little effect on the position of the group centre. This can distort the process if they are, in fact, natural groups and the observed numbers in each group are not necessarily representative of their relative frequency. The 'median' is a reference to the median of a triangle, the line joining a vertex to the mid-point of the opposite side, and not to the median in the statistical sense. Again, the rule given by the table is more general than the geometrical argument, which holds when d stands for a squared Euclidean distance.

The group average, or mean distance, rule defines the distance between two groups A and B as the average of all the $n_A n_B$ distances between the points in the groups.

Ward's method (Ward, 1963) merges groups to give the smallest possible increase in the sum of squares of within-group distances, or equivalently in the sum of squares of distances of individual items from the group centroid. This seems a natural criterion to statisticians accustomed to analysis of variance techniques, but it has some potential drawbacks. It can be rather sensitive to the effect of outliers, and it has a tendency to merge small, rather than large, groups even when the former are further apart.

Lance and Williams (1967b) discuss the general application of the four parameter model, and conclude that the best choice of parameters for most purposes would be $\alpha_A = \alpha_B$, with no size weighting, $\gamma = 0$, and $\alpha_A + \alpha_B + \beta = 1$. Different choices of β could then vary the characteristics of the clustering; they suggest that a small negative value would often be most appropriate.

Equation (10.6) gives rise to a range of hierarchical agglomerative clustering methods, probably covering all the sensible methods based on dissimilarities, although some similarity-based methods do not fit the equation in any simple way. The properties of the different choices of parameters have been extensively debated; see, for example, Cormack (1971), Everitt (1974), Romesberg (1984), Kaufman and Rousseeuw (1990) or Vach and Degens (1991). The differences are important when there is little structure in the data. When there is a well-defined grouping, they become less important, and the difference between methods is probably less critical than the choice of dissimilarities among individuals. If the aim is dissection, methods that tend to give equal groups are usually to be preferred.

The minimum spanning tree

10.12 The minimum spanning tree associated with a distance (dissimilarity) matrix is the shortest connected graph through all the points. It is a *tree*, with no loops, because clearly if there is a loop it can be made shorter by removing one segment of the loop. It is a structure well known in operational research, where it has applications to problems involving minimum distances. Gower and Ross (1969) first pointed out that the construction is precisely that of single linkage clustering, in the sense that if the $g-1$ longest links are broken, the resulting g groups are precisely those given by single linkage clustering after $n-g$ steps. Topologically, the tree is equivalent to the single linkage dendrogram, and it is easy to construct the dendrogram from the minimum spanning tree.

There are two major applications of the minimum spanning tree in multivariate statistics. In the first place, it may be used to carry out single linkage cluster analysis and to construct the dendrogram. There are efficient algorithms for doing this; see Gower and Ross (1969) and Ross (1969). Furthermore, one of the main constraints on cluster analysis of large data sets has been the storage and manipulation of the $\frac{1}{2}n(n-1)$ distinct elements of the dissimilarity matrix. Roger and Carpenter (1971) and Roger (1971) give an algorithm to *update* a minimum spanning tree. Given a tree with $n-1$ items, the tree for n items when a single item is added needs only the dissimilarities of the new item from those already in the tree. Thus, the tree can be constructed 'cumulatively'; the complete set of dissimilarities need not be stored, but may be calculated as they are needed to modify the tree step by step. This was a major advance in the methodology, but is perhaps less important now that data storage is cheaper and more readily available.

Secondly, the tree may be superimposed on any two- or three-dimensional representation of the p-dimensional data. The simplest example of this sort of plot uses the first two principal components. The distances will all be greater than their projections on the plane of the two components, but will agree fairly well if the two-dimensional representation accounts for most of the variability. The resulting graph makes it possible to judge which points are not well represented in the plot. (See Part 1, **5.11** and Example 5.3).

This idea has been developed by Friedman and Rafsky (1981), who give examples of the minimum spanning tree superimposed on other two-dimensional scatter plots of the data.

Comparing partitions

10.13 The problem of comparing different partitions of the same set of data is a difficult one, and will be only briefly discussed here. Day (1986) gives an introduction to the problem; this paper is the first of a complete issue of the *Journal of Classification* devoted to the subject.

In general, two partitions of the same data set may suggest some similarities, and it is tempting to match classes and count the numbers of objects in common. This approach, however, is not satisfactory. Unless there is close agreement, the matching is largely arbitrary, and any results are subjective and unreliable.

Suppose two partitions U and V divide the data into R and C classes respectively. The two partitions may be set out in the form of a $R \times C$ contingency table, in which n_{ij} represents the number of objects in class i of U and in class j of V. The form of the table is shown in Table 10.3.

Probably the best measure of conformity between two partitions is the Rand index, or one of the variations on it. Rand (1971) describes the index; see Hubert and Arabie (1985) for an account of other authors who discuss the index or variations of it. The account that follows is based closely on Hubert and Arabie (1985).

The $\binom{n}{2}$ pairs among the n objects are of four types.

(1) The objects are in the same class in U and in the same class in V.
(2) The objects are in different classes in U and in different classes in V.
(3) The objects are in different classes in U and in the same class in V.
(4) The objects are in the same class in U and in different classes in V.

Here, types 1 and 2 represent agreements between the classifications, and types 3 and 4 represent disagreements. Writing A for the number of pairs in types 1 and 2, and D for those in types 3 and 4, so that $A + D = \binom{n}{2}$, the Rand

Table 10.3 Comparison of two partitions

		Partition V				
	Class	v_1	v_2	...	v_C	Sums
	u_1	n_{11}	n_{12}	...	n_{1C}	$n_{1.}$
	u_2	n_{21}	n_{22}		n_{2C}	$n_{2.}$
	.				.	.
Partition U	.				.	.
	.				.	.
	u_R	n_{R1}	n_{R2}	...	n_{RC}	$n_{R.}$
	Sums	$n_{.1}$	$n_{.2}$	...	$n_{.C}$	$n_{..} = n$

index is defined as $A/(A + D)$. In terms of the entries in the contingency table

$$A = \binom{n}{2} + \sum_{i=1}^{R}\sum_{j=1}^{C} n_{ij}^2 - \frac{1}{2}\left(\sum_{i=1}^{R} n_{i.}^2 + \sum_{j=1}^{C} n_{.j}^2\right) \tag{10.7}$$

$$= \binom{n}{2} + 2\sum_{i=1}^{R}\sum_{j=1}^{C} \binom{n_{ij}}{2} - \left(\sum_{i=1}^{R} \binom{n_{i.}}{2} + \sum_{j=1}^{C} \binom{n_{.j}}{2}\right) \tag{10.8}$$

(Brennan and Light, 1974; Hubert and Arabie, 1985).

The Rand index, in this form, is not easy to interpret. The maximum value is 1 only if $R = C$, and the row and column totals match exactly. Hubert and Arabie (1985) suggest a modified form of the index:

$$\text{Corrected Index} = \frac{\text{Index} - \text{Expected Index}}{\text{Maximum Index} - \text{Expected Index}}$$

Here, it is natural to interpret the 'expected index' as the value expected when the partitions are independent and the numbers n_{ij} follow a hypergeometric distribution. The 'corrected index' is thus nearly zero for this situation. The choice of 'maximum index' is less clear. Hubert and Arabie (1985) consider various options. They go on to discuss significance tests of the null hypothesis that two partitions are unrelated, and to consider extensions and modifications to the Rand index.

When clustering is based on a hierarchical procedure leading to a dendrogram, the set of distances at which pairs of items join the ultrametric distances associated with the dendrogram, suggest various alternative methods of comparing clustering methods. Gower and Banfield (1975) consider various statistics relating the ultrametric distances, say d_{ij}^+, with the dissimilarities δ_{ij} on which the clustering was based. The discrepancies between these distances can be regarded as a measure of the distortion caused by imposing the hierarchical structure. Suitable measures include:

(1) the sum of squares of differences;
(2) the weighted sum of squares, $\sum(\delta_{ij} - d_{ij}^+)^2/\delta_{ij}^2$;
(3) the correlation between d_{ij}^+ and δ_{ij}, known as the cophenetic correlation;
(4) rank correlations between d_{ij}^+ and δ_{ij}.

Gower and Banfield (1975) examine the distributions of these, and other, criteria when the original data are drawn from a single multivariate normal distribution.

Similar criteria may be used to compare two hierarchical classifications, or to compare an ultrametric distance matrix with a dissimilarity matrix from which it was not derived. In particular, the correlation between two dissimilarity matrices may be used to test whether there is any relationship between them. The null distribution of this statistic is unknown—it is not, of course, that of the correlation between independent normal variables—and Monte Carlo type tests are necessary.

Finally, a distance matrix, either ultrametric or ordinary, may be related to a partition. The relationship is measured by some statistic that reflects the relationship between the distances among the items in the same cluster, and between those in different clusters.

All these questions are related to the space–time clustering problems familiar in epidemiology. There, a significant relationship suggests a relationship with an infection, and tests of independence are important in establishing the existence of such an element in the aetiology. In the general context of comparing clustering, such tests are seldom of much interest: it is more important to define statistics describing the closeness of the relationship between partitions.

Example 10.3

Lapointe and Legendre (1994) carried out a hierarchical cluster analysis, using Ward's method, of 109 Scotch whiskies. The data were taken from Jackson (1989), and consisted of five categorical variables, *colour, nose, body, palate and finish* scored as 68 binary variables. The resulting dendrogram was then used to give 2, 3, 6 and 12 classes.

Further data on the whiskies give the geographical position of the distillery, and hence the geographical distance matrix. They can also be classified by district, and a three-region model (Highlands, Lowlands and Islay) was used to give a binary dissimilarity matrix (0 for distilleries in the same region, 1 otherwise).

The geographical distance matrix and the regional dissimilarity matrix were compared with the cophenetic matrix—the matrix of ultrametric distances giving the dendrogram—and with the binary matrices representing the grouping, with four different numbers of classes. The null hypothesis that the classification was unrelated to geographical distance or regions was then investigated. Two tests were used: Mantel's (1967) test, and the double permutation test (DPT) developed by Lapointe and Legendre (1990, 1991, 1992). Both are based on the matrix correlation coefficient.

Tables 10.4 and 10.5 show the results. There are clearly, as might be expected, relationships between the tasters' descriptions on the one hand, and the geographical distance or region on the other. The relationship is not apparent for two or three groups, but shows up clearly for the dendrogram defined by the ultrametric distances, and for six or more groups. The significance is judged by a randomisation test. Notice that the correlation coefficients—based, of course, on a very large number of interdependent entries—are small.

Table 10.4 Test of ultrametric dissimilarities and grouping against geographical distance

Test	Dendrogram	2 groups	3 groups	6 groups	12 groups
Mantel	0.031(***)	−0.027(NS)	0.007(NS)	0.065(***)	0.064(***)
DPT	0.031(**)				

(NS) $P > 0.05$ (*) $P < 0.05$ (**) $P < 0.01$ (***) $P < 0.001$

Table 10.5 Test of ultrametric dissimilarities and grouping against the three-region classification

Test	Dendrogram	2 groups	3 groups	6 groups	12 groups
Mantel	0.042(***)	−0.025(NS)	0.019(NS)	0.066(***)	0.076(***)
DPT	0.042(**)				

(NS) $P > 0.05$ (*)$P < 0.05$ (**) $P < 0.01$ (***) $P < 0.001$

Optimisation methods of cluster analysis

10.14 A large group of clustering algorithms is based on the idea of optimising some criterion for a subdivision of the data into a given number of clusters. These methods are not hierarchical. The calculations are carried out for different numbers of clusters, and an appropriate choice of the final number is made at the end. The clusters are not nested, and do not allow the construction of a dendrogram.

These methods are sometimes referred to as k-means clustering, a name suggested by MacQueen (1967). This seems to be the first full published description of such an algorithm, although the author cites a number of earlier research reports. Friedman and Rubin (1967) discuss the problem further, and specifically investigate minimising $|\boldsymbol{W}|$ and $\text{tr}(\boldsymbol{T}^{-1}\boldsymbol{W})$, where $\boldsymbol{T}$ and $\boldsymbol{W}$ are the 'total' and 'within' matrices in a multivariate analysis of variance (see **7.4**, Part 1).

The standard procedure is to start with a partition into k groups, and then scan through the data set transferring each point to a different group if that would reduce the criterion. This scan continues until no further transfers take place. The criterion has then converged to an optimal value, although typically a local rather than global optimum.

With some algorithms, a group may become empty; the transfer of the last member of the group to another group improves the criterion. If, for example, all the relevant statistics are not updated whenever a point is transferred, a single point may appear further from the centre of its own group than from the centre of another group.

If a global minimum is required, a single run of this type of algorithm is inadequate. Traditionally, either the procedure is repeated with different starting points, or single point transfers are supplemented with exchanges and block-transfers (Friedman and Rubin, 1967). Banfield and Bassill (1977) give an algorithm that alternates between single-point transfers and two-point swapping, but still repetition is needed. Simulated annealing (Part 1, Chapter 4) is a more effective recent approach to the problem.

Clustering criteria

10.15 Most of the criteria suggested have been based on multivariate analysis of variance, and their use implies that clusters are expected to be roughly normally distributed, or at least elliptically contoured. Ward's method (Ward,

1963; see **10.11**) is a hierarchical precursor of this idea. It is based on minimising $\mathrm{tr}(\boldsymbol{W})$ at each step, and MacQueen's k-means clustering in its simplest form minimises $\mathrm{tr}(\boldsymbol{W})$ for a sequence of values of k. It depends on standardisation of variables before clustering, and tends to produce spherical clusters of similar size in the standardised space. The scale dependence is a major disadvantage, leading to the predisposition to form spherical clusters. The tendency to produce equal clusters is rather more subtle. In general, if a *test* criterion is optimised, the numbers in the clusters will be related to the optimal design. Thus, if a common covariance matrix within groups is assumed, the optimal design for comparing means would have equal numbers in the groups, and the corresponding clustering procedure tends to produce roughly equal groups.

Minimising $|\boldsymbol{W}|$ is equivalent to minimising the test criterion $|\boldsymbol{W}|/|\boldsymbol{T}|$. It avoids the scale dependence of $\mathrm{tr}(\boldsymbol{W})$, and can detect elliptically contoured clusters, but still has a tendency to produce equal clusters, even when the data are derived from a mixture, in unequal proportions, of normal distributions with equal covariance matrices. This criterion was proposed by Friedman and Rubin (1967). Scott and Symons (1971) show that it could be regarded as a maximum likelihood estimate if the data were a random sample from a mixture of normal distributions with equal covariance matrices. This derivation, however, is potentially misleading. The parameters estimated are the means, the elements of the covariance matrix, and an indicator variable for each observation indicating its cluster. This is maximum likelihood estimation with more parameters than observations, and does not have the usual consistency properties associated with the method. Neither the parameters of the underlying distributions nor the optimal divisions between them—the discriminant functions—are consistently estimated. The group of techniques is sometimes referred to as 'classification maximum likelihood'.

Maximising $\mathrm{tr}(\boldsymbol{B}\boldsymbol{W}^{-1})$ (or equivalently $\mathrm{tr}(\boldsymbol{T}\boldsymbol{W}^{-1})$) is also based on a test statistic, and has the same tendency to produce clusters of equal size and shape. It strongly favours elliptically contoured clusters with their centres roughly collinear. This is to be expected, since the associated test statistic gives much weight to the largest eigenvalue.

All the criteria so far discussed search for, and so tend to find, clusters of similar shape. Two criteria that allow for different within-cluster covariance matrices are $\prod |\boldsymbol{W}_s|^{n_s}$ and $\sum |\boldsymbol{W}_s|^{1/p}$. These criteria do not constrain the clusters to be of similar shape. There is, however, still a restriction on the cluster sizes; there is a tendency for small values of $|\boldsymbol{W}_s|$ to correspond to small cluster sizes. This characteristic is again related to the efficiency of sampling schemes or experimental design. Efficient schemes involve larger samples from the more variable populations and, correspondingly, the minimisation of these criteria involves a relationship between size and variability.

The criterion $\prod |\boldsymbol{W}_s|^{n_s}$ can be based on the same sort of likelihood argument as $|\boldsymbol{W}|$ (Scott and Symons, 1971), but for practical use must be constrained, by imposing a minimum value of either n_s or of $|\boldsymbol{W}_s|$. Otherwise, the absolute minimum of the criterion is achieved whenever one of the covariance matrices is singular.

The restrictive effect of these criteria on cluster sizes was pointed out by Symons (1981). He modified the argument in Scott and Symons (1971) by assuming a prior size distribution, and so including size explicitly in the optimisation. This led to two new criteria, $n \log|\boldsymbol{W}| - 2\sum n_s \log n_s$ and $\sum(n_s \log|\boldsymbol{W}_s| - 2n_s \log n_s)$. These criteria, modifications of $|\boldsymbol{W}|$ and $\prod|\boldsymbol{W}_s|^{n_s}$ respectively, avoid the constraint on sample size by the inclusion of the second term in the expressions. For the second criterion, it is still necessary to ensure that $\boldsymbol{W}_s$ cannot be singular. The modification certainly avoids the difficulties inherent in optimising test statistics. At the same time it must be realised that it does not lead to consistent estimates of the population parameters, or of the optimal divisions between populations, when the data are, in fact, independently sampled from a normal mixture.

Local optima for all these criteria can be found by starting with an arbitrary configuration and reassigning the data, point by point, until a stable configuration is reached. Marriott (1982) gives for each criterion the effect of assigning a given point $\boldsymbol{x}$ to cluster s.

First, define

$$\boldsymbol{d}_s = (\boldsymbol{x} - \bar{\boldsymbol{x}}_s)\left(\frac{n_s}{n_s + 1}\right)^{1/2} \tag{10.9}$$

The second term in this expression makes appropriate allowance for the change in $\bar{\boldsymbol{x}}_s$ when $\boldsymbol{x}$ is attached to group s.

Table 10.6 then gives the change in each criterion caused by adding $\boldsymbol{x}$ to group s. The optimisation procedure is then simply to assign $\boldsymbol{x}$ to the group for which the change is numerically smallest. This procedure converges to at least a local minimum of the criterion considered (apart from criterion 3, which is maximised).

Banfield and Raftery (1993) suggest other criteria in this general class. They also avoid the worst effects of outliers by including, as an additional component, a Poisson process of low density over the whole space covered by the data.

Table 10.6 Clustering criteria and changes on adding $\boldsymbol{x}$ to group s

	Criterion	Change
1.	$\mathrm{tr}(\boldsymbol{W})$	$+\boldsymbol{d}_s'\boldsymbol{d}_s$
2.	$\lvert\boldsymbol{W}\rvert$	$\times(1 + \boldsymbol{d}_s'\boldsymbol{W}^{-1}\boldsymbol{d}_s)$
3.	$\mathrm{tr}(\boldsymbol{B}\boldsymbol{W}^{-1})$	$-\boldsymbol{d}_s'\boldsymbol{W}^{-2}\boldsymbol{d}_s/(1 + \boldsymbol{d}_s'\boldsymbol{W}^{-1}\boldsymbol{d}_s)$
4.	$\prod\lvert\boldsymbol{W}_s\rvert^{n_s}$	$\times(1 + \boldsymbol{d}_s'\boldsymbol{W}_s^{-1}\boldsymbol{d}_s)^{n_s+1}\lvert\boldsymbol{W}_s\rvert$
5.	$\sum\lvert\boldsymbol{W}_s\rvert^{1/p}$	$+\lvert\boldsymbol{W}_s\rvert^{1/p}\{(1 + \boldsymbol{d}_s'\boldsymbol{W}^{-1}\boldsymbol{d}_s)^{1/p} - 1\}$
6.	$n \log\lvert\boldsymbol{W}\rvert - 2\sum n_s \log n_s$	$+\log\lvert\boldsymbol{W}\rvert + (n+1)\log(1 + \boldsymbol{d}_s'\boldsymbol{W}^{-1}\boldsymbol{d}_s)$ $-2(n_s+1)\log(n_s+1) + 2n_s \log n_s$
7.	$\sum(n_s \log\lvert\boldsymbol{W}_s\rvert - 2n_s \log n_s)$	$+\log\lvert\boldsymbol{W}_s\rvert + (n_s+1)\log(1 + \boldsymbol{d}_s'\boldsymbol{W}_s^{-1}\boldsymbol{d}_s)$ $-2(n_s+1)\log(n_s+1) + 2n_s \log n_s$

Choosing the number of clusters is a difficult problem. Engelman and Hartigan (1969) investigated by simulation the distribution of $|\boldsymbol{W}|/|\boldsymbol{T}|$ for a single normal distribution in the univariate case. This seems a natural test statistic, but the distribution for multivariate data is unknown. Marriott (1971) pointed out that the theoretical value of $g^2|\boldsymbol{W}|$ was constant for an optimal division of a multivariate *uniform* distribution into g groups, and suggested the minimum value of this criterion as an indication of the number of clusters. This suggestion was based on the rather dubious idea that the uniform distribution can be regarded as a transitional distribution between unimodal and multimodal distributions. Banfield and Raftery (1993) discuss a Bayesian variation of Wolfe's (1971) test (see **10.16**). This seems hard to justify. The distribution of the likelihood ratio has been investigated for maximum likelihood mixture estimation, not for 'classification maximum likelihood'.

In conclusion, these optimisation methods work well. This is particularly true of the last two, Symons' (1981) suggestions, which avoid difficulties that arise when the clusters are of very different sizes. They are, in origin, based on analysis of variance techniques applied to normal distributions, and are suitable only for observations of continuous variables. They are thus less flexible than distance-based methods of cluster analysis. They are not particularly sensitive to non-normality, but the fact that they do not perform ideally even when the data are drawn from a mixture of multivariate normal distributions suggests that direct estimation of the parameters of mixtures might be a better option. This is considered in the next section.

A recent development of these optimisation techniques is the program PAM (Partitioning Around Medoids) (Kaufman and Rousseeuw, 1990). This is similar to other k-means procedures in some respects, but is designed to be more robust against outliers. The 'medoids' are not centroids, but representative objects in each cluster. These objects may change as the clusters change, but are always actual data points. The other data points are attached to the representative points to which they are closest, using a L_1 dissimilarity measure.

Estimation for distribution mixtures

10.16 The idea of fitting distribution mixtures goes back to Pearson (1894), who fitted mixtures of (univariate) normal distributions using moments. Rao (1948) first suggested maximum likelihood estimation for univariate normal mixtures. Maximum likelihood fitting of multivariate normal distributions was first discussed by Wolfe (1965, 1967, 1970, 1971) and Day (1969b). This approach is quite different from that of Scott and Symons (1971) discussed in the last section. The data points are not assigned to components of the mixture with probability 1. Rather, the means and covariances, and the probabilities, of each component are estimated by maximum likelihood. Each point then has a probability of arising from each component of the mixture, estimated as the proportion of the density at that point accounted for by each component. For a detailed comparison of the two approaches, regarded as alternative methods of penalised maximum likelihood estimation, see Windham (1986) and Bryant (1988, 1991).

The end product, then, is not a grouping of the points into fixed clusters, but estimates of the parameters of the mixture components, and the posterior probabilities of membership of each component for each point. Of course, if there are well-separated clusters, most points will be assigned to one component with probability close to 1, and a formal clustering can be achieved by assigning each point to the component for which it has the highest posterior probability. If, in fact, the data constitute a random sample from a mixture of multivariate normal distributions, the method is consistent, in the sense that fitting a mixture with the right number of components will give consistent estimates of the parameters. Further, as sample size increases, the optimum divisions between the components, the contours of equal probability as defined in discriminant analysis, are correctly estimated.

In multivariate statistics, maximum likelihood estimation seems to be confined to normal mixtures. Other distributions have been used in univariate statistics, particularly for fitting populations of different age-groups; see McLachlan and Basford (1988) and Macdonald (1987).

The method is not hierarchical, and does not give a dendrogram. To fit a given number of groups, it is necessary to start with a set of guesses for the parameters and update them iteratively. If, as is usually the case, the number of groups is not known *a priori*, it is usual to fit one normal distribution, then a mixture of two, three, and so on, until it is decided that the fit is satisfactory. The choice of the appropriate number of components, as in other methods of cluster analysis, is a difficult problem and will be discussed in the next section.

The mathematics of maximum likelihood estimation of multivariate normal mixtures is straightforward. Suppose

$$f(\boldsymbol{x}) = \sum_{s=1}^{g} \lambda_s h_s(\boldsymbol{x}; \boldsymbol{\mu}_s, \boldsymbol{V}_s) \tag{10.10}$$

is the probability density of a mixture of g p-variate normal distributions with probabilities λ_s, $s = 1, \ldots, g$, where

$$h_s(\boldsymbol{x}; \boldsymbol{\mu}, \boldsymbol{V}) = (2\pi)^{-p/2} |\boldsymbol{V}_s|^{-1/2} \exp\left\{ -\frac{1}{2}(\boldsymbol{x} - \boldsymbol{\mu}_s)' \boldsymbol{V}_s^{-1} (\boldsymbol{x} - \boldsymbol{\mu}_s) \right\} \tag{10.11}$$

$$\sum_{s=1}^{g} \lambda_s = 1. \tag{10.12}$$

Now the probability that $\boldsymbol{x}$ belongs to component s is

$$\Pr(s|\boldsymbol{x}) = \frac{\Pr(s)\Pr(\boldsymbol{x}|s)}{\Pr(\boldsymbol{x})} = \frac{\lambda_s h_s(\boldsymbol{x}; \boldsymbol{\mu}_s, \boldsymbol{V}_s)}{f(\boldsymbol{x})} \tag{10.13}$$

It is now straightforward to derive the maximum likelihood equations based on a sample $\boldsymbol{x}_i$, $i = 1, \ldots, n$. Note, however, that if unequal dispersion matrices are used, they may not be singular.

$$\hat{\lambda}_s = \frac{1}{n}\sum_{i=1}^{n} \hat{\Pr}(s|\boldsymbol{x}_i) \tag{10.14}$$

$$\hat{\boldsymbol{\mu}}_s = \frac{1}{n\hat{\lambda}_s}\sum_{i=1}^{n} \boldsymbol{x}_i \hat{\Pr}(s|\boldsymbol{x}_i) \tag{10.15}$$

$$\hat{v}_{s(jk)} = \frac{1}{n\hat{\lambda}_s}\sum_{i=1}^{n} (x_i(j) - \hat{\mu}_j)(x_i(k) - \hat{\mu}_k).\hat{\Pr}(s|\boldsymbol{x}_i) \tag{10.16}$$

$$\hat{\Pr}(s|\boldsymbol{x}_i) = \frac{\hat{\lambda}_s h_s(\boldsymbol{x}; \hat{\boldsymbol{\mu}}_s, \hat{\boldsymbol{V}}_s)}{f(\boldsymbol{x}_i)} \tag{10.17}$$

In the full model, there are p parameters for the mean of each group, $\frac{1}{2}p(p+1)$ for each covariance matrix, and $g - 1$ for the vector of proportions—a total of $\frac{1}{2}g(p+1)(p+2) - 1$ parameters. For moderate-sized data sets, estimation of a separate covariance matrix for each group may mean that the model is overparametrised. Often it may be reasonable to assume either a common covariance matrix

$$\boldsymbol{V}_s = \boldsymbol{V},$$

or a common correlation matrix

$$\boldsymbol{V} = \boldsymbol{D}_s \boldsymbol{C} \boldsymbol{D}_s,$$

where $\boldsymbol{D}_s$ is a diagonal matrix of standard deviations and $\boldsymbol{C}$ is a correlation matrix assumed common for all groups.

The mixture model described has many attractions. In the first place, it is a genuine maximum likelihood estimate, with the usual asymptotic properties, when the data are actually sampled from a multivariate normal mixture. It is invariant under linear transformations. In particular, if the original data are replaced by principal components, the grouping is unaffected. If a subset of principal components is used, again the conclusions are practically unaffected if the discarded components contain no information about the grouping.

There is also a link with discriminant analysis. If certain observations are *known* to belong to different groups, they can have their prior (and posterior) probabilities of group membership fixed, so that they effectively define the groups to which they are assigned. When the number of groups with known members is equal to g, the model is that of the linear or quadratic discriminant functions with additional unclassified data.

This is an important application of the method of mixture analysis. In classical discrimination problems, data on a few unclassified cases provide negligible information about the distributions, or about discriminant rules. In certain situations, however, a very large proportion of the observations may be unclassified. This can happen, for example, with remotely sensed data. Often, the true classification is available only for a few parts of the whole image. Again, in medical statistics, if diagnosis is available only post mortem, data may be collected on a large number of patients before many diagnoses are available. In these situations, if it is reasonable to assume that the data within groups are approximately normally distributed, most of the information about the groups is derived from the unclassified data.

The algorithm is exactly the same as that for fitting mixtures, except that the classified data have the vector of group membership probabilities fixed, with one unit value and zeros elsewhere. The use of unclassified data in discrimination was discussed by Hosmer (1973), and the discrimination problem when the number of groups is known but no data are classified was discussed by O'Neill (1978) and Ganesalingam and McLachlan (1978).

The numerical problem of fitting multivariate normal mixtures was first solved by Wolfe (1965). A version of his program NORMIX is available in the CLUSTAN package. It can be regarded as an application of the E-M algorithm, for maximum likelihood estimation with the classification variables 'missing'. This contrasts with the maximum likelihood approach to optimisation methods of classification, in which they are treated as parameters.

The assumption of normality may appear restrictive, but in practice the method is reasonably robust, and works well with distributions with roughly elliptical contours. Often a transformation can give better results, particularly if the variance appears to increase with the mean.

Experience with NORMIX applied to simulated normal mixtures shows the following.

(1) When there are g reasonably well separated groups, NORMIX with g components gives good estimates of the parameters and the appropriate proportion of correct classifications.
(2) When NORMIX is used with fewer than g components, convergence is quite quick, but often local minima are found, depending on the starting values. This is not surprising, particularly if the distributions have a simple geometrical pattern with more than one theoretical optimum solution to the problem of fitting a mixture of fewer components.
(3) If there are more than g components specified in NORMIX, convergence is extremely slow, and the solution found typically depends on the starting values. The maxima of the likelihood depend entirely on 'noise' in the data, and convergence time is quite an effective indicator of whether too many components are being fitted.

Everitt (1988) suggested the extension of the normal mixture model to mixed discrete and continuous data by regarding the discrete variable values as corresponding to thresholds of an underlying normal variable—a latent variable model. The idea is developed in Everitt and Merette (1990); see Everitt (1993).

Choosing the number of mixture components

10.17 The procedure is to fit the distribution with $g = 1, 2, \ldots$ and adopt the solution with the value of g that is judged best. It is tempting to consider a formal significance test based on the likelihood ratio. Each increase in g increases the number of parameters fitted, by p for the extra mean, 1 for the extra proportion, and further parameters if the covariance matrices are allowed to differ. The likelihood correspondingly increases, and the change in log likelihood is an indication of whether the extra component is necessary.

Unfortunately, this does not lead to a suitable significance test. The null distribution of the likelihood ratio is unknown. The usual asymptotic form of the likelihood ratio test is not valid here, because the regularity conditions break down. The null distribution is 'on the edge of the parameter space', in the sense that when two components coincide, their proportions become unidentifiable, or if a proportion tends to zero, the corresponding parameters of the distribution become unidentifiable.

If the likelihood ratio is a suitable test criterion, its distribution could be found by simulation. Wolfe (1971), on the basis of a limited simulation, suggested that the usual χ^2 statistic, say $2\{\log(L_{g+1}) - \log(L_g)\}$, approximately followed a chi-squared distribution with $2f - 2$ degrees of freedom, where f is the number of extra parameters; that is, the degrees of freedom are double the number of extra parameters, ignoring the mixing proportion parameter. Later authors (e.g. Thode *et al*, 1989) have examined this suggestion. The consensus of opinion is that it is a reasonable approximation.

Everitt (1981) and Hernandez Avila (1979) both found fair agreement with Wolfe's suggested test. In a detailed investigation of the univariate case with a single normal distribution as the null hypothesis, and two normal distributions with equal variance as the alternative, Mendell *et al* (1991) showed that the Wolfe approximation was satisfactory for large samples, but that convergence to a limiting value was slow, giving a conservative test in moderate samples. McLachlan (1987) discussed a bootstrap test for the problem; his results supported Wolfe's test for homoscedastic univariate normal distributions, but suggested that it might be misleading for the heteroscedastic case. McLachlan (1992) cites an unpublished paper by McLachlan, Basford and Green as showing that the test is not satisfactory in general.

Mendell *et al* (1993) discuss the power of the likelihood ratio test, using the distribution calculated by Mendell *et al* (1991), and the same conditions. As might be expected, the likelihood ratio test performed reasonably well for all combinations of proportions in the mixture of two homoscedastic normal distributions, but other tests were better for special cases when the proportions were known—for example, if the proportions are equal, the Engelman–Hartigan test, based on minimising $|\boldsymbol{W}|$, which is strongly biased towards producing symmetric partitions, performs better, while for a 9-1 ratio a test of skewness is slightly more powerful.

It must be realised, though, that the difficulty of maximising the likelihood for $g + 1$ groups means that the test depends on the efficiency of the algorithm used. Most of the simulations cited took some steps to avoid local maxima, usually by trying different initial estimates. There is no guarantee, however, that the global maximum is found, and it is important that the simulations employ the same searching strategy as the practical example to be tested. At the least, the approximation shows how seriously the naive likelihood ratio test can break down.

The most serious objection, however, is that the null hypothesis being tested is unrealistic. The test of $g = 1$, for example, is a test of multivariate normality, powerful against a mixture, but capable of detecting any departure from normality. The fitting procedure is reasonably robust, but the test is not, and is of little practical value unless there are strong theoretical reasons for supposing that distributions really are multivariate normal.

Fuzzy clusters

10.18 The concept of 'fuzzy sets' was introduced by Zadeh (1965). In contrast to ordinary set theory, each item is characterised by a vector of *membership coefficients*, associating it with each of the sets. The membership coefficients sum to unity, and have some of the properties of probabilities of group membership, although set theorists avoid this interpretation.

Fitting distribution mixtures leads to a fuzzy clustering in which probabilities can be interpreted as membership coefficients. The algorithm involves optimising a criterion that depends on these probabilities—namely, the likelihood. In general, fuzzy clustering implies optimising a criterion involving membership coefficients. This allows standard forms of cluster analysis to be modified so that definite assignments to groups are replaced by fuzzy assignments involving membership coefficients. If there are well-defined clusters, most of the items will have one membership coefficient close to unity, but a few borderline cases will have coefficients that imply that they belong to one of two or more clusters.

A fuzzy k-means algorithm was proposed by Dunn (1974) and Bezdek (1974). The aim is to minimise the criterion

$$\sum_{k=1}^{g}\sum_{i=1}^{n} u_{ik}^2 ||\boldsymbol{x}_i - \boldsymbol{\mu}_k||$$

where u_{ik} is the membership coefficient of the ith item in the kth group, and $||\boldsymbol{x}_i - \boldsymbol{\mu}_k||$ is a distance measure, typically squared Euclidean distance, between $\boldsymbol{x}_i$ and the group mean $\boldsymbol{\mu}_k$, which is also a parameter to be estimated. For further discussion see Zadeh (1977).

Kaufman and Rousseeuw (1990) give a fuzzy clustering procedure FANNY that is based on dissimilarities. The criterion

$$\sum_{k=1}^{g} \frac{\sum_{i,j=1}^{n} u_{ik}^2 u_{jk}^2 d(i,j)}{2\sum_{j=1}^{n} u_{jk}^2}$$

is minimised, $d(i,j)$ being a dissimilarity measure.

When the model has been fitted, the coefficients u_{ij} give an idea of the sharpness of the clustering. *Dunn's partition coefficient* (Dunn, 1976) is defined as

$$F_g = \sum_{i=1}^{n}\sum_{k=1}^{g} u_{ik}^2/n$$

This coefficient approaches unity when all items are unequivocally classified, and has a minimum value $1/g$ when all items are equidistant from all the cluster centres.

Direct search methods

10.19 Wishart (1969) developed the first practical cluster analysis method based on a direct search for modes. His 'mode analysis', in an updated version, is incorporated in the CLUSTAN package (Wishart, 1987). It involves a search for 'dense points' by scanning the data with a hypersphere. The dense points become the centres of clusters, and the remaining points are associated with them. If this cluster building brings cluster centres close together, they can merge. The method is not hierarchical, and the number of clusters is not predetermined, but is decided by the parameters chosen for the search.

The method is adapted to search for spherical clusters and, like most methods not specifically distribution-based, is inefficient at finding strongly elliptical clusters. Estimation of modes is notoriously difficult in univariate data, and much more so in multivariate problems. A further weakness of the method is that the density at the centre of a large cluster is likely to be much higher than that at the centre of a small cluster, and the latter is therefore likely to be missed.

Wong and Lane (1983) describe a related method based on k nearest neighbours. This is a *hierarchical* technique, based on single linkage clustering applied using a distance that is finite only if either of the points is within the kth nearest neighbourhood of the other. The definition of neighbours is based on Euclidean distance.

Overlapping clusters

10.20 A different generalisation of the grouping of data into disjoint sets involves the idea of overlapping clusters. The space is divided into subspaces that may overlap. The data are then assigned to clusters corresponding to these subspaces. Each individual belongs to a cluster, but may well be a member of more than one. The concept is quite different from fuzzy clustering; multiple membership is a recognition that individuals may belong to more than one group, not a doubt about group assignment.

The first suggestion of an overlapping cluster analysis was made by Jardine and Sibson (1968, 1971). They suggested a modification of single linkage clustering in which some overlap is allowed. The B_k algorithm allows the overlap between clusters at a given threshold level to contain at most $k-1$ elements. The method was introduced mainly to avoid the 'chaining' effect often regarded as a weakness of single link clustering.

Rohlf (1974) examines the graph-theoretic structures associated with overlapping clusters. Figure 10.3 shows a '2-dendrogram', based on a B_2 cluster analysis. If the threshold is chosen to give three clusters, they are ABC, BD, and DE.

Other overlapping cluster techniques have been suggested, but practical examples were extremely rare until Shepard and Arabie (1979) proposed the ADCLUS model. An $n \times n$ similarity matrix $\boldsymbol{S}$, symmetric, with unit diagonal elements and off-diagonal elements in (0,1), is approximated by $\hat{\boldsymbol{S}}$, given by

$$\hat{\boldsymbol{S}} = \boldsymbol{P}\boldsymbol{W}\boldsymbol{P}'. \qquad (10.18)$$

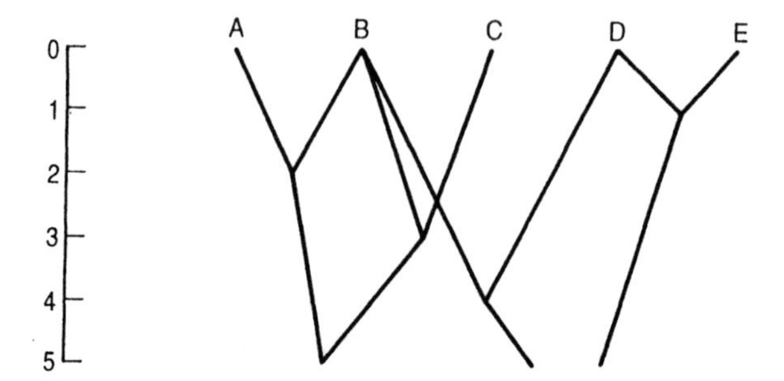

Figure 10.3 A 2-dendrogram for objects ABCDE. Reproduced from Rohlf (1974), *Journal of the American Statistical Society* with permission from the American Statistical Association.

Here, $\boldsymbol{P}$ is a $n \times m$ matrix of binary (0,1) variables. Each column corresponds to a cluster, and has a unit element for those of the n individuals in that cluster, with zero otherwise. In this formulation, one of the columns is a vector of units, corresponding to the complete set of individuals, and the remaining columns define $m - 1$ clusters. There is no constraint restricting membership to one of these clusters.

The matrix $\boldsymbol{W}$ is a diagonal matrix of weights. The aim is to approximate $\boldsymbol{S}$ as closely as possible, and the weights represent the 'tightness' of the clusters, the difference between the average similarity within the clusters and overall. The weights are positive; negative values are meaningless, but can possibly appear in the numerical fitting of $\hat{\boldsymbol{S}}$ unless precautions are taken to ensure otherwise.

The fraction of variance accounted for is

$$1 - \frac{\sum_{i>j}(s_{ij} - \hat{s}_{ij})^2}{\sum_{i>j}(s_{ij} - \bar{s})^2}$$

The components of $\boldsymbol{P}$ and $\boldsymbol{W}$ are chosen to maximise this fraction, for fixed values of m. Shepard and Arabie (1979) describe a method of estimation, but the alternative approach given by Arabie and Carroll (1980) seems a considerable improvement. The algorithm, called MAPCLUS, uses a combination of mathematical programming and alternating least squares for the optimisation.

Example 10.4

Miller and Nicely (1955) published a set of data on confusions between 16 consonant sounds. Table 10.7 shows the symmetric similarity matrix derived from these data (Shepard, 1972) in the upper triangle.

Table 10.8 shows eight clusters found by MAPCLUS, from Arabie and Carroll (1980). The weights in this table, together with the weight for the complete set, which is 0.049, define the model. They are not, however, the elements of the matrix $\boldsymbol{W}$. For computational convenience, the first step in the algorithm is to scale the *off-diagonal* elements of the similarity matrix

Table 10.7 Similarity matrix based on the Miller–Nicely data, from Shepard (1972), reproduced with permission

	p	t	k	f	th	s	sh	b	d	g	v	dh	z	zh	m	n
p	1	0.229	0.432	0.101	0.124	0.052	0.038	0.022	0.025	0.013	0.016	0.028	0.025	0.019	0.025	0.017
t		1	0.241	0.057	0.079	0.050	0.050	0.013	0.022	0.016	0.022	0.016	0.023	0.017	0.022	0.018
k			1	0.077	0.084	0.063	0.047	0.018	0.020	0.030	0.020	0.018	0.025	0.019	0.021	0.020
f				1	0.423	0.066	0.030	0.046	0.025	0.015	0.035	0.032	0.018	0.007	0.016	0.012
th					1	0.157	0.048	0.045	0.041	0.039	0.040	0.031	0.033	0.017	0.019	0.018
s						1	0.115	0.024	0.031	0.033	0.023	0.026	0.035	0.022	0.017	0.013
sh							1	0.012	0.033	0.021	0.020	0.018	0.017	0.136	0.021	0.016
b								1	0.058	0.069	0.210	0.145	0.055	0.027	0.016	0.030
d									1	0.342	0.059	0.094	0.106	0.089	0.024	0.151
g										1	0.054	0.120	0.139	0.125	0.032	0.030
v											1	0.338	0.080	0.029	0.030	0.022
dh												1	0.161	0.033	0.034	0.028
z													1	0.136	0.021	0.016
zh														1	0.016	0.030
m															1	0.151
n																1

Table 10.8 MAPCLUS solution for consonant confusions. Reproduced with permission from Arabie and Carroll (1980)

Cluster	Weight	Subset
1	0.814	f,th
2	0.729	v,dh
3	0.577	d,g
4	0.487	p,t,k
5	0.428	b,v
6	0.348	p,k
7	0.162	b,d,g,dh,z,zh
8	0.116	p,k,f,th,s,sh

to (0,1) by a linear transformation, in this instance of the form

$$s_{ij}^* = (s_{ij} - 0.007)/0.432.$$

To calculate the elements of $\hat{\boldsymbol{S}}$, the weights are added for all the groups in which both occur, and the reverse transformation is applied. Thus, for [d,g], calculate $\hat{S}_{dg} = (0.577 + 0.162 + 0.049)0.432 + 0.007 = 0.347$. This may be compared with the corresponding element in $\boldsymbol{S}$, 0.342.

The fitted matrix accounts for 89.6% of the variance. For a full discussion of the meaning of the clusters in the context of linguistics and phoneme recognition, see Arabie and Carroll (1980). Notice how the additive model accounts for the similarity structure; for example, the high similarities among p, t and k define *two* clusters, the second representing the extra similarity between p and k, above that already accounted for by the group of three.

10.21 The ADCLUS model has obvious affinities with the classical additive model of factor analysis, and with the latent class model for categorical data (Lazarsfeld, 1950; Lazarsfeld and Henry, 1968; Chapter 12, this book). It is based on a two-way data matrix, individuals by variables, or on a similarity matrix between individuals.

An important extension to the model is based on a three-way data array, individuals by individuals by variables, or on a set of similarity matrices. This situation can arise when sets of observations are made at different times or places, or when similarity matrices are constructed by different judges. The INDCLUS model (Carroll and Arabie, 1983) is the generalisation of ADCLUS for this problem. ADCLUS gives a set of possibly overlapping clusters with associated weights. INDCLUS (INdividual Differences CLUStering) gives a single set of clusters with a different set of associated weights for each of the underlying similarity matrices. See also Carroll *et al* (1984) for further methods.

Conclusions

10.22 Cluster analysis is generally regarded as a branch of exploratory data analysis, rather than of statistical inference. The concept of 'natural clusters' is not clearly defined.

For continuous variables, it seems reasonable to associate clusters with modes and, if a probability density were known, cluster boundaries and cluster membership could be determined. It is notoriously difficult to identify the modes, or determine their number, for even a univariate data set. Attempts to establish objective criteria for the number of clusters are generally based on the multivariate normal distribution. As a null hypothesis, multivariate normality is unreasonably restrictive; test criteria, powerful against mixtures, may be used, but eventually *any* departure from normality is interpreted as evidence of clustering.

For discrete variables, multimodality may be defined in terms of two or more vertices with higher probability than all their neighbours, but this occurs whenever two variables are sufficiently closely dependent. The only statistical models that have been worked out for this case, the latent class models, postulate clusters within which all variables are independent.

The commonest situation in practice involves continuous and discrete variables. No rational definition of what constitutes natural clustering seems possible.

The problem of significance testing in cluster analysis has attracted a great deal of attention. The simplest null hypothesis is that the data are homogeneous, with no evidence of natural clusters. The question of the number of clusters can also be investigated by successively testing models postulating one cluster, two clusters and so on until a satisfactory fit is found. A detailed discussion of the subject is given by Bock (1984, 1985): see also Ling (1973) and Gordon (1994).

The null hypothesis of homogeneity can be made specific in various ways.

- The data come from a single, specified, unimodal distribution. In practice, this is almost always the multivariate normal distribution. Maximum likelihood fitting of multivariate normal mixtures and the problems of associated tests have been discussed in **10.16**.
- The data are uniformly distributed over some specified region. The concept of clusters as modes, with a uniform distribution as the most extreme type of distribution that is not multimodal is an old one; its development is chiefly due to Hartigan (1975a,b, 1977, 1985).
- The data come from some unspecified unimodal distribution, and the null hypothesis is investigated by a suitable test of unimodality. Wishart (1969) based his modal analysis on this idea, though he did not develop a formal significance test. Tests for univariate unimodality have been proposed; Silverman (1981) gives a test based on normal kernels, and Hartigan and Hartigan (1985) introduce the 'dip' test. These tests can theoretically be extended to the multivariate case, but the sparseness of data in high-dimensional space means that they are scarcely practicable in general.

- The set of distances d_{ij} are consistent with unstructured data. This may be investigated using the theory of random graphs. Bock (1984) considers two forms of the null hypothesis.
  (1) The random permutation hypothesis, that all $\binom{n}{2}!$ rankings of the distances $d_{ij}, i < j$ are equally probable.
  (2) The random graph hypothesis, that all $\binom{n/2}{N}$ graphs with N edges are equally probable.

  The second hypothesis is implied by the first, and can be used stepwise to test different grouping structures. Bock (1984) describes a number of tests of these two hypotheses. This type of test can be applied to mixed, continuous and categorical, data if a suitable dissimilarity measure is chosen. It is, however, hardly a realistic hypothesis in practical situations.

Finally, it is worth noting that neural networks (**9.61**) have recently been investigated in the context of both cluster analysis and ordination (Chapter 5, Part 1). For example, Kohonen (1990) used the techniques to form 'self-organising maps'. The data are clustered by a k-means algorithm, and the inter-cluster distances are used to arrange them in a 2-dimensional map. This process is iterated with increasing numbers of smaller clusters, until the individuals are all located on the map.

Exercises

10.1 A distribution is given by a mixture of two multivariate normal distributions, with common covariance matrix, in proportions α, $1 - \alpha$. Show that the condition for bimodality is that the Mahalanobis distance Δ exceeds Δ_0, where the critical value Δ_0 is found by solving the equations

$$k = \frac{1-y}{1+y}\exp\left(\frac{2y}{1-y^2}\right)$$

$$\text{and } y = \frac{\sqrt{\Delta_0^2 - 4}}{\Delta_0}$$

where

$$k = \frac{1-\alpha}{\alpha} \geqslant 1$$

10.2 Carry out a nearest-neighbour cluster analysis, using Euclidean distance, on the data in Table 1.5 (tooth measurements) *after taking logs of all variables*. Construct the dendrogram, and compare with Fig. 1.4. Notice that the dogs link up in a similar way, but that the wolves join together before linking with the dogs, and the jackal joins last of all.

11

Covariance and Interaction Structures

Introduction

11.1 As the number of variables, say p, in a multivariate distribution increases, the number of parameters to be estimated also increases. Unless the number of observations, n, is very large, estimation is often inefficient, and models with many parameters are, in general, difficult to interpret. In many situations reduced models, with constraints on the parameters, are easier to understand and more useful for purposes of prediction. Two familiar instances are the selection of subsets of regressor variables, and the fitting of parsimonious time-series models.

The present chapter is concerned with constraints on the relationships among variates. There are three special cases in which reduced models have been extensively studied.

(1) The multivariate normal distribution involves $\frac{1}{2}p(p+1)$ elements in the variance-covariance matrix. It is often reasonable to assume that some variates are related only indirectly, through other variates. There may be prior information suggesting cause and effect relationships among the variates, and implying conditional independence among some sets of variables; this is the basis for 'path analysis' models, introduced by Wright (1921) in connection with problems in genetics, and widely used in econometrics. Alternatively, in the absence of such prior information, an investigation of conditional independence may give a clearer picture of the relationships among the variates. The importance of the multivariate normal distribution in this context is that a zero partial correlation coefficient implies conditional independence.
(2) Log-linear models are widely used for multi-way contingency tables. The 'saturated' model, including all possible interaction terms, fits the observed

counts exactly, and is often almost impossible to interpret. Reduced models, in which some of the interaction terms are set to zero, can be constructed and their fit tested, and they may throw light on the structure of the table.

(3) The conditional Gaussian distribution for mixed binary and continuous variables (Part 1, **2.25**) can be heavily overparametrised. If there are q binary and p continuous variables, there are, in the most general form, $2^{q-1}p(p+3)$ parameters (mean and covariance matrix for the p-variate normal distribution at each of the 2^q locations). The distribution can be parametrised to reveal interactions among the discrete variates, conditional covariances among the continuous variates, and interactions between the two types, and reduced models can be studied. The first systematic studies of this problem are fairly recent (Lauritzen and Wermuth, 1984, 1989; Edwards, 1990), but were foreshadowed by work on the location model in discriminant analysis (Olkin and Tate, 1961; Krzanowski, 1975, 1980, 1982).

In the first part of this chapter we consider each of these reduced models in turn, and then go on to discuss methods for handling constraints on the relationships among variates in regression models.

Conditional independence in the multivariate normal distribution

11.2 For the multivariate normal distribution the following statements are equivalent.

(1) The variates x_i and x_j are independent, conditional on $\boldsymbol{x}_{\{k\}}$, where $\boldsymbol{x}_{\{k\}}$ stands for any subset of the p variates $\boldsymbol{x}$—often for all the variates except x_i and x_j.
(2) The partial correlation (Part 1, **2.6**) $\rho_{ij.\{k\}} = 0$.
(3) $E(x_i|x_j, \boldsymbol{x}_{\{k\}}) = E(x_i|\boldsymbol{x}_{\{k\}})$, or the partial regression coefficient $\beta_{ij.\{k\}} = 0$, with a similar relationship for the regression of x_j on x_i, $\boldsymbol{x}_{\{k\}}$. This shows a relationship between conditional independence and regressor selection, but the practical problem, based on a data set rather than a theoretical distribution, is much more complex.
(4) If $\boldsymbol{\Sigma}$ is the covariance matrix, the element ω_{ij} of the inverse matrix $\boldsymbol{\Omega} = \boldsymbol{\Sigma}^{-1}$ is zero. (For the relationship between the inverse covariance matrix and the variances and partial correlations, see Part 1, **2.6**. The inverse covariance matrix is also known as the *concentration matrix*.)

Conditional independence holds when x_i is affected by x_j only by way of $\boldsymbol{x}_{\{k\}}$. For example, the statement 'Exercise produces increased carbon dioxide tension and reduced oxygen tension in the blood, and these changes cause increased ventilation' implies that exercise and ventilation are independent conditionally upon blood O_2 and CO_2, or that the connection is entirely through the blood. Of course, conditional independence does not imply this sort of causal relationship, but, coupled with knowledge of the physical mechanisms involved, it often suggests such a model.

Independence graphs

11.3 Conditional independence can be represented by a *graph*, in the sense of graph theory. The variables correspond to the nodes of the graph, and the nodes are joined by edges when the variables are not conditionally independent (conditionally, that is, upon all the other variables in the model). The graph is constructed by joining those pairs of nodes that correspond to variates having non-zero values in the inverse covariance matrix.

A p-variate concentration matrix has $\frac{1}{2}p(p-1)$ off-diagonal elements, and there are thus $2^{\frac{1}{2}p(p-1)}$ possible graphs, ranging from the unreduced model with no conditional independences and all nodes joined, to the model of complete independence with no joins.

A convenient notation for conditional independence was introduced by Dawid (1979). The statement

$$A \perp \{B, C\} | \{D, E\}$$

may be read 'A is independent of B and of C (each of the elements in the first braces) conditional on D and E (the set of elements in the second braces)'.

Example 11.1 (Whittaker, 1990)
The graph in Fig. 11.1 shows the conditional independences among five variates. Of the ten possible joins, four are missing; (1,3), (1,4), (1,5) and (3,5). The independences can be expressed as:

$$1 \perp \{3, 4, 5\} | 2, \quad 3 \perp \{1, 5\} | \{2, 4\}, \quad 4 \perp 1 | \{2, 3, 5\}, \quad 5 \perp \{1, 3\} | \{2, 4\}$$

These independence relationships imply

$$\omega_{13} = \omega_{14} = \omega_{15} = \omega_{35} = 0.$$

The graph can be derived from these conditional independence relationships: see Dawid (1979) or Whittaker (1990). The independence structure is, however, much clearer from the graph than from the set of zero elements or from the list of independences.

Notice, too, that there is some redundancy in the expressions above. $1 \perp 3 | 2$ obviously implies $3 \perp 1 | \{2, 4\}$, but it is convenient to list them in that form.

11.4 The simple type of independence graph discussed above is the only type to be considered in this chapter. Various other forms of graphical representa-

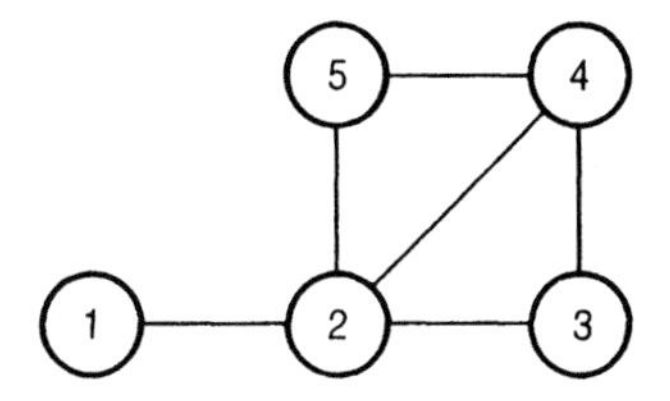

Figure 11.1 An independence graph with five vertices. Reproduced with permission from Whittaker (1990).

tion are possible, and have been used to represent relationships among variates, but they require more detailed knowledge of graph theory for their construction and interpretation. Some examples will be briefly mentioned.

It seems natural to use *directed graphs* to represent causal relationships among variates. Indeed, the path analysis introduced by Wright (1921) was based on causal relationships and used directed graphs as illustrations. There are, however, certain difficulties; the idea of putting arrows on the joins of an independence graph to indicate the direction of causation is an oversimplification.

In the first place, there is a problem with cycles. A situation of the type $A \rightarrow B \rightarrow C \ldots \rightarrow A$ is a possible causal chain, but gives little useful information about the joint distribution of the variables. Secondly, there is a question of whether a directed graph without cycles implies that the independence graph is the same figure without arrows. The answer, in general, is no; a directed graph in which A and B affect C but in which there is no causal relationship between A and B does not necessarily imply distributional independence of A and B conditional on C. Wermuth (1980), however, shows that if a directed graph contains no subgraph of this type, the independence graph is obtained by removing the arrows from the directed graph.

Another application of independence graphs involves independences among variates, some of which are unobservable. Latent variable models (see Chapter 12) are usually based on the assumption that the manifest variables are independent conditionally on the latent variables. Models of this type include classical factor analysis, and neural network models used in classification.

The success of graphical models in representing conditional independences between pairs of variables suggests the possibility of extending them to represent more complex interactions, particularly the interaction structure in log-linear models in which high-order interactions may be set to zero. This would involve some form of 'polygraph', with multiple connections, but at present there is no accepted method of deriving a suitable plot.

The subject of graphical models is large, and constantly growing. Further details may be found in the books by Whittaker (1990) and Lauritzen (1993), or in the statistical package MIM and its manual (Edwards, 1993).

Fitting conditional independence models

11.5 The conditional independence models discussed above provide a complete and unique description of the pattern of zero elements in the inverse covariance matrix. The practical problem is to identify a model of this type that gives a satisfactory fit to a set of data, and can be interpreted, often in terms of causal relationships. The problem is not unlike that of selecting a subset of regressor variables, and some of the difficulties of that procedure also arise here. In particular, if two variates are strongly correlated, the partial correlation of another variate with each conditional upon the other is likely to be small, but that does not at all imply that *both* partial correlations can reasonably be set to zero.

The procedure is to postulate a pattern of conditional independences, fit the parameters conditional upon these dependences by maximum likelihood, and test the null hypothesis that the reduced model is adequate against the alternative represented by the unrestricted model. This can be done by a likelihood ratio test; if k zeros are postulated, the usual likelihood ratio statistic has an asymptotic χ^2 distribution with k degrees of freedom on the null hypothesis. Models are examined until one is found that cannot be reduced further, and this can be regarded as a suitable approximation to the true conditional independence structure. This procedure was first described by Dempster (1972) under the name of covariance selection.

This description skates over a number of difficulties.

(1) Fitting reduced models is not trivial; the procedure for the multivariate normal distribution will be outlined in the next section.
(2) The likelihood ratio test is valid for a single reduced model—one, for example, suggested by theory about the variates. As a procedure for finding a model that can subsequently be interpreted, it involves multiple testing—see Part 1, **7.7**—and can really be justified only as an exploratory method without a basis in statistical inference.
(3) There may well be a number of 'adequate' models of conditional independences, suggesting quite different interpretations of the relationships among the variables.
(4) When the number of variates is large, examination of all possible patterns may be impracticable. A stepwise approach may be used, but has the same drawbacks as stepwise methods of variable selection—see Part 1, **7.9**.

11.6 The estimation problem consists of finding maximum likelihood estimates of the parameters of the multivariate normal distribution, subject to the constraint that certain specified off-diagonal elements in the concentration matrix are zero. The sums, and sums of squares and products, of the observations are sufficient statistics for the unconstrained parameter estimates, and this is obviously still true when the constraints are imposed. Further, the sample mean vector is still the maximum likelihood estimate of the mean, and the estimate of the constrained variance-covariance matrix, say $\hat{\boldsymbol{\Sigma}}$, is based on the sums of squares and products about the sample mean.

Maximum likelihood estimation then reduces to the following rule (Dempster, 1972).

(1) Suppose that I denotes a subset of the index pairs (i, j), with $1 \leqslant i < j \leqslant p$. This is the subset of variate pairs that are to be conditionally independent. Let J denote the remaining set of pairs.
(2) Let

$$\boldsymbol{S} = \frac{1}{n}\sum_{l=1}^{n}(\boldsymbol{x}_l - \bar{\boldsymbol{x}})(\boldsymbol{x}_l - \bar{\boldsymbol{x}})'$$

This is the maximum likelihood estimator of $\boldsymbol{\Sigma}$ in the unconstrained case. Note that Dempster actually used the divisor $n-1$, giving the usual unbiased estimates.

(3) Choose $\hat{\Sigma}$ to be the positive definite symmetric matrix such that S and $\hat{\Sigma}$ are identical for index pairs (i, j) in J, while $\hat{\Sigma}^{-1}$ is identically zero for index pairs (i, j) in I.

Dempster shows that there is a unique matrix defined in this way, positive definite provided that S is positive definite, that it is the constrained maximum likelihood estimate of Σ, and further that it has maximum entropy, defined as $-E[\log f(x)]$. In a few simple cases it is possible to find explicit expressions for the elements of $\hat{\Sigma}$ that are not equal to those of S but, in general, it requires an iterative calculation. The method generally used is iterative proportional scaling, first described by Haberman (1974) and Kiiveri and Speed (1982), and implemented in the computer program MIM (Edwards, 1993). Whittaker (1990) gives a clear informal account of the procedure. The same technique can be used for estimation in log-linear models, and in conditional Gaussian models. Quasi-Newton methods, used in GLIM, have the advantage of giving direct estimates of the variances and covariances of parameter estimates, but are, as a rule, slower.

11.7 The likelihood ratio test statistic for a model with k elements of the concentration matrix set to zero takes the form

$$T = -n \log |S\hat{\Sigma}^{-1}| - \sum_{i=1}^{n} (x_i - \bar{x})'(S^{-1} - \hat{\Sigma}^{-1})(x_i - \bar{x}) \qquad (11.1)$$

which, with some manipulation, reduces to

$$T = n(\operatorname{tr} S\hat{\Sigma}^{-1} - \log |S\hat{\Sigma}^{-1}| - p) \qquad (11.2)$$

Under the null hypothesis, T has asymptotically a χ^2 distribution on k degrees of freedom, and this gives a test, based on the sufficient statistics, for the adequacy of any postulated conditional independence model. A significant value of T indicates that the model is unsatisfactory; when S and $\hat{\Sigma}$ are consistent with each other, the test statistic is small.

A special case throws further light on the properties of the test. Suppose x is partitioned into (x_A, x_B, x_C), with q, s and t elements respectively, so that $q + s + t = p$. Consider the model in which x_A and x_B are independent conditionally upon x_C. The independence graph shows the elements of x_A and x_B completely separated from each other by those of x_C, with the qs edges between them deleted. Whittaker (1990) calls this the 'butterfly model'; a simple neural network model has this structure with x_C representing latent variables.

For this model, the likelihood ratio test becomes

$$-n \log(1 - R^2_{AB|C}) \sim \chi^2_{qs} \qquad (11.3)$$

where $R^2_{AB|C}$ is the squared multiple correlation between x_A and x_B after regression on x_C.

When $q = s = 1$ with $x_A = x_i$ and $x_B = x_j$, the test reduces to a test of the single partial correlation $r_{ij.\{k\}}$,

$$-n \log(1 - r^2_{ij.\{k\}}) \sim \chi^2_1 \qquad (11.4)$$

where $\{k\}$ stands for the $p-2$ variables omitting i and j. Notice that this is almost equivalent to testing a partial correlation coefficient as a normally distributed statistic with standard error $1/\sqrt{n}$. The chi-squared statistic is known as the *edge exclusion deviance*.

The procedure for fitting conditional independence models can now take various forms. When the number of variates is small, it may be reasonable to examine all possible models, generated by deleting all possible subsets of edges, and selecting from among the most parsimonious models giving a satisfactory fit. Even though some models can be rejected without actual fitting, as being either bad fits or reduceable to more parsimonious models that fit well, the computations become very heavy as the number of variates increases. The obvious remedy is to fit models stepwise. Stepwise fitting can be done starting from complete independence and adding, term by term, non-zero elements in the concentration matrix corresponding to the largest partial correlations; this approach was suggested by Dempster (1972). The alternative is to start with the full covariance matrix fitted, and start by deleting the terms corresponding to the smallest edge exclusion deviances. This approach is usually preferred when using a specialised modelling program, such as MIM.

11.8 The likelihood ratio tests described above can be justified as a form of exploratory data analysis when many different subsets of edges are considered. An alternative is to examine those subsets with high values of the Akaike information criterion (AIC) (Akaike, 1973, 1974). This is defined as

$$\text{AIC} = -2\log(\text{maximised likelihood}) + 2\,(\text{number of parameters fitted}).$$

It is an estimate of the Kullback–Leibler information measure. Obviously, the relationship with the likelihood ratio test is very close; in fact

$$AIC = -n(\operatorname{tr} \boldsymbol{S}\hat{\boldsymbol{\Sigma}}^{-1} - \log|\boldsymbol{S}\hat{\boldsymbol{\Sigma}}^{-1}| - p) - 2k + C$$

where C is constant for all models. AIC and the likelihood ratio order the models in the same way for equal k, but among models with different k values, AIC tends to be less parsimonious.

Example 11.2
Mardia *et al* (1979) give a data set consisting of the marks of 88 students in five mathematics examinations; mechanics (ME), vectors (VE), algebra (AL), analysis (AN) and statistics (ST). It is used as an illustration of graphical models both in the MIM guide (Edwards, 1993) and by Whittaker (1990).

The correlation matrix has the form:

	ME	VE	AL	AN	ST
ME	1.000	0.553	0.547	0.409	0.389
VE		1.000	0.610	0.485	0.436
AL			1.000	0.711	0.665
AN				1.000	0.607
ST					1.000

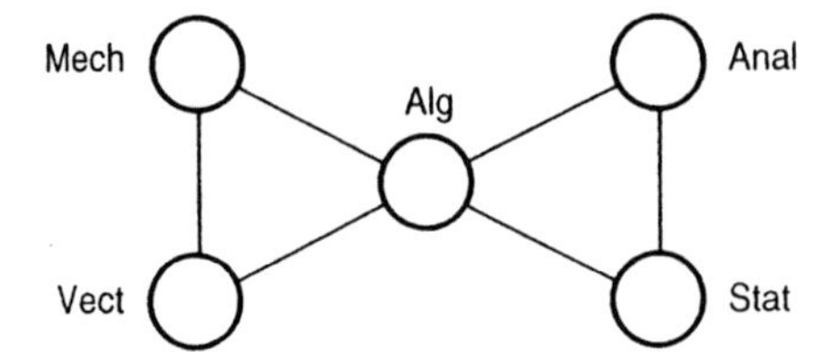

Figure 11.2 Graphical model of butterfly type for the mathematics marks.

Apart from the fact that all correlations are positive, this is not very informative. The inverse of this matrix, scaled to have unit elements on the diagonal, is:

	ME	VE	AL	AN	ST
ME	1.000	−0.329	−0.230	0.002	−0.025
VE		1.000	−0.281	−0.078	−0.020
AL			1.000	−0.432	−0.357
AN				1.000	−0.253
ST					1.000

The off-diagonal elements of this matrix are the partial correlations (see Part 1, **2.6**). Four of them are well below their standard errors (about 0.11), and this suggests a model with conditional independence for these pairs of variables. When deleting several edges simultaneously, it is always wise to check the associated likelihood ratio test. Here, $\chi^2_{(4)} = 0.895$, and the conditional independence model is clearly acceptable. The correlation matrix for this model shows small changes in the four corresponding elements and no changes elsewhere, while the scaled inverse correlation matrix has zeros for the four elements and only small changes elsewhere. The scaled inverse matrix is now:

	ME	VE	AL	AN	ST
ME	1.000	−0.332	−0.235	0.000	0.000
VE		1.000	−0.327	0.000	0.000
AL			1.000	−0.451	−0.364
AN				1.000	−0.256
ST					1.000

This leads to a butterfly model (Fig. 11.2), in which statistics and analysis on one side are independent of mechanics and vectors on the other, conditionally upon algebra. It is tempting to think of a general factor accounting for the correlations with algebra, with specific factors for the two pairs of connected subjects, but a factor model with five variates and three factors is unidentifiable (see **12.18**).

Log-linear models

11.8 The interpretation of multi-way contingency tables using log-linear models is treated in another monograph in this series, and will be mentioned only briefly here. Everitt (1977) is an excellent introduction to the subject; Bishop *et al* (1975) is a comprehensive treatise with many examples.

The full model for the three-way table may be written

$$\log f_{ijk} = u + u_{1(i)} + u_{2(j)} + u_{3(k)} + u_{12(ij)} + u_{13(ik)} + u_{23(jk)} + u_{123(ijk)} \quad (11.5)$$

where f_{ijk} is the frequency in the cell (ijk), and the u terms represent the mean, main effects, two-factor interactions and so on in the same way as the usual parametrisation of models for factorial designs. A suitable set of constraints must be imposed to make the parameters identifiable; for example, the sum of each parameter over its indices may be defined to be zero (the usual parametrisation for factorial designs) or each may be defined to be zero when all its subscripts are 1 (the GLIM parametrisation). The extension to more than three dimensions is obvious. The restriction to binary factors is an important simplification, often appropriate in medical and psychological applications.

This model, the saturated model, can be fitted exactly to any set of frequencies. It is therefore uninformative, and the main purpose of fitting log-linear models is to find simpler models, in which some of the parameters are set to zero, that are easier to interpret.

One important class of models embodies the *hierarchy principle* (Bishop, 1971). This states that if any parameter is set to zero, all higher order interactions involving the same suffixes must also be set to zero. The corresponding rule in factorial designs is the *marginality principle* (Nelder, 1977). For example, a model with $u_{12} = 0$ in all cells but with non-zero values of u_{123} is non-hierarchical.

In most situations, it is common sense that a factor that has no effect cannot interact with another, and here it will be assumed that the hierarchy principle holds. Models in which two factors are completely ineffective on their own, but do have an effect in combination, known as synergic models, are not impossible but rarely occur in practice.

If one or more of the two-factor u-parameters are set to zero, a corresponding independence graph can be constructed. The graph, however, does not uniquely define the model. A *graphical model* is one in which certain two-factor terms are set to zero, implying further zeros by the hierarchy principle, but in which no other parameters are set to zero.

Example 11.3

Consider a five-way table in which factors 1 and 2 are independent. The graph has no edge {12}. The parameters $u_{12} = 0$ for all levels, and by the hierarchy principle the same is true for u_{123}, u_{124}, u_{125}, u_{1234}, u_{1235}, u_{1245} and u_{12345}. It may also be reasonable to postulate $u_{345} = 0$ (entailing $u_{1345} = 0$ and $u_{2345} = 0$), but this is not a graphical model.

Most statistical computer packages have facilities for fitting log-linear models. These are usually based on a quasi-Newton algorithm, as used in GLIM, or on iterative proportional scaling (Haberman, 1974). These are essentially the same methods as used for the multivariate normal models and are discussed in **11.6**. The program CoCo (Badsberg, 1991, 1992, 1993) fits hierarchical log-linear models, and displays the corresponding independence graphs.

Hierarchical conditional Gaussian models

11.9 The conditional Gaussian distribution (see Part 1, **2.25**) is an appropriate model for many situations involving both categorical and continuous variates. Standard linear and quadratic discriminant analysis is based on a model of this type, with a single categorical variable representing the classes. Many data sets involve both types of variable, and the 'location model' for discrimination (Olkin and Tate, 1961; Krzanowski, 1980, 1993a) is an example of the conditional Gaussian model in which one of the categorical variables is regarded as dependent, and is to be predicted from the combination of the continuous variables and the remaining categorical variables (see **9.13**).

The general conditional Gaussian model is flexible and plausible in many situations, but the number of parameters becomes very large as the number of categories increases. The development of hierarchical independence models for this distribution has been an important advance. Edwards (1993) has written the program MIM for fitting these models; the theory is set out in Edwards (1990), Wermuth and Lauritzen (1990), and the discussion following these two papers.

The following discussion follows the notation of Lauritzen (1993).

Suppose $\boldsymbol{x} = (\boldsymbol{i}, \boldsymbol{y})$ represents a vector of p discrete variates $\boldsymbol{i}$, defining a set of locations, and q continuous variates $\boldsymbol{y}$. The continuous variates are assumed to follow a multivariate normal distribution with means and covariance matrices dependent on the location. If each of the discrete variables is binary, there are 2^p locations, and $2^{p-1}q(q+3)$ parameters associated with the normal distributions. Suppose $\Pr(\boldsymbol{i})$ is the probability of location $\boldsymbol{i}$. Then the conditional Gaussian distribution has the form:

$$\begin{aligned} f(\boldsymbol{x}) &= f(\boldsymbol{i}, \boldsymbol{y}) \\ &= \Pr(\boldsymbol{i})(2\pi)^{-q/2}|\boldsymbol{\Sigma}(\boldsymbol{i})|^{-1/2}\exp\left[-\frac{1}{2}\{\boldsymbol{y}-\boldsymbol{\mu}(\boldsymbol{i})\}'\boldsymbol{\Sigma}(\boldsymbol{i})^{-1}\{\boldsymbol{y}-\boldsymbol{\mu}(\boldsymbol{i})\}\right] \end{aligned} \tag{11.6}$$

In this equation, $\boldsymbol{\mu}(\boldsymbol{i})$ and $\boldsymbol{\Sigma}(\boldsymbol{i})$ are the mean and covariance matrix of $\boldsymbol{y}$ at location $\boldsymbol{i}$.

This is the familiar parametrisation of the multivariate normal distribution at each location. In discriminant analysis with mixed discrete and continuous variables, it is often assumed that the covariance matrix is independent of location and population. In the conditional Gaussian formulation, this implies $\boldsymbol{\Sigma}(\boldsymbol{i}) = \boldsymbol{\Sigma}$. This is referred to as the homogeneous conditional Gaussian distribution.

The parametrisation above is not the most convenient for constructing conditional independence models, and may be replaced by the *canonical* parametrisation, as in the parametrisation of the exponential family,

$$\begin{aligned} f(\boldsymbol{x}) &= f(\boldsymbol{i}, \boldsymbol{y}) \\ &= \exp\left\{\alpha(\boldsymbol{i}) + \boldsymbol{\beta}'(\boldsymbol{i})\boldsymbol{y} - \frac{1}{2}\boldsymbol{y}'\boldsymbol{\Omega}(\boldsymbol{i})\boldsymbol{y}\right\} \end{aligned}$$

The relationships between the two sets of parameters may be expressed as

$$\begin{aligned}
\Pr(\boldsymbol{i}) &= (2\pi)^{q/2}|\boldsymbol{\Omega}(\boldsymbol{i})|^{-1/2}\exp\left\{\alpha(\boldsymbol{i}) + \frac{1}{2}\beta(\boldsymbol{i})'\boldsymbol{\Omega}(\boldsymbol{i})\beta(\boldsymbol{i})\right\} \\
\mu(\boldsymbol{i}) &= \boldsymbol{\Omega}(\boldsymbol{i})^{-1}\beta(\boldsymbol{i}) \\
\boldsymbol{\Sigma}(\boldsymbol{i}) &= \boldsymbol{\Omega}(\boldsymbol{i})^{-1}
\end{aligned}$$

or as

$$\begin{aligned}
\alpha(\boldsymbol{i}) &= \log\Pr(\boldsymbol{i}) - \frac{1}{2}\log|\boldsymbol{\Sigma}(\boldsymbol{i})| - \frac{1}{2}\mu(\boldsymbol{i})'\boldsymbol{\Sigma}(\boldsymbol{i})^{-1}\mu(\boldsymbol{i}) - \frac{q}{2}\log(2\pi) \\
\beta(\boldsymbol{i}) &= \boldsymbol{\Sigma}(\boldsymbol{i})^{-1}\mu(\boldsymbol{i}) \\
\boldsymbol{\Omega}(\boldsymbol{i}) &= \boldsymbol{\Sigma}(\boldsymbol{i})^{-1}
\end{aligned}$$

The new parameters $\alpha(\boldsymbol{i})$, $\beta(\boldsymbol{i})$, and $\boldsymbol{\Omega}(\boldsymbol{i})$ are referred to as, respectively, the discrete, linear and quadratic canonical parameters of the conditional Gaussian distribution.

The next step is to express the canonical parameters, $\alpha(\boldsymbol{i})$ and the individual terms of $\beta(\boldsymbol{i})$ and $\boldsymbol{\Omega}(\boldsymbol{i})$, in terms of interactions among the discrete variables in the same way as in the log-linear model. Formally, this may be written

$$\begin{aligned}
\alpha(\boldsymbol{i}) &= \sum_{a\subseteq\Delta}\lambda(\boldsymbol{i}_a) \\
\beta(\boldsymbol{i}) &= \sum_{a\subseteq\Delta}\eta(\boldsymbol{i}_a) \\
\boldsymbol{\Omega}(\boldsymbol{i}) &= \sum_{a\subseteq\Delta}\psi(\boldsymbol{i}_a)
\end{aligned}$$

In these expressions, the summation is over all subsets a of the set Δ of discrete variates. As in the log-linear model, these represent a constant term, a set of main effects, a set of two-factor interactions, and so on. As in the log-linear model, also, a constraint is needed to ensure identifiability.

The density may then be written:

$$f(\boldsymbol{i}, \boldsymbol{y}) = \exp\left\{\sum_{a\subseteq\Delta}\lambda(i_a) + \sum_{r=1}^{q}\sum_{a\subseteq\Delta}\eta_r(i_a)y_r - \frac{1}{2}\sum_{r=1}^{q}\sum_{s=1}^{q}\sum_{a\subseteq\Delta}\psi_{rs}(i_a)y_r y_s\right\} \quad (11.7)$$

Now it is easy to specify hierarchical conditional independence models.

- Conditional independence of two discrete variates implies that all parameters (λ, η and ψ) involving a subset containing those variables are zero. The same rule extends to subsets of more than two discrete variables, giving hierarchical models that are not graphical.
- Conditional independence of a discrete and continuous variate implies that all parameters η and ψ involving the two variates are zero. Again, this extends to subsets of the discrete variates.
- Conditional independence of two continuous variates implies that all parameters ψ involving the two variates are zero; in other words, the corresponding element of $\boldsymbol{\Omega}$ is zero at every location.

Example 11.4 (Whittaker, 1990)
Consider the saturated model for a conditional Gaussian distribution with $p = 2$, $q = 2$ involving two binary variates i, j and two continuous variates, say (x, y). The distribution thus consists of four bivariate normal distributions at locations specified by (i, j) = (0,0),(0,1),(1,0) and (1,1). There are 24 parameters; in the original parametrisation, there are two means, two variances and a covariance at each location, plus four probabilities. There is one constraint, namely, that the four probabilities add to one.

In terms of the new parameters, the distribution may be written:

$$\begin{aligned}\log f(i,j,x,y) = \lambda_0 + \lambda_1 i + \lambda_2 j + \lambda_{12} ij \\ + \{\eta_{X,0} + \eta_{X,1} i + \eta_{X,2} j + \eta_{X,12} ij\}x \\ + \{\eta_{Y,0} + \eta_{Y,1} i + \eta_{Y,2} j + \eta_{Y,12} ij\}y \\ - \frac{1}{2}\{\psi_{XX,0} + \psi_{XX,1} i + \psi_{XX,2} j + \psi_{XX,12} ij\}x^2 \\ - \frac{1}{2}\{\psi_{YY,0} + \psi_{YY,1} i + \psi_{YY,2} j + \psi_{YY,12} ij\}y^2 \\ - \{\psi_{XY,0} + \psi_{XY,1} i + \psi_{XY,2} j + \psi_{XY,12} ij\}xy\end{aligned}$$

It is now possible to define reduced models by imposing constraints on these parameters. There are three types of conditional independence:

- $i \perp j | x, y$. The six parameters
  $\lambda_{12}, \eta_{X,12}, \eta_{Y,12}, \psi_{XX,12}, \psi_{YY,12}, \psi_{XY,12}$
  are set equal to zero.
- $i \perp x | j, y$. The six parameters
  $\eta_{X,1}, \eta_{X,12}, \psi_{XX,1}, \psi_{XX,12}, \psi_{XY,1}, \psi_{XY,12}$
  are set equal to zero.
- $x \perp y | i, j$. The four parameters
  $\psi_{XY,0}, \psi_{XY,1}, \psi_{XY,2}, \psi_{XY,12}$
  are set to zero.

Each of these constraints may be tested by a likelihood ratio test. Conditional independence constraints may be combined to give the complete range of graphical models for the conditional Gaussian distribution. These may be tested similarly, but care is needed in determining the degrees of freedom. Also, of course, the simultaneous testing aspect of the procedure means that interpretation as formal statistical inference is not possible.

Other hierarchical, non-graphical, models can be defined and tested. For example, setting $\eta_{X,12}$, $\psi_{XX,12}$ and $\psi_{XY,12}$ equal to zero gives a model without a three-factor interaction. The homogeneous model, with the same covariance matrix at each location, is given by setting the nine parameters

$$\psi_{XX,1}, \psi_{XX,2}, \psi_{XX,12}, \psi_{YY,1}, \psi_{YY,2}, \psi_{YY,12}, \psi_{XY,1}, \psi_{XY,2}, \psi_{XY,12}$$

equal to zero; this is not a hierarchical model.

Fitting these models is not straightforward. Edwards (1990) and Whittaker (1990, p357) describe how iterative proportional scaling can be used, and the program MIM (Edwards, 1993) fits conditional Gaussian models using this algorithm.

This concludes the discussion of graphical and hierarchical interaction models; in the final sections we discuss some of the problems that arise when covariance matrices are singular, or nearly singular.

Path analysis

11.10 Path analysis has been mentioned as a precursor of graphical modelling. The aim of constructing models is to find a simplified structure that fits the data and can be interpreted. Path analysis starts with a structure known *a priori*; variables are linked by a *directed* graph, representing cause and effect. This graph may be modified in the course of the analysis by deleting links found to be superflous, but not by adding new links. The aim is to find out the extent to which certain variables are affected by others in the model.

Path analysis was introduced by Wright (1921) as a tool for the analysis of data concerned with population genetics. Wright (1934) is a full and formal account of the method. Applications have also been found in econometrics, and in the social sciences, where the techniques are often referred to as 'structural equation models'. There are considerable conceptual difficulties involved, and some of the early work seemed to be intuitive, rather than based on a firm mathematical foundation. Wermuth (1980) and Kang and Seneta (1980) are expositions of the subject from a statistical point of view.

A set of $p + q$ variables

$$\begin{aligned} \boldsymbol{x}' &= (x_1, \ldots, x_p, x_{p+1}, \ldots, x_{p+q}) \\ &= (\boldsymbol{z}', \boldsymbol{y}'), \text{ say,} \end{aligned}$$

is divided into two groups. The p variables in the first group, $\boldsymbol{z}$, act on the system from outside; no attempt is made to model their behaviour. They are known as 'exogenous' variables. The remainder, the 'endogenous' variables, may be affected by the exogenous variables, and by each other. The underlying model defines a subset of the complete set of variables that may affect each of the endogenous variables. These subsets define the 'path diagram', in which directed lines link each of the exogenous variables to one or more of the endogenous variables, and also link certain pairs of endogenous variables. These lines define the variables regarded as direct causes of other variables. Variables may also have an indirect causal effect when a directed path links them to other variables only through intermediate variables.

This structure is represented by a linear model of the form

$$\boldsymbol{y}_{(q\times 1)} = \boldsymbol{B}_{(q\times q)}\boldsymbol{y}_{(q\times 1)} + \boldsymbol{\Gamma}_{(q\times p)}\boldsymbol{z}_{(p\times 1)} + \boldsymbol{u}_{(q\times 1)} \tag{11.8}$$

In this equation, $\boldsymbol{B}$ and $\boldsymbol{\Gamma}$ represent matrices of unknown coefficients, with zero elements wherever there is not a causal link in the path diagram, and $\boldsymbol{u}$ is a vector of error terms. The form of the equation immediately suggests that there may be problems of identifiability. This is the first point that has concerned statisticians in the past. In fact, if there are directed cycles in the path diagram, so that some endogenous variables are regarded as indirectly causing themselves, the model is said to be recursive. That, fairly obviously, implies that unique estimates do not exist.

If the model is not recursive, $\boldsymbol{I}-\boldsymbol{B}$ is non-singular, and (11.8) may be written

$$y = (\boldsymbol{I}-\boldsymbol{B})^{-1}\boldsymbol{\Gamma}z + \boldsymbol{v} \qquad (11.9)$$

Equation (11.9) has the form of a multivariate regression of $\boldsymbol{y}$ on z, and a possible method of estimating the non-zero elements of $\boldsymbol{B}$ and $\boldsymbol{\Gamma}$ would be to treat $\boldsymbol{v} = (\boldsymbol{I}-\boldsymbol{B})^{-1}\boldsymbol{u}$ as an error term and estimate the non-zero elements to minimise the generalised variance of $\boldsymbol{v}$ subject to appropriate constraints on the matrix $(\boldsymbol{I}-\boldsymbol{B})^{-1}\boldsymbol{\Gamma}$ of regression coefficients (at least, when the terms are identifiable). This method is awkward, because of the nature of the constraints, and most methods try to fit (11.8) directly. For details of the various approaches, see Wermuth (1980) and Kang and Seneta (1980).

Path analysis remains controversial, but Wright's model has had an enormous influence on modern statistical methods. Apart from graphical modelling, neural networks (**9.61**) and the design of Expert Systems owe much to the concept. Like the factor analysis model, it was introduced before the statistical methods and computing power needed for its efficient fitting were available, and in consequence was, for a long time, neglected by statisticians.

Example 11.5 (Wright, 1960)
Figure 11.3 shows a complex path diagram, representing the effects of various factors on weight of guinea pigs at 33 days (W). W is directly affected by birth weight B and gain G after birth; G is also affected by B. The gestation period A affects both B and G, and litter size L affects G directly, and through A. For details of the other variables see Wright (1960).

The arrowed lines represent the causal effects assumed in constructing the model, and the correlation coefficients beside them are estimated from the data. The dotted lines represent effects that were assumed in constructing the model, but found to be unimportant.

Regression models with multicollinearity

11.11 The problem of multicollinearity in regression is not new, but methods of automatic data collection have given rise to a number of problems in which it appears in an extreme form. Particularly in chemistry, the routine use of spectrophotometry gives a set of variables $\boldsymbol{x}$ representing transmission or absorption at different wavelengths. The values actually correspond to a continuous absorption curve, so the number of variables is determined by the setting of the recording device; typically, it will be hundreds or even thousands. The aim is to use these variables in prediction of a variable y, which may be a single continuous variable, such as toxicity or the concentration of a chemical component, or a vector of continuous variables, such as the concentration of a number of components, or a discrete classificatory variable in a discrimination

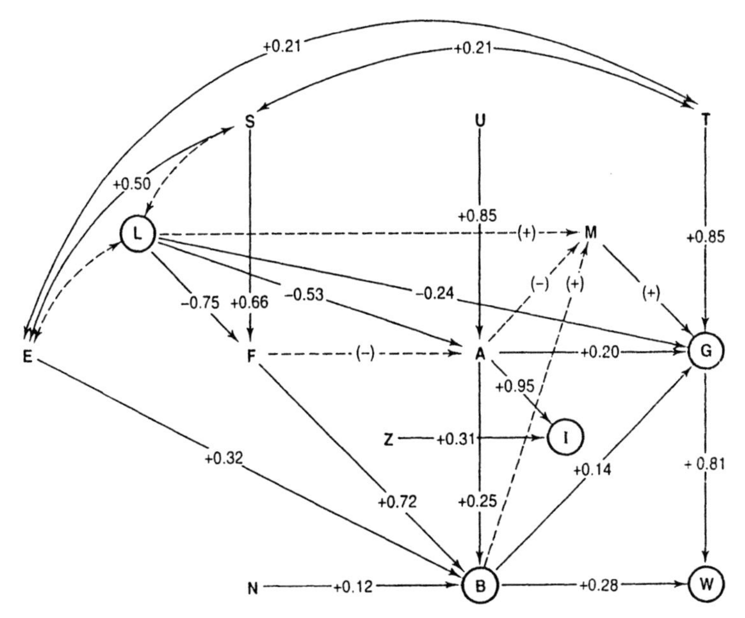

Figure 11.3 Path diagram showing the relationships among variables relating to the development of guinea pigs. Reproduced from Wright (1968) with permission from the University of Chicago Press.

problem, such as whether the substance has a particular useful activity. Generally, y is difficult to determine, involving a full chemical analysis or some form of bioassay, while the variables $\boldsymbol{x}$ are easy to obtain. This leads to problems in which the number of x variables p is much larger than the number of observations, n.

Generally, in these problems y is known accurately, while the x variables are subject to random variation, owing to minor contaminants, instrumental errors, and so on. The problem is therefore one of multivariate calibration (Part 1, **7.3** and **14.11**), rather than of regression in the classical sense. Nevertheless, it is usually dealt with by standard regression procedures rather than the reverse regression of $\boldsymbol{x}$ on y. For a full discussion of calibration problems, see Brown (1982, 1993) or Martens and Naes (1989).

In this situation, $p \gg n$, variable selection is not an attractive option. The important features in the data, for example peaks in an absorption curve, are probably represented by contrasts among groups of the x variables, and stepwise selection of single variables is unlikely to pick out meaningful combinations. Since any set of $n - 1$ variables will, with the constant term, give a perfect representation of the observed y, there is a serious risk that the chosen subset will be artefactual, with no predictive value.

11.12 A more attractive alternative is to reduce the x variables to $r < n$ linear combinations, chosen to represent the 'information' in them as well as possible, and to carry out regression of y on these combinations. This is sometimes called 'bilinear modelling' (Martens and Naes, 1989). The choice of suitable linear combinations may or may not make explicit assumptions about latent variables associated with $\boldsymbol{x}$ and y, but there is in any case a relationship with networks involving a hidden layer of cells.

The general bilinear calibration model may be put in the form:

$$\underset{n \times p}{\boldsymbol{X}} = \underset{n \times r}{\boldsymbol{T}} \; \underset{r \times p}{\boldsymbol{\Theta}'} + \underset{n \times p}{\boldsymbol{E}}$$

$$\underset{n \times q}{\boldsymbol{Y}} = \underset{n \times r}{\boldsymbol{T}} \; \underset{r \times q}{\boldsymbol{\Phi}'} + \underset{n \times q}{\boldsymbol{F}}$$

$$\underset{n \times r}{\boldsymbol{T}} = \underset{n \times p}{\boldsymbol{X}} \; \underset{p \times r}{\boldsymbol{V}}$$

In these equations, $\boldsymbol{X}$ and $\boldsymbol{Y}$ are centred data matrices, and the columns of $\boldsymbol{T}$ are the derived vectors intended to contain the relevant information in $\boldsymbol{X}$. $\boldsymbol{E}$ and $\boldsymbol{F}$ are matrices of residuals; in the standard regression model $\boldsymbol{E}$ is assumed to be zero. The matrix $\boldsymbol{V}$ determines the linear transformation of $\boldsymbol{X}$; it may be based on $\boldsymbol{X}$, or on $\boldsymbol{X}$ and $\boldsymbol{Y}$.

An obvious choice for $\boldsymbol{T}$ is the set of the first r principal components of $\boldsymbol{X}$; see Part 1, Chapter 4. The value of r may be chosen arbitrarily, or using a scree plot or similar criterion. Principal component regression has several advantages:

(1) The components are orthogonal, and the regression analysis is thus particularly simple.
(2) The components are combinations of the regressor variables, depending on the complete set or on groups rather than single variables.
(3) Interpretation of the components may be possible; for example, in spectrophotometry certain components may correspond to peaks in particular positions.
(4) Exact statistical inference is possible. The chosen components may be treated in the same way as a set of r regressor variables, provided r is chosen without reference to the relationships of the components with the variables in $\boldsymbol{Y}$.

There is, however, one serious drawback. There is no reason why the components with highest variance should be useful for predicting the dependent variables. This point has been emphasised by Chang (1983), Jolliffe (1986) and others, and many real and artificial data sets show that the information relevant to the prediction of $\boldsymbol{Y}$ may be contained in low ranked principal components, which might be left out of the chosen subset, or at least might be masked by the inclusion of irrelevant components of higher variance.

Krzanowski (1992) suggested the use of principal components ranked, not by their variances, but rather by their correlations (canonical correlations in the case of multiple dependent variables) with $\boldsymbol{y}$. This is equivalent, since the components are orthogonal, to stepwise selection from the complete set of n principal components. The new ranking ensures that the most important components for prediction are not dropped. Statistical inference is no longer possible, but that is probably a small loss compared with the likely gain in predictive power.

11.13 H. Wold introduced the method of partial least squares in the 1970s; see Wold (1975, 1985). The earliest, and many of the most important, applications involve a single dependent variable, and the assumption that y is scalar simplifies the argument and the algorithms. The extension to multiple y variables will be deferred. The underlying idea is that the relationship between y and $\boldsymbol{x}$ depends on a small number of latent factors. Both observable sets of variables are subject to random variation, but latent variables that are not related to *both* sets are not considered relevant. It is assumed that relationships with the latent variables are linear, and that unexplained variation is reasonably regular, so that least squares is a suitable method of estimation. The practical algorithm involves maximising the *covariance* of y with a constrained linear function of the x variables. Regression and canonical analysis of course maximise correlation.

There are many variations and extensions to the basic partial least squares algorithm, and the method is obviously closely related to other approaches based on linear modelling, with or without explicit reference to latent variables.

(1) Principal components are derived entirely from the $\boldsymbol{x}$ variables, and, unlike partial least squares, do not postulate latent variables (**12.43**). They are merely an attempt to condense the information in $\boldsymbol{X}$, and when ordered by their variances, may be quite inefficient for prediction of y.

(2) Classical factor analysis (Chapter 12) does postulate latent factors, but again there is no mechanism for picking out the factors that are relevant to y.

(3) Structural analysis considers both y and $\boldsymbol{x}$ as functions of unobservable 'causal' variables, and the unobservable sets are deterministically related. The best known model of this type was introduced by Jöreskog (1973a), and forms the basis of the LISREL (LInear Structural RELationships) computer program (**12.37**). The algorithm was developed at about the same time as partial least squares, and Wold (1985) gives a detailed comparison of the two approaches. LISREL, like factor analysis, estimates parameters using either maximum likelihood or unweighted least squares on the elements of covariance matrices. In maximum likelihood form it is severely restricted in the number of variables it can handle, whereas partial least squares works well even with very large numbers of variables.

(4) Path analysis covers a variety of techniques describing relationships among sets of (manifest or latent) variables. Wright (1921) introduced the idea of graphs representing causal relationships in the context of genetics. Since then, it has been widely used for modelling econometric and sociological structures, in particular. Partial least squares can be regarded as one of the techniques in this group. Wermuth (1980) surveys the general field of linear recursive equations, including linear path analysis, partial least squares, and covariance selection.

(5) Redundancy analysis (see Part 1, **7.10**), is closely related and, like partial least squares, is scale dependent. It concentrates, however, on minimising the trace of the covariance matrix of the y variables, and depends on the covariance matrix of the explanatory variables being of full rank.

(6) Canonical analysis is scale independent and symmetrical in the two sets of variables, but of course cannot give a unique answer when there are more explanatory variables than observations.

11.14 The original partial least squares algorithm was developed by S. Wold; see, for example, Wold *et al* (1983). The variables y and $\boldsymbol{x}$ are assumed to be related through A latent variables $t_1, \ldots, t_A$, which are defined to be orthogonal. These factors are estimated in order. The number of latent variables A is unknown, but the calculation starts with a value $A_{\max}$ as the largest acceptable value of A—usually rather higher than the number of latent variables expected—and rejects those that are judged to be insufficiently correlated with y. Finally, y is predicted by linear regression on $\boldsymbol{t}$, or equivalently as a linear function of $\boldsymbol{x}$.

The algorithm given below is based on Martens and Naes (1989); see also Garthwaite (1994) for further elucidation.

Step 1. Scale the observed vector $y' = (y_1, \ldots, y_n)$ and the data matrix X, $n \times p$, typically by scaling each variable to unit variance, and centre each variable to zero mean, to give starting values $\boldsymbol{y}_0$ and $\boldsymbol{X}_0$. Choose $A_{\max}$.
For $a = 1$ to $A_{\max}$ carry out Steps 2 to 6.

Step 2. Regress each variable x_{a-1} on y_{a-1}, giving

$$\boldsymbol{X}_{a-1} = \boldsymbol{y}_{a-1}\boldsymbol{w}'_a + \boldsymbol{E}$$

and scale the vector of weights to length $\boldsymbol{w}'_a\boldsymbol{w}_a = 1$. This gives

$$\hat{\boldsymbol{w}}_a = c\boldsymbol{X}'_{a-1}\boldsymbol{y}_{a-1}$$

where

$$c = (\boldsymbol{y}'_{a-1}\boldsymbol{X}_{a-1}\boldsymbol{X}'_{a-1}\boldsymbol{y}_{a-1})^{-1/2}$$

Step 3. Regress each variable x_{a-1} on $\hat{\boldsymbol{w}}_a$, giving

$$\boldsymbol{X}_{a-1} = \boldsymbol{t}_a\hat{\boldsymbol{w}}'_a + \boldsymbol{E}$$

This gives the estimated scores $\hat{\boldsymbol{t}}_a$ as

$$\hat{\boldsymbol{t}}_a = \boldsymbol{X}_{a-1}\hat{\boldsymbol{w}}_a$$

since

$$\hat{\boldsymbol{w}}'_a\hat{\boldsymbol{w}}_a = 1$$

Step 4. Estimate the x loadings $\boldsymbol{p}_a$ by regressing each variable x_{a-1} on $\hat{\boldsymbol{t}}_a$, giving

$$\boldsymbol{X}_{a-1} = \hat{\boldsymbol{t}}_a\boldsymbol{p}'_a + \boldsymbol{E}$$

or

$$\hat{\boldsymbol{p}}_a = \boldsymbol{X}'_{a-1}\hat{\boldsymbol{t}}_a/\hat{\boldsymbol{t}}'_a\hat{\boldsymbol{t}}_a$$

Step 5. Estimate the y loading q_a by regressing $\boldsymbol{y}_{a-1}$ on $\hat{\boldsymbol{t}}_a$ giving

$$\boldsymbol{y}_{a-1} = \hat{\boldsymbol{t}}_a q_a + \boldsymbol{f}$$

or

$$\hat{q}_a = \boldsymbol{y}'_{a-1}\hat{\boldsymbol{t}}_a/\hat{\boldsymbol{t}}'_a\hat{\boldsymbol{t}}_a$$

Step 6. Define new variables

$$\boldsymbol{X}_a = \boldsymbol{X}_{a-1} - \hat{\boldsymbol{t}}_a \hat{\boldsymbol{p}}'_a$$
$$\boldsymbol{y}_a = \boldsymbol{y}_{a-1} - \hat{\boldsymbol{t}}_a \hat{q}_a$$

and set $a = a + 1$.

The last step defines new variables as the residuals at each stage in the calculation. This makes it possible to check for outliers and to examine the change in the prediction error corresponding to $\boldsymbol{y}_{a-1}$ and $\boldsymbol{y}_a$. On the basis of these calculations, the value of A, the number of factors to retain in the final model, is decided.

Prediction of a new value $\hat{y}_i$ from a vector of observations $\boldsymbol{x}_i$ is then given by the linear predictor

$$\hat{y}_i = \hat{b}_0 + \boldsymbol{x}'_i \hat{\boldsymbol{b}}$$

where

$$\hat{\boldsymbol{b}} = \hat{\boldsymbol{W}}(\hat{\boldsymbol{P}}'\hat{\boldsymbol{W}})^{-1}\hat{\boldsymbol{q}}$$
$$\hat{b}_0 = \bar{y} - \bar{\boldsymbol{x}}'\hat{\boldsymbol{b}}.$$

In these equations, $\hat{\boldsymbol{W}}$ and $\hat{\boldsymbol{P}}$ are $A \times n$ matrices formed from the vectors $\hat{\boldsymbol{w}}_a$ and $\hat{\boldsymbol{p}}_a$ with $a = 1, \dots, A$.

Alternatively, prediction may proceed by estimating the factors $\hat{t}_{i,a}$ and using them directly. Since one of the purposes of partial least squares is to find a meaningful factor for the calibration, this method may give more insight into the characteristics of $\boldsymbol{x}_i$ that lead to the prediction of y_i.

The method of partial least squares has two objectives, first to find common latent variables associated with both y and $\boldsymbol{x}$ and, if possible, to interpret them, and secondly to find a prediction formula for y from $\boldsymbol{x}$ that will be valid for observations not in the training set. In fact, the first partial least squares factor $\hat{\boldsymbol{t}}_1 = \boldsymbol{X}\boldsymbol{w}_1$ is the linear combination of x-variables that has maximum covariance with $\boldsymbol{y}$, conditional on $\boldsymbol{w}'_1\boldsymbol{w}_1 = 1$, and the other factors have the same property conditional on orthogonality (Frank, 1987; Höskuldsson, 1988; Helland, 1990). This seems a reasonable choice, and in practice it often works well, but Exercises 11.4 and 11.5 below show that it may not always give sensible results.

11.15 The extension to multivariate $\boldsymbol{y}$ is straightforward, although the computations are more elaborate. Linear combinations $\boldsymbol{v}'\boldsymbol{y} = \boldsymbol{u}$ and $\boldsymbol{w}'\boldsymbol{x} = \boldsymbol{t}$ are chosen to maximise the covariance cov$(\boldsymbol{u}, \boldsymbol{t})$, subject to $\boldsymbol{w}'_1\boldsymbol{w}_1 = 1$ and $\boldsymbol{v}'_1\boldsymbol{v}_1 = 1$. Further components are defined as in the case of a single y variable, so that $\boldsymbol{t}_i, \boldsymbol{t}_j$ are orthogonal. Components are calculated until most of the variability in $\boldsymbol{y}$ is accounted for.

This extension, however, is of limited value. The aim of partial least squares is prediction of $\boldsymbol{y}$. Further, the $\boldsymbol{t}$ variables may be interpretable as latent variables affecting $\boldsymbol{y}$. This extension effectively assumes that the underlying factors affecting the different components are *the same*. If this is so, the components of $\boldsymbol{y}$ are likely to be highly correlated, and combined estimation is best. Otherwise, it is easier to carry out the partial least squares procedure for each component separately, and if in fact the latent variables are different, the estimates of $\boldsymbol{t}$ will be easier to interpret.

Partial least squares, which maximises a covariance, is in a sense intermediate between ordinary least squares, which maximises a correlation, and principal component regression, which maximises a variance. It is a compromise, aimed at finding meaningful components that are also good predictors. Stone and Brooks (1990) describe a continuous range of prediction procedures ('continuum regression') with principal components and ordinary least squares at the extremes and partial least squares intermediate.

In conclusion, partial least squares is an exploratory technique that works best with explanatory variables of a similar type—spectral absorbance or reflectance in chemistry, economic indices in econometric applications, and scores on the same scale in psychological statistics. Practical experience suggests that in these situations it often works well, although attempts to fit it in a framework of formal statistical inference seem unconvincing. There seems to have been no serious attempt to compare partial least squares with Krzanowski's (1992) use of reordered principal components.

11.16 There are many other approaches to the general problem of multicollinearity. Krzanowski *et al* (1995) discuss and compare them; see also **14.15** in this volume.

Another area in which the modelling of covariance structures is important is in the analysis of (univariate and multivariate) growth curves and experiments with repeated measures. Chapter 13 of this volume deals with data of this type.

Exercises

11.1 $\boldsymbol{S}$ is the estimated variance-covariance matrix of a 3-variate normal distribution. Show that the maximum likelihood estimate of the covariance matrix for the model with x_1, x_2 conditionally independent is the matrix $\boldsymbol{S}$ with s_{12} replaced by $s_{13}s_{23}/s_{33}$.

11.2 Derive equations (11.2), (11.3), and (11.4) as special cases of equation (11.1).

11.3 Show that the first partial least squares factor $\hat{\boldsymbol{t}}_1 = \boldsymbol{X}\boldsymbol{w}_1$, as defined in the algorithm in **11.14**, is the linear combination of x-variables that has maximum covariance with $\boldsymbol{y}$, conditional on $\boldsymbol{w}_1'\boldsymbol{w}_1 = 1$, as stated in **11.14.**

11.4 Suppose $n = 3$, $p = 2$, $\boldsymbol{y}' = (-1, 0, 1)$, $\boldsymbol{x}_1' = (49, -100, 51)$, $\boldsymbol{x}_2' = (50, -100, 50)$. Show that $\boldsymbol{t}_1 = \boldsymbol{x}_1$. This variable has correlation 0.01 with $\boldsymbol{y}$, while in fact $\boldsymbol{y} = \boldsymbol{x}_1 - \boldsymbol{x}_2$.

(G.Weatherhill, cited in Helland, 1990)

11.5 Suppose $n = 3$, $p = 700$, $\boldsymbol{y}' = \boldsymbol{x}'_1 = (-1, 0, 1)$, and $\boldsymbol{x}'_2 = \ldots = \boldsymbol{x}'_{700} = (0, -1, 1)$. Show that $\boldsymbol{w}_1' = (2, 1, \ldots, 1)/\sqrt{703}$, giving a correlation with $\boldsymbol{y}$ of approximately 0.5.

(G.Weatherhill, cited in Helland, 1990)

12

Latent Structure Models

Introduction and historical background

12.1 While much of statistical inference and design of experiments owes its beginnings to agricultural science, and to those pioneering workers employed in it who provided the necessary mathematical rigour and logical reasoning, so the now-large body of techniques involving latent variables has arisen and grown almost exclusively within the framework of the social and behavioural sciences. A broad survey of the impact of sociological methodology on the development of statistical methodology has been provided by Clogg (1992). This article, and the accompanying discussion by a set of prominent researchers in the area, gives testimony to the wealth of statistical ideas that have emanated from this quarter — some of which, indeed, are still relatively unfamiliar outside their own areas of application. In this chapter, we focus on just one such idea, that of a latent variable, and outline the rich battery of techniques that are now founded on it.

12.2 A *latent* variable is simply a variable that cannot be measured directly. Many concepts in the social and behavioural sciences, such as *social class*, *personality*, *intelligence* and *ambition*, are of this type. In order to obtain information on such variables researchers are thus forced to consider other variables, which *can* be measured (i.e. are *manifest* variables) and which are related to the latent quantities of interest, but which may contain additional noise or error. For example, we may attempt to measure intelligence by a battery of tests including an I.Q. test, an arithmetic test, a comprehension test and a spelling test, but of course none of these tests is a pure measure of intelligence; arithmetic may additionally involve *numerical ability*, spelling will depend to some extent on *memory*, comprehension on *verbal facility*, and all the tests will be subject to sampling fluctuation as well as unpredictable measurement errors. Thus, an individual's intelligence will need to be *estimated* in some way from that individual's test scores, and to do this a model will first need to be formulated linking the latent and manifest variables. Latent variables can be explicitly built into a statistical analysis in this way; we shall see how this is done later in the chapter.

12.3 However, latent variables may also be employed as convenient descriptive tools by extracting them from an analysis, and this is the use that we examine first. Much research in the social and behavioural sciences involves the measurement of a large number of variables, often highly related ones. It is then tempting to seek just a few (uncorrelated) latent variables that explain, in some way, all the associations existing among the initial manifest variables. The most natural way in which associations can be explained is via the concept of conditional independence outlined in the previous chapter (see also Kiiveri and Speed, 1982), and this in turn leads to a *structural* model for the manifest variables in terms of a set of hypothesised latent variables. Now it turns out that this structural model is precisely the same mathematically as the one introduced in **12.2** above, but the important practical difference is that any extracted latent variables are purely hypothetical quantities that have no *a priori* labels attached to them. The model here simply provides a convenient descriptive summary of the data. It is, of course, entirely possible for the investigator to seek a *post hoc* interpretation of these latent variables (in the manner, say, of a principal component analysis, as in Part 1, Chapter 4), and many investigators expend much energy and ingenuity in such a pursuit. However, even if successful, there is no guarantee that the identified latent variables are 'real' phenomena; this is an area that has given rise to no little controversy and conflicting interpretations, and it may often be better not to attempt spurious interpretation but just to treat the results as a convenient parsimonious description of a complex data set. Certainly it is prudent, if interpreting a set of latent variables in this way and making substantive claims about them, to verify such claims on further independent data sets.

12.4 In practice, observed (i.e. manifest) variables can be of various types but the fundamental dichotomy is between *continuous* and *categorical* ones. Latent variables can also be conceived in either of these forms, thus giving rise to four possible combinations. Each of these possibilities has been studied, and methods of analysis have been developed for all of them. The oldest technique is *factor analysis*, in which both manifest and latent variables are continuous; *latent class analysis* deals with the case when both manifest and latent variables are categorical, *latent trait analysis* when the manifest variables are categorical while the latent variables are continuous, and *latent profile* analysis when the manifest variables are continuous and the latent variables categorical. Each of these techniques is discussed below, as are various extensions that allow ordering among the classes of any categorical variable. More comprehensively, recent work has extended the search for a parsimonious description of, and inference from, covariance and correlation matrices by allowing models that mix manifest and latent variables in various ways. Techniques that arise from this approach are grouped under the general title 'covariance structure modelling', and these techniques are also discussed below.

12.5 On an historical note, the origins of latent variable modelling date from the start of the twentieth century, specifically from the study of human abilities conducted by Spearman. The early studies were almost exclusively focused on the simplest factor analysis model that included just a single factor, but further impetus came in the 1930s with the work of Thurstone and colleagues on multiple-factor models and simple-structure solutions. The chief problems at this time arose from poor estimation procedures, but a start on improved techniques was made by Lawley in the 1940s with his development of the maximum-likelihood approach. However, efficient and reliable estimation was not guaranteed until the computing developments of the 1960s led to the routines in common use today. These same computing advances led directly to the development of covariance structure modelling through the 1970s and 1980s. In the meantime, Lazarsfeld, Anderson, Henry and others developed the latent class model in the early 1950s, and gradually added the other latent structure models. Early estimation was done by a variant of the method of moments known as the determinantal method, but formulation of the E-M algorithm in the late 1970s opened the way to the efficient procedures now in common use.

A good, simple, introduction to the whole area is provided by Everitt (1984a), while Bartholomew (1987) gives a synthesised account at a more mathematical level. Lazarsfeld and Henry (1968) is the classic text for latent class modelling, but of course predates the important computational advances on which modern estimation methods are founded. A similarly classic text on factor analysis is that by Thurstone (1947); more recent views are given by Lawley and Maxwell (1971), Harman (1976) and Cureton and D'Agostino (1983). Covariance structure modelling is covered thoroughly by Bollen (1989).

The nature of the factor analysis model, and some of the estimation procedures for it, has meant that factor analysis and principal component analysis have become inextricably interlinked in many people's minds, and much confusion has arisen between the two techniques over the years. It is hoped to try and dispel some of this confusion below; we return to this point at the end of the chapter. Close connections also exist between latent structure models and other statistical techniques. The idea of a manifest variable being modelled by a latent variable (or combination of latent variables) plus an error term renders latent variable models similar to measurement error models (Fuller, 1987), while the fact that individuals are assumed to fall into distinct latent classes establishes a connection with finite mixture separation (Everitt and Hand, 1981; Titterington *et al*, 1985; **10.16**). The latter connection also relates latent class analysis to cluster analysis (Chapter 10), and all these connections will be drawn on as appropriate below.

Foundations

12.6 To motivate the concepts in latent variable modelling, consider the following simple example given by Everitt (1984a). Table 12.1 shows data relating survival of infants to amount of pre-natal care received in a clinic. Conducting a chi-squared test of association on these data yields a test statistic

Table 12.1 Contingency table relating survival of infants to amount of pre-natal care received. From Everitt (1984a) with permission

Amount of care	Infants: Died	Survived
Less	20	373
More	6	316

value of 5.26 on 1 degree of freedom, showing that survival is significantly related to the amount of care received.

However, these data were, in fact, aggregated from two separate clinics; the full set of data is given in Table 12.2. Analysing the data for the two clinics separately, we obtain two near-zero test statistic values. Thus, the association between survival and care received has disappeared on disaggregating the data into the two clinics, or in other words the relationship between the two variables is *explained* by the relationship of each to the third variable (place of birth).

If continuous variables had been used instead of the categorical ones in each clinic, then a correlation coefficient would need to replace the chi-squared statistic as a measure of association, but otherwise the reasoning would remain the same. On the other hand, if a continuous variable (e.g. weight at birth) were to replace the categorical 'place of birth' as a conditioning variable then it might be difficult to disaggregate the data into reasonably sized subgroups in order to carry out a process as above. In this case we would need to *estimate* the effect of holding 'weight' constant on the correlation between the other variables, and this is done via partial correlations (Part 1, **2.6**). In all cases, however, the relationships between specified variables have been 'explained' by other variables if the associations between the former variables disappear on holding the latter variables fixed, i.e. if the former variables are *conditionally independent* given the latter.

12.7 In practice, situations often arise where we wish to account for the associations among a set of manifest variables but do not have access to a subsidiary set of explanatory variables, and so have to introduce latent variables in their place. The basic reasoning, however, remains as above. Given

Table 12.2 Contingency table relating survival of infants to amount of pre-natal care received in two separate clinics. From Everitt (1984a) with permission

	Infants			
	Died		Survived	
Clinic	Less Care	More Care	Less Care	More Care
A	3	4	176	293
B	17	2	197	23

a set of p manifest variables $\boldsymbol{x}' = (x_1, \ldots, x_p)$, we can explain the observed associations among them by finding a set of m latent variables $\boldsymbol{y}' = (y_1, \ldots, y_m)$ such that the x_i are conditionally independent given the values of the y_i. If we write $f(.)$ for the (marginal) density function of any manifest variables, $\phi(.)$ for the (prior, marginal) density function of any latent variables, and $g(.|.)$ for the conditional density function of any manifest variables given values of the latent variables (and if density functions are replaced by mass functions for categorical variables), then the joint density function of the manifest variables is given in general by

$$f(\boldsymbol{x}) = \int \phi(\boldsymbol{y}) g(\boldsymbol{x}|\boldsymbol{y}) d\boldsymbol{y}, \tag{12.1}$$

with summation replacing integration in the case of categorical latent variables. Thus if the latent variables explain the associations among the manifest variables then

$$f(\boldsymbol{x}) = \int \phi(\boldsymbol{y}) \prod_{i=1}^{p} g(x_i|\boldsymbol{y}) d\boldsymbol{y}. \tag{12.2}$$

This is the central assumption of all latent variable modelling, and the different forms of analysis mentioned in **12.4** arise directly on taking specific functional forms for f, ϕ and g. We will now consider the various possibilities in turn, looking first at the models with categorical latent variables and then at the models with continuous latent variables.

Latent class models, binary manifest variables

12.8 This is probably the simplest structural situation for latent variable modelling. The p manifest variables are all binary, so that $x_i = 0$ or 1 for $i = 1, \ldots, p$, and we attempt to explain the associations among them by a single categorical latent variable which has k classes. Let θ_{ij} be the probability that x_i takes the value 1 for an individual in latent class j, and let α_j be the prior probability that a randomly chosen individual belongs to latent class j ($i = 1, \ldots, p; j = 1, \ldots, k$ and $\sum_{j=1}^{k} \alpha_j = 1$). Thus, the conditional densities $g(.|.)$ in (12.2) are Bernoulli, the prior density $\phi(.)$ is the discrete set α_j, and integration is replaced by summation to give

$$f(\boldsymbol{x}) = \sum_{j=1}^{k} \alpha_j \prod_{i=1}^{p} \theta_{ij}^{x_i}(1 - \theta_{ij})^{1-x_i} \tag{12.3}$$

The posterior probability that an individual with response $\boldsymbol{x}$ belongs to latent class j is thus

$$\phi(j|\boldsymbol{x}) = \alpha_j \prod_{i=1}^{p} \theta_{ij}^{x_i}(1 - \theta_{ij})^{1-x_i} / f(\boldsymbol{x}) \tag{12.4}$$

The principal tasks of latent class analysis are to estimate the parameters of the model, to test its goodness of fit, to assign each individual to one of the latent classes, and to interpret these classes in terms meaningful to the particular study.

12.9 Popular early estimation methods, as set out by Lazarsfeld and Henry (1968), were based essentially on the method of moments. If we write π_i for the marginal probability that $x_i = 1$, π_{ij} for the marginal probability that $x_i = 1$ and $x_j = 1$, π_{ijl} for the marginal probability that $x_i = 1$, $x_j = 1$ and $x_l = 1$, and so on, then we have the series of relationships: $\sum_{s=1}^{k} \alpha_s = 1$, $\sum_{s=1}^{k} \alpha_s \theta_{is} = \pi_i$, $\sum_{s=1}^{k} \alpha_s \theta_{is} \theta_{js} = \pi_{ij}, \ldots$. Taking as many of these relationships as there are unknown parameters α_j and θ_{ij}, replacing the $\pi_{ij..}$ on the right-hand sides by their corresponding observed proportions in the sample, and solving the resulting *accounting equations* then yields estimates of the unknown parameters.

Another early estimation method was the one proposed by Madansky (1959), in which the data individuals were partitioned into k classes in such a way as to minimize a numerical measure of the extent to which a particular partition fails to meet the conditional independence requirement of the latent class model. This method is thus closely related to the optimisation methods of cluster analysis discussed in Chapter 10; for details see Everitt (1984a, pp. 77–78).

12.10 However, since the computing advances of the 1970s, and in particular since the development of the E-M algorithm (Dempster *et al*, 1977), all these methods have been replaced by the maximum likelihood approach. This approach was first proposed by McHugh (1956, 1958) using Newton-Raphson techniques, and recast into E-M form by Goodman (1978) and Everitt (1984c). In addition to its providing efficient estimates, maximum likelihood allows calculation of standard errors and evaluation of goodness of fit of models.

We suppose that $\boldsymbol{x}_1, \ldots, \boldsymbol{x}_n$ is a random sample of n vectors of manifest variables, where $\boldsymbol{x}_h' = (x_{h1}, \ldots, x_{hp})$ for $h = 1, \ldots, n$. Then from (12.3) the log-likelihood of the sample is

$$L = \sum_{h=1}^{n} \log \left\{ \sum_{j=1}^{k} \alpha_j \prod_{i=1}^{p} \theta_{ij}^{x_{hi}} (1 - \theta_{ij})^{1 - x_{hi}} \right\} \tag{12.5}$$

This has to be maximised subject to $\sum \alpha_j = 1$, which is done by finding the unconstrained maximum of

$$L' = L - \lambda \sum_{j=1}^{k} \alpha_j \tag{12.6}$$

where λ is a Lagrange multiplier. Straightforward differentiation yields

$$\frac{\partial L'}{\partial \alpha_j} = \sum_{h=1}^{n} \phi(j|\boldsymbol{x}_h)/\alpha_j - \lambda \tag{12.7}$$

and

$$\frac{\partial L'}{\partial \theta_{ij}} = \sum_{h=1}^{n} (x_{hi} - \theta_{ij}) \phi(j|\boldsymbol{x}_h)/[\theta_{ij}(1 - \theta_{ij})] \tag{12.8}$$

where $\phi(j|\boldsymbol{x})$ is given by (12.4) and (12.3). Setting these equations equal to zero, summing the first over j and remembering that $\sum \alpha_j = 1$ shows that the maximum likelihood estimates $\hat{\alpha}_j$ and $\hat{\theta}_{ij}$ satisfy the equations

$$\hat{\alpha}_j = \sum_{h=1}^{n} \hat{\phi}(j|\boldsymbol{x}_h)/n \; (j = 1, \ldots, k) \tag{12.9}$$

and

$$\hat{\theta}_{ij} = \sum_{h=1}^{n} x_{hi}\hat{\phi}(j|\boldsymbol{x}_h)/n\hat{\alpha}_j \; (i = 1, \ldots, p; j = 1, \ldots, k) \tag{12.10}$$

where

$$\hat{\phi}(j|\boldsymbol{x}_h) = \hat{\alpha}_j \prod_{i=1}^{p} \hat{\theta}_{ij}^{x_{hi}}(1 - \hat{\theta}_{ij})^{x_{hi}} \Big/ \left\{ \sum_{s=1}^{k} \hat{\alpha}_s \prod_{i=1}^{p} \hat{\theta}_{is}^{x_{hi}}(1 - \hat{\theta}_{is})^{x_{hi}} \right\} \tag{12.11}$$

Using the E-M algorithm, these equations can be solved in iterative fashion as follows.

(1) Choose an initial set of posterior probabilities $\{\hat{\phi}(j|\boldsymbol{x}_h)\}$.
(2) Use (12.9) and (12.10) to obtain approximations to $\{\hat{\alpha}_j\}$ and $\{\hat{\theta}_{ij}\}$.
(3) Substitute these values into (12.11), giving new probabilities $\{\hat{\phi}(j|\boldsymbol{x}_h)\}$.
(4) Continue the cycle 2–3 until convergence is attained.

A simple way of starting the process is to allocate individuals arbitrarily to latent classes (e.g. on the basis of total manifest score) and then take $\hat{\phi}(j|\boldsymbol{x}_h) = 1$ if $\boldsymbol{x}_h \in$ class j, and 0 otherwise. The process is usually a fast one, and yields posterior probabilities of belonging to each latent class for each individual as well as the parameter estimates. However, there is a danger of converging to a local rather than global maximum, so the process should perhaps be run several times from different starting values to check whether a global maximum has been reached. Aitkin *et al* (1981) briefly discuss the problem of multiple maxima in their latent class analysis of data on teaching styles.

Note that maximisation of the log-likelihood (12.5) can be effected in a variety of ways. This form arises directly in finite mixture separation, where the Newton-Raphson method has been studied extensively (Redner and Walker, 1984; Everitt, 1984a, 1987), while other possibilities include the Fletcher-Reeves and simplex methods as well as direct numerical optimisation. Everitt (1984b) has discussed and compared all these approaches in a simple case.

12.11 Writing $z_{hij} = x_{hi} - \theta_{ij}$, $\psi_{ij} = \theta_{ij}(1 - \theta_{ij})$, and δ_{ij} for the Kronecker delta, the following expressions are obtained (Exercise 12.1) for the second derivatives of the log-likelihood:

$$\frac{\partial^2 L}{\partial \alpha_s \partial \alpha_t} = -\sum_{h=1}^{n} \phi(s|\boldsymbol{x}_h)\phi(t|\boldsymbol{x}_h)/\alpha_s\alpha_t \tag{12.12}$$

$$\frac{\partial^2 L}{\partial \theta_{us} \partial \theta_{vt}} = \sum_{h=1}^{n} z_{hus}z_{hvt}\phi(s|\boldsymbol{x}_h)[\delta_{st}(1 - \delta_{uv}) - \phi(t|\boldsymbol{x}_h)]/\psi_{us}\psi_{vt} \tag{12.13}$$

and

$$\frac{\partial^2 L}{\partial \alpha_s \partial \theta_{ut}} = \sum_{h=1}^{n} z_{hus}\phi(s|\boldsymbol{x}_h)[\delta_{st} - \phi(t|\boldsymbol{x}_h)] \tag{12.14}$$

for $s, t = 1, \ldots, k$ and $u, v = 1, \ldots, p$. The inverse of the matrix comprising the expected values of the negatives of these derivatives thus provides the asymptotic variance-covariance matrix of the parameter estimates, so that standard errors of the estimates are given by the square roots of the diagonal elements of this matrix. Unfortunately, calculation of the expected values involves summations over the 2^p possible response patterns $\boldsymbol{x}$ for each element of the matrix, so is very time consuming. However, use of observed, in place of expected, values is very quick and generally gives an excellent approximation. Indeed, Efron and Hinkley (1978) and Skovgaard (1985) have argued that, in general, using the observed information matrix conditions on the ancillary information and provides the better inference. These observed values follow directly from the above second derivatives. Siddiqi (1991) has provided some alternative procedures for calculation of standard errors of parameters in the latent class model, two of which derive from direct use of the likelihood function, one of which uses a result for the E-M algorithm provided by Louis (1982), and one of which attempts a model-robust inference based on a result by Royall (1986).

12.12 An individual with manifest response pattern $\boldsymbol{x}_h$ is allocated to that latent class for which its estimated posterior probability $\hat{\phi}(j|\boldsymbol{x}_h)$ is greatest, following the standard methods of cluster analysis based on finite mixture separation. To interpret the latent classes we need to identify the characteristics that individuals allocated to a particular class have in common, and what it is that distinguishes them from individuals in other classes.

If p is small then goodness-of-fit of the model can usually be tested by comparing observed response pattern frequencies with those predicted by the model via a standard chi-squared or likelihood ratio (i.e. deviance) goodness-of-fit test; the predicted value for pattern $\boldsymbol{x}$ is obtained by substituting parameter estimates into (12.3) and multiplying by the sample size n. For large p, or more generally sparse data, there does not as yet appear to be a satisfactory procedure; see Aitkin *et al* (1981) for a simulation-based approach. Determining the 'best' model (i.e. the best choice of k, number of latent classes) is somewhat akin to determination of the number of groups in cluster analysis. An informal procedure is to plot deviance against number of classes, and to choose k as the number of classes after which the drop in deviances is comparatively small. See Hinde (1985) for some further discussion.

Finally, it is worth noting that models with more than one latent variable can be accommodated within the same scheme as above but with the imposition of some extra constraints. For example, if we postulate a model with two independent latent variables, each with two classes, then we can recast it as a model having one latent variable with four classes whose parameters are of the form $\alpha_1 = B\gamma, \alpha_2 = B(1-\gamma), \alpha_3 = (1-B)\gamma, \alpha_4 = (1-B)(1-\gamma)$ with a similar structure for the θ_{ij}.

Example 12.1

Table 12.3 gives a set of data reported by Macready and Dayton (1977), showing the results of four arithmetic tasks carried out by 142 subjects. Each task involved the multiplication of a two-digit number by either a three-digit or a four-digit number, and is scored 1 or 0 according to whether the subject obtained the right or the wrong answer. Macready

Table 12.3 Data from Macready and Dayton (1977), with added expected frequencies and posterior class probabilities on fitting a model with two latent classes

Response pattern	Frequency	Expected frequency	$\hat{\phi}(2\|x)$
1111	15	15.0	1.00
1110	7	6.0	1.00
1101	23	19.7	1.00
1100	7	8.9	0.91
1011	1	4.2	1.00
1010	3	1.9	0.90
1001	6	6.1	0.91
1000	13	12.9	0.18
0111	4	4.9	1.00
0110	2	2.1	0.97
0101	5	6.6	0.98
0100	6	5.6	0.47
0011	4	1.4	0.97
0010	1	1.3	0.43
0001	4	4.0	0.45
0000	41	41.0	0.02

and Dayton analysed the data by assuming a model in which any subject was either a 'master' or a 'non-master' of the relevant arithmetic operation, with probability η that a randomly chosen individual was a 'master'.

Bartholomew (1987) fitted a latent class model having two classes to these data, using the above theory. There were thus nine parameters (since $\alpha_1 + \alpha_2 = 1$, so that we can take $\alpha_1 = \alpha$ and $\alpha_2 = 1 - \alpha$). Estimates of these parameters, and their standard errors obtained from the observed information matrix as above, are given in Table 12.4. Using these fitted values, we obtain the expected frequencies shown in Table 12.3 and hence a chi-squared goodness-of-fit value of 2.77 on 3 degrees of freedom (since the 16 expected frequencies are reduced to 12 after grouping to ensure all expected frequencies are greater than 5, and there are nine parameters to be estimated). The fit is thus a good one, and the two-class model explains the associations among the four items well. To interpret the two classes, we can compute the posterior probabilities of class membership, $\hat{\phi}(j|\boldsymbol{x})$, for an individual with response vector $\boldsymbol{x}$. These probabilities are shown in

Table 12.4 Parameter estimates and standard errors (in parentheses) for a two-class latent structure model fitted to the data of Table 12.3

Class j	$\hat{\theta}_{1j}$	$\hat{\theta}_{2j}$	$\hat{\theta}_{3j}$	$\hat{\theta}_{4j}$	$\hat{\alpha}$
1	0.21	0.07	0.02	0.05	0.59
	(0.07)	(0.06)	(0.03)	(0.05)	(0.06)
2	0.75	0.78	0.43	0.71	
	(0.06)	(0.06)	(0.06)	(0.07)	

Table 12.3 (only class 2 probabilities are tabulated, as class 1 probabilities are simply their complements). We see that individuals are allocated to class 2 if they obtain two or more correct answers out of four, and to class 1 otherwise. Class 2 thus aligns with the 'masters' of Macready and Dayton, and class 1 with the 'non-masters'; in fact, closer inspection shows Bartholomew's model to be a reparametrised version of Macready and Dayton's model.

Siddiqi (1991) also analysed the same set of data, recomputing the standard errors of Table 12.4 by each of the four methods mentioned in **12.11**. All methods gave 0.06 instead of 0.07 for $\hat{\theta}_{41}$, and the method derived from Royall (1986) gave 0.07 instead of 0.06 for $\hat{\theta}_{21}$ and 0.06 instead of 0.05 for $\hat{\theta}_{42}$. Apart from these small discrepancies, all other values agreed with those in Table 12.4 to two decimal places.

Latent class models, multicategorical manifest variables

12.13 In the most general case, we suppose that p categorical manifest variables are observed, the ith having c_i possible categories ($i = 1, \ldots, p$). Thus, we can denote the ith variable this time by the *vector* $\boldsymbol{x}_i' = (x_{i1}, \ldots, x_{ic_i})$: if the response on variable i is in the jth category, then $x_{ij} = 1$ and $x_{is} = 0$ for $s \neq j$ (so that $\sum_{u=1}^{c_i} x_{iu} = 1,\ i = 1, \ldots, p$). The complete response vector is $\boldsymbol{x}' = (\boldsymbol{x}_1', \ldots, \boldsymbol{x}_p')$. The theory for binary manifest variables in **12.8** now extends in the obvious way: θ_{isj} is the probability that the response on the ith variable is in category s for an individual in latent class j, and α_j is again the prior probability that a randomly chosen individual belongs to latent class j ($s = 1, \ldots, c_i; i = 1, \ldots, p; j = 1, \ldots, k$), but the constraints $\sum_{s=1}^{c_i} \theta_{isj} = 1$ for $i = 1, \ldots, p$ and $j = 1, \ldots, k$ are added to the original constraint $\sum_{j=1}^{k} \alpha_j = 1$. The conditional densities $g(.|.)$ in (12.2) are now multinomial, the prior density $\phi(.)$ is again the discrete set α_j, and the joint probability function of $\boldsymbol{x}$ becomes

$$f(\boldsymbol{x}) = \sum_{j=1}^{k} \alpha_j \prod_{i=1}^{p} \prod_{s=1}^{c_i} \theta_{isj}^{x_{is}} \tag{12.15}$$

The posterior probability that an individual with response $\boldsymbol{x}$ belongs to latent class j is thus

$$\phi(j|\boldsymbol{x}) = \alpha_j \prod_{i=1}^{p} \prod_{s=1}^{c_i} \theta_{isj}^{x_{is}} / f(\boldsymbol{x}) \tag{12.16}$$

Writing $\boldsymbol{x}^h$ for the observed response vector on individual h in a sample of size n, and mindful of the extra constraints introduced above, maximum likelihood estimates of the unknown parameters α_j and θ_{isj} are obtained by maximising

$$L' = \sum_{h=1}^{n} \log f(\boldsymbol{x}^h) - \lambda \sum_{j=1}^{k} \alpha_j - \sum_{j=1}^{k} \sum_{i=1}^{p} \mu_{ij} \sum_{s=1}^{c_i} \theta_{isj} \tag{12.17}$$

where λ and $\{\mu_{ij}\}$ are Lagrange multipliers. Straightforward differentiation, and summation over categorical states where necessary to eliminate Lagrange multipliers, yields the estimating equations

$$\hat{\alpha}_j = \sum_{h=1}^{n} \hat{\phi}(j|\boldsymbol{x}^h)/n \; (j = 1, \ldots, k) \tag{12.18}$$

and

$$\hat{\theta}_{isj} = \sum_{h=1}^{n} x_{his}\hat{\phi}(j|\boldsymbol{x}^h)/n\hat{\alpha}_j \; (s = 1, \ldots, c_i; i = 1, \ldots, p; j = 1, \ldots, k), \tag{12.19}$$

where x_{his} is the value of x_{is} for individual h and

$$\hat{\phi}(j|\boldsymbol{x}^h) = \hat{\alpha}_j \prod_{i=1}^{p}\prod_{s=1}^{c_i} \hat{\theta}_{isj}^{x_{his}} \Big/ \left\{ \sum_{u=1}^{k} \hat{\alpha}_u \prod_{i=1}^{p}\prod_{s=1}^{c_i} \hat{\theta}_{isu}^{x_{his}} \right\} \tag{12.20}$$

These equations are solved by an iterative E-M scheme of exactly the same form as in **12.10**; asymptotic standard errors can be computed in the usual way from second derivatives of L, which are obtained as easily as were those in **12.11** for binary variables; goodness-of-fit of any model is tested as described in **12.12**; allocation of any specified individual with response vector $\boldsymbol{x}^h$ is to the latent class with highest value of $\hat{\phi}(j|\boldsymbol{x}^h)$; and interpretation of latent classes is done in the same way as before.

Example 12.2
Siddiqi (1991) has fitted latent class models to the 'chest pain' data described by Emerson *et al* (1989) and previously analysed by various other techniques including correspondence analysis (Part 1, **5.19**; Crichton and Hinde, 1989). The data concerned patients arriving at the Accident and Emergency department of Westminster Hospital, London, complaining of anterior chest pain, the aim of the analysis being to develop a method of identifying those patients who need treatment in the Coronary Care Unit of the hospital. The data set comprised 295 cases, on each of which 18 indicants were recorded. Two of the indicants (age and diastolic blood pressure) were continuous variables categorised into four and three categories respectively, one (site of pain) was categorical with five categories, and the remaining 15 were simple presence/absence binary symptoms (e.g. did the patient arrive by ambulance?; is there visible sweating?; is the pain knife-like? etc). However, 130 patients had less than their full set of values (106 had one missing response, 19 had two missing responses, three had three missing responses and one each had five and eight missing responses respectively); missing responses were identified and corresponding terms were removed from estimating equations.

Using the above approach, Siddiqi (1991) fitted a k-class latent structure model for k ranging from 1 to 10, obtaining the deviances shown in Table 12.5. (Note that every extra latent class in the model thus requires estimation of 25 additional parameters.) From Table 12.5, and balancing requirements of parsimony with those of adequate fit, it seems as if either a 3- or 4-class model will provide the best summary of the data. Looking

Table 12.5 Deviances for different latent class models fitted to the chest pain data

Classes	Deviance	Decrease	% Decrease
1	6712.629	—	—
2	6381.156	331.473	4.938
3	6153.992	227.164	3.560
4	6035.719	118.273	1.921
5	5980.641	55.078	0.911
6	5933.326	47.315	0.792
7	5891.614	41.712	0.702
8	5819.492	72.122	1.224
9	5798.965	20.527	0.353
10	5725.199	73.766	1.272

at the 3-class model, the indicants whose presence characterised class 1 (and their posterior probabilities) were age over 45 years (0.94); arrival by ambulance (0.57); pain mainly at central site (0.65); pain radiates to arm (0.51); pain very severe (0.80); pain is constricting (0.85); pain provoked by exertion and relieved by rest (0.59); history of pain due to angina (0.62); and history of pain due to myocardial infarction (0.58). This class seemed to correspond most with the group of patients identified independently by consultants as the high-risk group requiring attention in the Coronary Care Unit, while latent classes 2 and 3 were characterised by indicants which identified patients in them as low-risk patients not requiring attention. A further 133 patients were subsequently seen at the Accident and Emergency department and classified as high- or low-risk by the same consultants; applying the fitted 3-class model with the above class identification to these cases gave an overall correct classification rate of 82%.

Fitting a 4-class model, and matching the fitted classes to the consultants' assessments as above, gave poorer results with the percentage of correct classification of the extra patients falling to 77.4%. Models with more than four classes similarly gave poorer results, so the 3-class model seems to be satisfactory for data summary and description.

Latent class models, continuous manifest variables

12.14 If the manifest variables are continuous but the latent variables are categorical, then the resulting analysis is known as *latent profile analysis*. The problem now is that there are many possible choices for the conditional distribution of the manifest variables, whereas previously we have had only one possibility. Consequently, we have to be content with merely sketching out the general approach here; each specific choice of distribution will then lead to its own set of equations to be solved.

Denote, again, the manifest variables by $\boldsymbol{x}' = (x_1, \ldots, x_p)$ and the latent class probabilities by α_j for $j = 1, \ldots, k$ (with $\sum_{j=1}^{k} \alpha_j = 1$). Then assumption of local independence means that the latent profile model has the form

$$f(\boldsymbol{x}) = \sum_{j=1}^{k} \alpha_j \prod_{i=1}^{p} g_i(x_i|j) \tag{12.21}$$

and the posterior probability that an individual with response $\boldsymbol{x}$ belongs to latent class j is

$$\phi(j|\boldsymbol{x}) = \alpha_j \prod_{i=1}^{p} g_i(x_i|j)/f(\boldsymbol{x}) \tag{12.22}$$

where $g_i(x_i|j)$ is the conditional probability density of x_i in latent class j $(i = 1, \ldots, p; j = 1, \ldots, k)$. This probability density will usually be assumed to be a (univariate) parametric function, depending on a parameter θ_{ij} for variable i and class j, so that we can write $g_i(x_i|j)$ as $g(x_i|\theta_{ij})$.

Now if $\boldsymbol{x}^h = (x_{h1}, \ldots, x_{hp})'$ denotes the value of $\boldsymbol{x}$ for individual h in a random sample of size n, then the log-likelihood of the sample is

$$L = \sum_{h=1}^{n} \log \sum_{j=1}^{k} \alpha_j g(x_{hi}|\theta_{ij}) \tag{12.23}$$

Thus

$$\frac{\partial L}{\partial \alpha_s} = \sum_{h=1}^{n} g(x_{hi}|\theta_{is}) \Bigg/ \left\{ \sum_{j=1}^{k} \alpha_j g(x_{hi}|\theta_{ij}) \right\} \tag{12.24}$$

and

$$\frac{\partial L}{\partial \theta_{ij}} = \sum_{h=1}^{n} \alpha_j \frac{\partial g}{\partial \theta_{ij}} / g(x_{hi}|\theta_{ij}) \tag{12.25}$$

Maximum likelihood estimates are obtained by setting these derivatives equal to zero. The equations derived from (12.24) are the same for all choices of $g(.|.)$, namely

$$\hat{\alpha}_j = \sum_{h=1}^{n} \hat{\phi}(j|\boldsymbol{x}^h)/n \quad (j = 1, \ldots, k) \tag{12.26}$$

where $\hat{\phi}(j|\boldsymbol{x}^h)$ has the usual interpretation. (Note that this same equation for $\hat{\alpha}_j$ occurs for all latent class models.) However, the equations derived from (12.25) will depend on the particular choice of $g(.|.)$. For example, if we assume a normal distribution with mean θ_{ij} and variance 1 for x_i in class j then we obtain the simple form

$$\hat{\theta}_{ij} = \sum_{h=1}^{n} x_{hi} \hat{\phi}(j|\boldsymbol{x}^h) \Bigg/ \left\{ \sum_{h=1}^{n} \hat{\phi}(j|\boldsymbol{x}^h) \right\} \quad (i = 1, \ldots, p; j = 1, \ldots, k) \tag{12.27}$$

(which is also the same form as in previous sections), but other choices will lead to other, more complicated, equations. Whatever form is chosen, solution of the equations follows from application of the E-M algorithm in the usual manner and an individual with responses $\boldsymbol{x}^h$ is allocated to the latent class for which $\hat{\phi}(j|\boldsymbol{x}^h)$ is maximum.

Ordered latent classes

12.15 In all the foregoing, the latent classes have been simply taken as distinct and subject to no constraints. If the manifest variables are either metrical or ordered categorical, however, it may be sensible to look for a solution in which the latent classes have some ordering imposed on them. Lazarsfeld and Henry (1968) propose two possible solutions, which they term the *latent distance model* and the *located classes model* respectively. However, these models suffer from two drawbacks: they are only defined for binary manifest variables, and they are both essentially discretised versions of continuous latent variable models (see below). The latent distance model assumes that the items may be ordered along a latent continuum and the different latent classes correspond to contiguous regions along this continuum, while in the located classes model the ordering of the classes is obtained by assuming that the classes may be represented by a coordinate on this latent continuum. Similar drawbacks apply to the subsequent methods developed by Rost (1985; 1988).

A more general solution has been proposed by Croon (1990), who obtains a maximum likelihood latent class solution for categorical manifest variables, subject to appropriate monotonicity constraints on the latent classes. If the latent classes are considered to be ordered from 'low' to 'high' and if, following **12.13**, θ_{isj} is the probability that the response on the ith variable is in category s for an individual in latent class j, then by 'monotonicity' Croon requires the probability of a 'positive' response to be an increasing function of the latent class number. This implies that θ satisfies

$$\sum_{s=g}^{c_i} \theta_{isj} \leqslant \sum_{s=g}^{c_i} \theta_{is,j+1} \tag{12.28}$$

for $1 \leqslant j \leqslant k-1$ and $2 \leqslant g \leqslant c_i$ and $i = 1, \ldots, p$. Since for each latent class j and each variable i we have $\sum_{s=1}^{c_i} \theta_{isj} = 1$, this condition is equivalent to

$$\sum_{s=1}^{g} \theta_{isj} \geqslant \sum_{s=1}^{g} \theta_{is,j+1} \tag{12.29}$$

for $1 \leqslant g \leqslant c_i - 1$, $1 \leqslant j \leqslant k-1$ and $i = 1, \ldots, p$. If, further, cumulative response probabilities F_{igj} are defined by

$$F_{igj} = \sum_{s=1}^{g} \theta_{isj} \tag{12.30}$$

then the condition is equivalent to

$$F_{igj} \geqslant F_{ig,j+1} \tag{12.31}$$

for $1 \leqslant g \leqslant c_i - 1$, $1 \leqslant j \leqslant k-1$ and $i = 1, \ldots, p$.

Croon thus tackles the maximisation of (12.17), but in addition imposes the inequality constraints (12.31). By identifying a sequence of random variables $Z_1, \ldots, Z_k$ whose distribution functions are given by the F above, he shows that this problem is equivalent to the maximum likelihood estimation of a sequence of stochastically ordered distribution functions. This latter problem was solved by Feltz and Dykstra (1985), and Croon gives a computational scheme based on their method.

Example 12.3
To illustrate his method, Croon (1990) analysed data from a Dutch survey on religious and secular beliefs (Felling *et al* 1987). He focused on six of the questionnaire items.

(1) People can be divided into two classes, the weak and the strong.
(2) Familiarity breeds contempt.
(3) Young people sometimes have rebellious ideas, but as they grow up they ought to get over them and settle down.
(4) Most of our social problems would be solved if we could somehow get rid of immoral, crooked and feeble-minded people.
(5) What this country needs most, more than laws and political programmes, is a few courageous, fearless, devoted leaders in whom the people can put their faith.
(6) A person who has bad manners, habits and breeding can hardly expect to get along with decent people.

For the purpose of his analysis, Croon recoded the responses into the three (ordered) categories: disagree with, am neutral to, and agree with the item content, and included in the analysis only the 1581 interviewees who responded to all six items.

Croon fitted, in turn, models containing from one to six latent classes. By examining the deviances (log-likelihood ratio statistics) and the Akaike Information Criterion (AIC; Akaike, 1987) for each model, he concluded that the five-class model provided the best fit. Table 12.6 gives the probabilities of responding with category 1 ('disagree with item'), while Table 12.7 gives the probabilities of responding with either categories 1 or 2 ('disagree with' or am neutral to the item').

It is clear from these tables that the five latent classes can be ordered along the latent continuum, with latent class 1 representing the most 'negative' and latent class 5 the most 'positive' class within this set of classes. Estimated proportions of respondents falling in these classes were 0.125, 0.223, 0.238, 0.215 and 0.199 from class 1 to class 5 respectively. The distribution is thus nearly symmetric, the majority of respondents

Table 12.6 Probabilities of responding with category 1, Dutch survey data. From Croon (1990) with permission

	Class 1	Class 2	Class 3	Class 4	Class 5
Item 1	0.70	0.65	0.32	0.18	0.08
Item 2	0.46	0.46	0.11	0.08	0.06
Item 3	0.73	0.70	0.17	0.16	0.07
Item 4	0.59	0.31	0.07	0.07	0.01
Item 5	0.84	0.60	0.28	0.28	0.00
Item 6	0.78	0.65	0.25	0.25	0.00

Table 12.7 Probabilities of responding with categories 1 or 2, Dutch survey data. From Croon (1990) with permission

	Class 1	Class 2	Class 3	Class 4	Class 5
Item 1	0.87	0.72	0.72	0.27	0.27
Item 2	0.80	0.63	0.63	0.21	0.20
Item 3	1.00	0.80	0.72	0.39	0.24
Item 4	0.89	0.44	0.40	0.16	0.08
Item 5	1.00	0.74	0.74	0.50	0.14
Item 6	0.97	0.76	0.76	0.61	0.08

occupying a moderate or central position on the continuum with only a relative minority at the two extremes. Also, clear differences emerge among the six items with respect to the way in which their response and cumulative response probabilities vary as a function of the ordinal latent variable. Croon singled out two aspects of this relationship, namely overall level and steepness. Comparing pairs of items across latent classes in the two tables we see that, for example, each of items 1 and 3 are responded to in a more negative way than item 2, and similarly that item 5 is more negatively responded to than item 4. This overall level can thus be identified with the 'popularity' of a given item. The second feature is the rate at which the probabilities move from 0.0 to 1.0 along the latent continuum. For example, the change in probabilities for item 3 is more drastic (steeper) than for item 2. Steepness is thus associated with discriminatory power: item 3 is, in a sense, a better indicator of class membership than is item 2.

Latent class models: other

12.16 There have been various other developments and applications of latent class analysis in the social sciences. Clogg (1981) reviews the use of latent structure models for mobility data, while de Leeuw *et al* (1990) and van der Heijden *et al* (1992) discuss *latent budget analysis.* This last technique is applicable to behavioural data collected in the form of a contingency table, in which the entry in row i and column j denotes the number of times that individual i was observed carrying out activity j. The latent budget model postulates that there are k latent time-budgets that determine the behaviour of all individuals, so that the theoretical conditional proportions π_{ij} (with $\sum_j \pi_{ij} = 1 \ \forall i$) have the form $\sum_{s=1}^{k} \alpha_{is} B_{js}$. The above references focus on the issues of identifiability of parameters, their estimation by maximum likelihood, and the testing of goodness-of-fit of the model. A similar model underlies archetypal analysis (Cutler and Breiman, 1994) and end-member analysis for compositional data (Renner, 1993). Further latent class models, for preference ratings and for two-way compositional data, have been described by De Soete and Winsberg (1993) and DeSarbo *et al* (1993) respectively. De Leeuw and van der Heijden (1991) explore equivalences among latent class, latent budget, canonical, and correspondence analyses and their reduced rank versions for the analysis of contingency tables.

This completes our look at categorical latent variable models, and we now turn to continuous latent variables.

Factor models, continuous manifest variables

12.17 Suppose first that the manifest variables $\boldsymbol{x}' = (x_1, \ldots, x_p)$ are continuous, and that we wish to explain the associations (i.e. correlations) among them by a set $\boldsymbol{y}' = (y_1, \ldots, y_m)$ of m continuous latent variables. We therefore need to formulate a model that satisfies (12.2) for these types of variables. Assume first that $\boldsymbol{x}$ and $\boldsymbol{y}$ have multivariate normal distributions. Then the joint distribution $f(\boldsymbol{x}, \boldsymbol{y})$ and the conditional distribution $g(\boldsymbol{x}|\boldsymbol{y})$ are also both normal. Moreover, the conditional mean of $\boldsymbol{x}$ given $\boldsymbol{y}$ is linear in $\boldsymbol{y}$ and the conditional dispersion matrix does not depend on $\boldsymbol{y}$ (Part 1, **2.8**). Thus we can write

$$\boldsymbol{x} = \boldsymbol{\mu} + \boldsymbol{\Gamma}\boldsymbol{y} + \boldsymbol{e}, \tag{12.32}$$

where $\boldsymbol{e}$ is a normal random vector with zero mean and constant dispersion matrix. If, additionally, we assume that elements of $\boldsymbol{e}$ are independent, then the conditional distribution $g(\boldsymbol{x}|\boldsymbol{y})$ will be of the form $\prod_{i=1}^{p} g(x_i|\boldsymbol{y})$, where the means of the $g(x_i|\boldsymbol{y})$ are given by the rows on the right-hand side of (12.32) and the variances are constant. Thus the basic condition (12.2) is satisfied. Furthermore, without loss of generality we can take the y_i to be independent standard normal deviates, in which case the *normal linear factor model* is given by (12.32) with $\boldsymbol{e} \sim N(\boldsymbol{0}, \boldsymbol{\Psi})$ and $\boldsymbol{y} \sim N(\boldsymbol{0}, \boldsymbol{I})$ where $\boldsymbol{\Psi}$ is a diagonal matrix. The latent variables $\boldsymbol{y}$ are commonly referred to as *factors*, the 'residuals' $\boldsymbol{e}$ as *specific variates* and the weights $\boldsymbol{\Gamma}$ as the *factor loadings*.

12.18 The model as specified above is indeterminate. First note that if $\boldsymbol{\Sigma}$ is the dispersion matrix of the manifest variables $\boldsymbol{x}$, then we have the fundamental relationship that

$$\boldsymbol{\Sigma} = \boldsymbol{\Gamma}\boldsymbol{\Gamma}' + \boldsymbol{\Psi}. \tag{12.33}$$

Our objective in factor analysis is to determine m and the elements of $\boldsymbol{\Gamma}$ and $\boldsymbol{\Psi}$, given a sample estimate $\boldsymbol{S}$ of $\boldsymbol{\Sigma}$. Let us assume for the time being that m is fixed and known. Then from $\frac{1}{2}p(p+1)$ elements of $\boldsymbol{S}$, we need to estimate $p(m+1)$ unknown parameters. However, if we consider a non-singular orthogonal transformation of $\boldsymbol{y}$ to new latent variables $\boldsymbol{z} = \boldsymbol{M}\boldsymbol{y}$, then the z_i will also be standard normal and

$$\boldsymbol{x} = \boldsymbol{\mu} + \boldsymbol{\Gamma}\boldsymbol{M}'\boldsymbol{z} + \boldsymbol{e},$$

so that

$$\boldsymbol{\Sigma} = \boldsymbol{\Gamma}\boldsymbol{M}\boldsymbol{M}'\boldsymbol{\Gamma}' + \boldsymbol{\Psi} = \boldsymbol{\Gamma}\boldsymbol{\Gamma}' + \boldsymbol{\Psi}.$$

Thus our y_i are indeterminate up to an orthogonal transformation. All we can hope to do, without further assumptions, is to estimate the *factor space*. We can determine a unique set of axes in that space by imposing further constraints, and this is done most easily through the condition

$$\boldsymbol{\Gamma}'\boldsymbol{\Gamma} = \boldsymbol{\Lambda}, \tag{12.34}$$

where $\boldsymbol{\Lambda}$ is a diagonal matrix with diagonal elements λ_i. Having identified the factor space in this way, we are then at liberty to *rotate* the axes into any position that provides better interpretation of the rotated weights for the given application. The question of factor rotation is discussed further in **12.26**.

More generally, if we drop the restriction that $\boldsymbol{M}$ be an orthogonal matrix and simply require it to be non-singular, it follows that (12.32) can be written as

$$\boldsymbol{x} = \boldsymbol{\mu} + \boldsymbol{\Gamma}\boldsymbol{M}^{-1}\boldsymbol{z} + \boldsymbol{e}. \tag{12.35}$$

Thus, weights $\boldsymbol{\Gamma}\boldsymbol{M}^{-1}$ and latent variables $\boldsymbol{z}$ provide the same linear model as weights $\boldsymbol{\Gamma}$ and latent variables $\boldsymbol{y}$; the difference is that the former variables are no longer independent in general. However, if dependent latent variables are permitted, then this equivalence forms the rationale behind *oblique rotation* of factors, also considered in **12.26**.

The practical implication of the above discussion is for the number m of latent variables that can be included in the model. We have $\frac{1}{2}p(p+1)$ items of information (elements of $\boldsymbol{S}$) from which to estimate the $p(m+1)$ parameters (pm factor loadings and p specific variances), but (12.34) contains $\frac{1}{2}m(m-1)$ constraints (off-diagonal elements of the symmetric matrix $\boldsymbol{\Gamma}'\boldsymbol{\Gamma}$ all zero). For estimability of parameters we thus need $\frac{1}{2}p(p+1) \geqslant p(m+1) - \frac{1}{2}m(m-1)$, so that m must satisfy $(p-m)^2 \geqslant p+m$. For example, with five manifest variables we cannot have more than two latent variables, with eight manifest variables we cannot have more than four latent variables, and with 15 manifest variables we cannot have more than 10 latent variables. Of course, in practice we aim to explain correlations among manifest variables by as few latent variables as possible, and we would not generally expect to approach these maxima in most applications.

Factor analysis: maximum likelihood approach

12.19 Having made assumptions of normality about the factors $\boldsymbol{y}$, the specific variates $\boldsymbol{e}$ and the manifest variables $\boldsymbol{x}$, it is natural to tackle estimation of the parameters $\boldsymbol{\Gamma}$ and $\boldsymbol{\Psi}$ in **12.17** by maximum likelihood. If $\boldsymbol{x}_1, \ldots, \boldsymbol{x}_n$ is a random sample of size n, then the likelihood is

$$L = (2\pi)^{-np/2}|\boldsymbol{\Sigma}|^{-n/2} \exp -\frac{1}{2}\sum_{i=1}^{n}(\boldsymbol{x}_i - \boldsymbol{\mu})'\boldsymbol{\Sigma}^{-1}(\boldsymbol{x}_i - \boldsymbol{\mu}) \tag{12.36}$$

where $\boldsymbol{\Sigma} = \boldsymbol{\Gamma}\boldsymbol{\Gamma}' + \boldsymbol{\Psi}$ from (12.33). Differentiating the log-likelihood with respect to $\boldsymbol{\mu}$, setting the result to zero and solving for $\boldsymbol{\mu}$ yields $\hat{\boldsymbol{\mu}} = \bar{\boldsymbol{x}}$. As this does not interact with the other model parameters, the latter can be estimated by maximising the profile log-likelihood

$$l = -\frac{1}{2}np\log(2\pi) - \frac{1}{2}n\log|\boldsymbol{\Sigma}| - \frac{1}{2}n\mathrm{tr}(\boldsymbol{\Sigma}^{-1}\boldsymbol{S}) \tag{12.37}$$

where $\boldsymbol{S} = \frac{1}{n}\sum_{i=1}^{n}(\boldsymbol{x}_i - \bar{\boldsymbol{x}})(\boldsymbol{x}_i - \bar{\boldsymbol{x}})'$. We thus need to maximise l with respect to $\boldsymbol{\Gamma}$ and $\boldsymbol{\Psi}$.

First, if $\boldsymbol{\Gamma} = (\gamma_{ij})$, $\boldsymbol{\Sigma} = (\sigma_{ij})$ and the diagonal elements of $\boldsymbol{\Psi}$ are ψ_i, then (12.33) gives

$$\sigma_{rs} = \sum_{h=1}^{m} \gamma_{rh}\gamma_{hs} + \delta_{rs}\psi_r$$

where δ_{rs} is the Kronecker delta. Next, use of the chain rule for differentiation produces

$$\frac{\partial l}{\partial \gamma_{ij}} = \sum_{h=1}^{p}\left(\frac{\partial l}{\partial \sigma_{ih}}\right)\left(\frac{\partial \sigma_{ih}}{\partial \gamma_{ij}}\right)$$

and

$$\frac{\partial l}{\partial \psi_i} = \frac{\partial l}{\partial \sigma_{ii}}\frac{\partial \sigma_{ii}}{\partial \psi_i}.$$

Consider the equation for $\frac{\partial l}{\partial \gamma_{ij}}$. From above, $\frac{\partial \sigma_{ih}}{\partial \gamma_{ij}} = (1 + \delta_{ih})\gamma_{hj}$, so that if we use the matrix differential form

$$\frac{\partial l}{\partial \boldsymbol{\Gamma}} = \left(\frac{\partial l}{\partial \gamma_{ij}}\right)$$

then

$$\frac{\partial l}{\partial \boldsymbol{\Gamma}} = \left(\frac{\partial l}{\partial \boldsymbol{\Sigma}} + \text{diag}\frac{\partial l}{\partial \boldsymbol{\Sigma}}\right)\boldsymbol{\Gamma}$$

Now

$$\frac{\partial \log|\boldsymbol{\Sigma}|}{\partial \boldsymbol{\Sigma}} = 2\boldsymbol{\Sigma}^{-1} - \text{diag}(\boldsymbol{\Sigma}^{-1})$$

and

$$\frac{\partial \text{tr}(\boldsymbol{\Sigma}^{-1}\boldsymbol{S})}{\partial \boldsymbol{\Sigma}} = -2\boldsymbol{\Sigma}^{-1}\boldsymbol{S}\boldsymbol{\Sigma}^{-1} + \text{diag}(\boldsymbol{\Sigma}^{-1}\boldsymbol{S}\boldsymbol{\Sigma}^{-1})$$

are standard matrix differential results (e.g. Mardia *et al*, 1979, p108), so that we obtain

$$-\frac{2}{n}\frac{\partial l}{\partial \boldsymbol{\Sigma}} = 2\boldsymbol{\Omega} - \text{diag}\boldsymbol{\Omega} \tag{12.38}$$

where $\boldsymbol{\Omega} = \boldsymbol{\Sigma}^{-1}(\boldsymbol{\Sigma} - S)\boldsymbol{\Sigma}^{-1}$. Hence

$$-\frac{2}{n}\frac{\partial l}{\partial \boldsymbol{\Gamma}} = 2\boldsymbol{\Omega}\boldsymbol{\Gamma}. \tag{12.39}$$

Turning to the equation for $\frac{\partial l}{\partial \psi_i}$, it follows that

$$-\frac{2}{n}\text{diag}\frac{\partial l}{\partial \boldsymbol{\Psi}} = \text{diag}\boldsymbol{\Omega} \tag{12.40}$$

since $\frac{\partial \sigma_{ii}}{\partial \psi_i} = 1$.

Maximum likelihood estimators of the unknown parameters are now obtained by setting the right-hand sides of (12.39) and (12.40) to zero and solving for $\boldsymbol{\Gamma}$ and $\boldsymbol{\Psi}$. Given relationship (12.33) connecting $\boldsymbol{\Sigma}$, $\boldsymbol{\Gamma}$ and $\boldsymbol{\Psi}$, it can readily be verified that

$$\boldsymbol{\Sigma}^{-1} = \boldsymbol{\Psi}^{-1} - \boldsymbol{\Psi}^{-1}\boldsymbol{\Gamma}(\boldsymbol{I} + \boldsymbol{\Theta})^{-1}\boldsymbol{\Gamma}'\boldsymbol{\Psi}^{-1} \tag{12.41}$$

where $\boldsymbol{\Theta} = \boldsymbol{\Gamma}'\boldsymbol{\Psi}^{-1}\boldsymbol{\Gamma}$. Substituting this expression into (12.39) and (12.40) in turn, and doing some algebraic manipulation, yields the two likelihood equations

$$(\hat{\boldsymbol{\Psi}}^{-\frac{1}{2}}\boldsymbol{S}\hat{\boldsymbol{\Psi}}^{-\frac{1}{2}})(\hat{\boldsymbol{\Psi}}^{-\frac{1}{2}}\hat{\boldsymbol{\Gamma}}) = (\hat{\boldsymbol{\Psi}}^{-\frac{1}{2}}\hat{\boldsymbol{\Gamma}})(\boldsymbol{I} + \hat{\boldsymbol{\Theta}}) \tag{12.42}$$

and

$$\hat{\boldsymbol{\Psi}} = \text{diag}(\hat{\boldsymbol{\Gamma}}\hat{\boldsymbol{\Gamma}}' - \boldsymbol{S}) \tag{12.43}$$

In principle, these two equations can be solved iteratively as follows. Starting with initial estimates $\hat{\psi}_i$, $i = 1, \ldots, p$, equation (12.42) defines an eigenvalue/eigenvector decomposition of $\hat{\boldsymbol{\Psi}}^{-\frac{1}{2}}\boldsymbol{S}\hat{\boldsymbol{\Psi}}^{-\frac{1}{2}}$ so that $\hat{\boldsymbol{\Gamma}}$ can be obtained from the eigenvectors corresponding to the m largest eigenvalues of this matrix. Substitution of $\hat{\boldsymbol{\Gamma}}$ into (12.43) yields an improved estimate of $\boldsymbol{\Psi}$, and so on.

This approach essentially dates back to the work of Lawley (1940, 1941). Unfortunately, the iterative process may converge very slowly (c.f. Howe, 1955), and the estimates of some ψ_j^2 may tend to zero. Although some improvements were made over the years, maximum likelihood estimation remained impracticable for all but the smallest problems until the work of Jöreskog (1967, 1969). His procedure replaces (12.43) by a numerical optimisation of the log-likelihood treated as a function of $\boldsymbol{\Psi}$, using the Fletcher-Powell method based on a second-degree approximation to l. For computational details see Lawley and Maxwell (1971, pp. 142–143); subsequent improvements include implementation of the Newton-Raphson approach (Jennrich and Robinson, 1969; Clark, 1970) and consideration of scoring methods to overcome poor initial estimates (Lee and Jennrich, 1979). Numerical optimisation also ensures that the solution is constrained to have $\hat{\psi}_i > 0 \; \forall i$. Thus, maximum likelihood now affords a fast and reliable method of estimation in factor analysis, and has been implemented on most standard statistical packages.

12.20 Two aspects of the above procedure require brief amplification: the choice of initial values for $\hat{\psi}_i$, $i = 1, \ldots, p$, to start the iterative process, and the tendency for some $\hat{\psi}_i$ to go to zero in practical applications.

Taking the first point, if we regard the model (12.32) as expressing each x_i in terms of a 'systematic' part (the common factors y_i) and an 'error' e_i, then we may regard each x_i as a linear function of all the other xs plus a residual. It can be shown (Exercise 12.3) that the *communality* of the ith variable, $h_i^2 = \text{var}(x_i) - \psi_i$, cannot be less than $R_i^2\text{var}(x_i)$ (where R_i^2 is the multiple correlation coefficient between x_i and all the other xs), and the two are often close in value. Thus, a reasonable starting value for $\hat{\psi}_i$ is $\text{var}(x_i)\{1 - R_i^2\}$, $i = 1, \ldots, p$. Note, however, that other simple starting values are also available. For example, Jöreskog (1963) suggests using $(1 - \frac{1}{2}m/p)(1/s^{ii})$, where s^{ii} is the ith diagonal element of $\boldsymbol{S}^{-1}$.

As regards the second point, the simple approach of setting partial derivatives to zero and solving for the unknown parameters leads (surprisingly often) to a solution with one or more negative $\hat{\psi}_i$ ('improper solutions', also called 'Heywood cases' after Heywood, 1931). Numerically, this situation is now handled by constraining the optimisation of the log-likelihood to be over $\psi_i \geqslant \epsilon$ for some pre-specified value $\epsilon > 0$ and then, possibly, reworking the analysis with $\psi_i = 0$ to explore the boundary solutions. Van Driel (1978) suggests three possible reasons for such behaviour. The movement to the boundary may be:

(1) due to sampling variation;
(2) because no interpretable factor model fits; or
(3) because of the indefiniteness of the model (e.g. too many true factor loadings are zero).

The boundary solution is appropriate in the first case, but in the other situations an inappropriate choice of variables for the model seems to be indicated. Either the offending variables (i.e. those with $\hat{\psi}_i \to 0$) should be dropped from the analysis, or new variables and/or additional factors should be included, or no factor model may fit the data.

Finally, an alternative approach to maximum likelihood estimation, via the E-M algorithm, has been explored by Rubin and Thayer (1982, 1983) and Bentler and Tanaka (1983), but this method does not appear to offer any advantages over the previously described one.

12.21 One problem that arises in factor analysis, in common with other multivariate techniques, is that the variates x_i $(i = 1, \ldots, p)$ may be measured in arbitrary or in non-commensurate units. In either case, it might be reasonable to consider standardising each variate by dividing each of its values by the sample standard deviation. The upshot of such standardisation is that the factor analysis is conducted on the sample correlation matrix $\boldsymbol{R}$ instead of the covariance matrix $\boldsymbol{S}$. Here, maximum likelihood has a major advantage over the other estimation methods described below, in that it is the only method for which the results on analysing $\boldsymbol{R}$ can be deduced in consistent fashion from those obtained on analysing $\boldsymbol{S}$: if s_{ii} is the sample variance of the ith manifest variable, and if $\hat{\lambda}_{ir}$, $\hat{\psi}_i$ are the loading and specific variance estimates from $\boldsymbol{S}$, then $\hat{\lambda}_{ir}/\sqrt{s_{ii}}$, $\hat{\psi}_i/s_{ii}$ are the corresponding estimates from $\boldsymbol{R}$. Indeed, it is easy to show that maximum likelihood factor analysis is independent of the scales of measurement (Exercise 12.4); see Krane and McDonald (1978) for further details.

12.22 The second major benefit of making normality assumptions about the data, and using maximum likelihood for estimation, is that the goodness-of-fit of the m-factor model can be judged using the likelihood ratio test (see Part 1, Appendix **A.6**). The null hypothesis here is that $\boldsymbol{\Sigma}$ has the form (12.33) while the alternative is that $\boldsymbol{\Sigma}$ is unconstrained. The test statistic reduces to

$$\omega = n(\mathrm{tr}\,\hat{\boldsymbol{\Sigma}}^{-1}\boldsymbol{S} - \log|\hat{\boldsymbol{\Sigma}}^{-1}\boldsymbol{S}| - p), \tag{12.44}$$

where $\hat{\boldsymbol{\Sigma}} = \hat{\boldsymbol{\Gamma}}\hat{\boldsymbol{\Gamma}}' + \hat{\boldsymbol{\Psi}}$ is the estimated dispersion matrix under the factor model and S is the usual sample dispersion matrix. If $\boldsymbol{\Psi} > \boldsymbol{0}$, ω has an asymptotic chi-squared distribution on $\nu = \frac{1}{2}[(p-m)^2 - (p+m)]$ degrees of freedom under the null hypothesis, which enables the test to be carried out. Bartlett (1954) showed that the approximation can be improved by replacing n in (12.44) by $n - 1 - \frac{1}{6}(2p+5) - \frac{2}{3}m$, and the behaviour of the test has been investigated by Gweke and Singleton (1980) and Schönemann (1981).

12.23 The large-sample distributions of the ML estimators have been examined by Anderson and Rubin (1956), who demonstrate asymptotic normality of $\hat{\boldsymbol{\Gamma}}$ and $\hat{\boldsymbol{\Psi}}$ if the manifest variables have a joint normal distribution, and Gill (1977), who demonstrates their consistency even when the manifest variables are non-normal. Asymptotic variances and covariances for the estimators can be obtained in the usual way, by computing the second derivative matrix. Formulae are given by Lawley and Maxwell (1971, Chapter 5), with a correction by Jennrich and Thayer (1973), but these expressions are only approximate if the parameters have been estimated from non-normal data. Bartholomew (1987, p56) discusses the use of bootstrapping to obtain standard errors in these circumstances.

It is worth noting that Young (1941) and Whittle (1952) also discuss maximum likelihood factor analysis, but for a different model to the one adopted above. In their model the $\boldsymbol{e}$ are the only random variates; the factors $\boldsymbol{y}$ now take fixed (unknown) values for each individual in the sample and are treated as parameters to be estimated. This leads to consistency problems, as the number of parameters thus tends to infinity with sample size, so the method is seldom used and remains of historic interest only.

Factor analysis: other methods of estimation

12.24 Equation (12.33) remains true whatever the distribution of the manifest variables (subject to standard regularity conditions). Thus, a reasonable approach if we do not wish to assume normality of these variables is to estimate $\boldsymbol{\Gamma}$ and $\boldsymbol{\Psi}$ in such a way as to make $\hat{\boldsymbol{\Sigma}} = \hat{\boldsymbol{\Gamma}}\hat{\boldsymbol{\Gamma}}' + \hat{\boldsymbol{\Psi}}$ as close to the sample dispersion matrix $\boldsymbol{S}$ as possible. To do this, we need to define some scalar measure of discrepancy (or distance) between $\boldsymbol{S}$ and $\boldsymbol{\Sigma} = \boldsymbol{\Gamma}\boldsymbol{\Gamma}' + \boldsymbol{\Psi}$, and then find the values $\hat{\boldsymbol{\Gamma}}$, $\hat{\boldsymbol{\Psi}}$ that minimise this measure.

A natural alternative to maximum likelihood is least squares, so that a simple criterion to minimise is

$$D_1 = \text{tr}[(\boldsymbol{S} - \boldsymbol{\Sigma})^2] \tag{12.45}$$

To obtain the estimates, we thus differentiate with respect to $\boldsymbol{\Gamma}$ and $\boldsymbol{\Psi}$ in turn and set the resulting expressions to zero, yielding the estimating equations

$$(\boldsymbol{S} - \hat{\boldsymbol{\Psi}})\hat{\boldsymbol{\Gamma}} = \hat{\boldsymbol{\Gamma}}(\hat{\boldsymbol{\Gamma}}'\hat{\boldsymbol{\Gamma}}) \tag{12.46}$$

and

$$\hat{\boldsymbol{\Psi}} = \text{diag}(\boldsymbol{S} - \hat{\boldsymbol{\Gamma}}\hat{\boldsymbol{\Gamma}}') \tag{12.47}$$

These two equations suggest a natural iterative scheme: starting with an initial estimate $\hat{\boldsymbol{\Psi}}$, the m eigenvectors of $\boldsymbol{S} - \hat{\boldsymbol{\Psi}}$ yield the columns of $\hat{\boldsymbol{\Gamma}}$ from (12.46), and an improved estimate $\hat{\boldsymbol{\Psi}}$ is then obtained by substituting $\hat{\boldsymbol{\Gamma}}$ into (12.47). This cycle is continued until convergence is achieved. This method is known as the *principal factor* method (sometimes qualified as being *with iteration*); it was the most commonly used method before the computing advances described above brought the maximum likelihood method to prominence in the 1970s. If $\boldsymbol{\Psi} = \boldsymbol{0}$, then the method reduces exactly (in one step) to principal component analysis (Part 1, Chapter 4), and principal component analysis is therefore sometimes offered as an option to principal factor analysis in statistical software packages. Note that principal factor analysis has the same lack of scale invariance as principal component analysis: if we use the correlation matrix as the starting point of the analysis, the derived loadings and specific variances will *not* be the same as those obtained from the covariance matrix and then standardised (as they were in the case of maximum likelihood).

The above estimates are the ordinary least squares ones. Various generalisations have also been considered by different authors; two simple variants are the weighted least squares criterion

$$D_2 = \mathrm{tr}[(\boldsymbol{S} - \boldsymbol{\Sigma})\boldsymbol{\Psi}^{-1}]^2 \tag{12.48}$$

and the generalised least squares criterion

$$D_3 = \mathrm{tr}[(\boldsymbol{S} - \boldsymbol{\Sigma})\boldsymbol{S}^{-1}]^2 \tag{12.49}$$

The former leads to exactly the same estimating equations (12.46) and (12.47), but with $\hat{\boldsymbol{\Gamma}}$ replaced by $\hat{\boldsymbol{\Psi}}^{-\frac{1}{2}}\hat{\boldsymbol{\Gamma}}$ and with $\boldsymbol{S}$ replaced by $\hat{\boldsymbol{\Psi}}^{-\frac{1}{2}}\boldsymbol{S}\hat{\boldsymbol{\Psi}}^{-\frac{1}{2}}$. The latter has the same first equation, (12.46), but replaces (12.47) by

$$\mathrm{diag}\{\boldsymbol{S}^{-1}[(\hat{\boldsymbol{\Gamma}}\hat{\boldsymbol{\Gamma}}' + \hat{\boldsymbol{\Psi}}) - \boldsymbol{S}]\boldsymbol{S}^{-1}\} = \mathrm{diag}\{\mathbf{0}\}. \tag{12.50}$$

For further details see Anderson (1984, Section 14.3.4). Note that the above criteria are all special cases of

$$D = \mathrm{tr}([\boldsymbol{S} - \boldsymbol{\Sigma}]\boldsymbol{V})^2 \tag{12.51}$$

which is the objective function considered by Browne (1982, 1984) and Tanaka and Huba (1985). However, use of this general function destroys the simplicity of the eigenvector-based iterative schemes described above.

Factor analysis: model selection

12.25 All the development so far has assumed that the number m of factors in the model is fixed and known, but in most practical applications this will not be the case: m will be another parameter of the model that has to be estimated from the data. Before widespread use of maximum likelihood factor analysis, this choice was generally made on the basis of one of the ad hoc criteria used to determine the number of 'important' components in principal component analysis (discussed in Part 1, Chapter 4). These criteria all involve the eigenvalues of the correlation matrix, the most common being the 'proportion of trace' (criterion 1, **4.6**), the number of eigenvalues greater than 1 (criterion 2, **4.6**), and the 'scree plot' (criterion 5, **4.6**). If maximum likelihood factor analysis is used, however, then the likelihood ratio test of **12.22** can be applied sequentially: starting with $m = 1$, we can fit successive values and test the goodness-of-fit using (12.44) until we obtain a non-significant result (indicating that the fit of the model is adequate). It is possible, of course, that no solution with $(p - m)^2 > p + m$ gives an adequate fit. Also, although this seems to be a suitably objective method for determining m, there are problems: it is not strictly valid as a hypothesis test, because no adjustment is made to the significance levels to allow for its sequential nature, and the mere fact that the test is not significant does not mean that we have found the optimum value of m. The larger is m, the more parameters are fitted, and the better will be the fit of the model to the data. Determining the 'best' model involves a trade-off between number of parameters and goodness-of-fit, and Akaike (1983) suggests choosing m to minimise $\omega - 2\nu$ (where these quantities are defined in **12.22**). The question of model choice has received very intensive attention recently in the psychometric literature. However, as it relates directly to the wider issue of covariance structure modelling, we will defer any further discussion until we have covered that topic; see **12.38–12.40**. We merely note here that some simulation studies of the criteria discussed in the present section have been conducted by Hakstian *et al* (1982) and Fachel (1986), but without any clear-cut recommendations.

Factor rotation and factor scores

12.26 We have already noted in **12.18** that the indeterminacy of the linear factor model means that we essentially estimate the factor *space* using one of the above methods, and we are then at liberty to select any *rotation* we like within this space. A rotation would usually be chosen so as to improve interpretability or understanding of the fitted model, and various authors have considered this aspect in more detail. In general, the more 'definite' the pattern of estimated loadings $\hat{\gamma}_{ij}$ the more easily will the analysis be understood. Thurstone (1947) considers various criteria that this pattern should satisfy if it is to have a 'simple structure', and the different rotations that are available in factor analysis software all aim to meet these criteria to a greater or lesser extent. The way they operate is to specify an index of 'simplicity' in terms of the factor loadings, and then to seek the transformation that optimises this index.

The two most popular orthogonal rotations are the *varimax* (Kaiser, 1958) and the *quartimax* (Neuhaus and Wrigley, 1954) rotations. The former seeks the orthogonal transformation that maximises

$$R_1 = \frac{1}{p^2}\sum_{j=1}^{m}\left\{m\sum_{i=1}^{p}\beta_{ij}^4 - \left(\sum_{i=1}^{p}\beta_{ij}^2\right)^2\right\} \tag{12.52}$$

where $\beta_{ij} = \gamma_{ij}/\left(\sum_{j=1}^{m}\gamma_{ij}^2\right)^{1/2}$, while the latter seeks the orthogonal transformation that maximises

$$R_2 = \sum_{i=1}^{p}\sum_{j=1}^{m}\gamma_{ij}^4 - \frac{1}{pm}\left(\sum_{i=1}^{p}\sum_{j=1}^{m}\gamma_{ij}^2\right)^2 \tag{12.53}$$

Both transformations thus aim to end up with solutions in which some loadings are 'small' (i.e. close to zero) while the others are 'large', thereby making it easier to label the factors in terms of the variables. Other criteria for orthogonal rotation are summarised in Harman (1976, Chapter 13).

If the factors can be identified with fundamental quantities in whose reality we have confidence (e.g. verbal and arithmetical abilities in educational tests), and if there is reason to suppose that such factors are correlated, then we shall not uncover them if we restrict ourselves to orthogonal factors only. The above ideas have thus been generalised to encompass more general non-singular transformations rather than just orthogonal ones, and various criteria have been proposed for obtaining such *oblique* rotations. Statistical packages typically include such oblique transformations as *promax* and *oblimin*; for details of these methods see, for example, Lawley and Maxwell (1971, Section 6.4) or Harman (1976, Chapter 14).

Finally, if we wish to obtain a rotation that matches our solution as closely as possible with some prior specification, we need to employ Procrustes analysis (Part 1, **5.22**); see Meredith (1977). Care should be taken in any interpretation after using this procedure, however, as Horn and Knapp (1973) and Huitson (1989) both highlight some problems with it. The former show that a structured matrix can be matched well with a random target, while the latter demonstrates that apparent structure can be extracted from a random matrix.

12.27 Having completed all the above steps of a factor analysis, and having successfully labelled some of the factors, it might then be of interest to estimate the ith individual's vector $\boldsymbol{y}_i$ of *scores* on a given factor. For example, if one of the factors has been identified as 'arithmetic ability', then it is clearly of interest to estimate each individual's value on this factor if we wish to select the half-dozen individuals best suited to an arithmetical assignment.

The two methods of factor score estimation in common use both have an intuitive basis and both provide estimates that are linear functions of the observed data. We suppose that the observed vector of manifest variables for the ith individual is $\boldsymbol{x}_i$, and that the sample mean vector is $\bar{\boldsymbol{x}}$. Then Bartlett (1937, 1938) used least squares on the linear factor model (12.32), and sought $\boldsymbol{y}_i$ to minimise

$$S = (\boldsymbol{x}_i - \bar{\boldsymbol{x}} - \hat{\boldsymbol{\Gamma}}\boldsymbol{y}_i)'\hat{\boldsymbol{\Psi}}^{-1}(\boldsymbol{x}_i - \bar{\boldsymbol{x}} - \hat{\boldsymbol{\Gamma}}\boldsymbol{y}_i) \tag{12.54}$$

Straightforward calculus yields the estimate

$$\hat{\boldsymbol{y}}_i = (\hat{\boldsymbol{\Gamma}}'\hat{\boldsymbol{\Psi}}^{-1}\hat{\boldsymbol{\Gamma}})^{-1}\hat{\boldsymbol{\Gamma}}'\hat{\boldsymbol{\Psi}}(\boldsymbol{x}_i - \bar{\boldsymbol{x}}) \tag{12.55}$$

Thomson (1951), on the other hand, assumes normality of $\boldsymbol{y}$ and $\boldsymbol{e}$ in (12.32), puts the manifest and latent variables into a single vector and uses standard properties of the multivariate normal distribution (**2.8**, Part 1) to obtain the expected value of $\boldsymbol{y}$ given that $\boldsymbol{x} = \boldsymbol{x}_i$ as

$$E(\boldsymbol{y}|\boldsymbol{x}_i) = \boldsymbol{\Gamma}'(\boldsymbol{\Gamma}\boldsymbol{\Gamma}' + \boldsymbol{\Psi})^{-1}(\boldsymbol{x}_i - \boldsymbol{\mu}) \tag{12.56}$$

An obvious estimate of the factor score vector is thus

$$\hat{\boldsymbol{y}}_i = \hat{\boldsymbol{\Gamma}}'(\hat{\boldsymbol{\Gamma}}\hat{\boldsymbol{\Gamma}}' + \hat{\boldsymbol{\Psi}})^{-1}(\boldsymbol{x}_i - \bar{\boldsymbol{x}}) \tag{12.57}$$

The set of factor scores is generally standardised before use with either method, to accord with the fundamental assumptions of the model. Both methods thus provide estimated factor scores as linear combinations $\boldsymbol{A}(\boldsymbol{x}_i - \bar{\boldsymbol{x}})$ of the mean-centred manifest variables, and the matrix $\boldsymbol{A}$ is often termed the *factor score estimation matrix* in computer package output.

Example 12.4 (Kendall *et al*, 1983, p357)
The rows of Table 12.8 contain the scores obtained by each of 48 applicants for a post, on a set of 15 characteristics. The following characteristics were used, each scored on a 10-point scale.

(1) Form of letter of application.
(2) Appearance.
(3) Academic ability.
(4) Likeability.
(5) Self-confidence.
(6) Lucidity.
(7) Honesty.
(8) Salesmanship.
(9) Experience.
(10) Drive.
(11) Ambition.
(12) Grasp.
(13) Potential.
(14) Keenness to join the company.
(15) Suitability.

A straightforward principal component analysis of the correlation matrix gave eigenvalues 7.50, 2.06, 1.46, 1.21, 0.74, 0.49, 0.35, 0.31, 0.26, 0.20, 0.15, 0.09, 0.08, 0.06 and 0.03. There are four eigenvalues greater than 1, accounting for 81.5% of the total variation, so using the eigenvalue criteria it seems as if four factors could be used to represent the data. A first approximation to the principal factors can be obtained simply by taking the eigenvectors resulting from the principal component analysis, but normalising them so that each has the sum of squared elements equal to the corresponding eigenvalue (the principal component *loadings*). These loadings are shown in the left-hand portion of Table 12.9, where rows correspond to original measured characteristics (i.e. variables) and columns to the successive components (ranked in decreasing order of eigenvalues).

If we read this table across so that, for example,

$$x_1 = 0.45y_1 + 0.62y_2 + 0.37y_3 - 0.12y_4 + e_1$$

we are, in effect, defining the residual and hence the communality. To interpret the analysis we read down the columns, to see which variables are most heavily weighted in each factor. We must also remember that we have only estimated the factor space of four dimensions, and that these particular factors are arbitrary within a rotation.

Taking the factors as they stand, the picture is not very clear. Factor 1 seems loaded on all variables except perhaps x_3 ('academic'). Factor 2 loads on 'letter' (x_1), 'honesty' (x_7), 'experience' (x_9) and 'suitability' (x_{15}); factor 3 on 'academic' (x_3), 'likeability' (x_4) and 'keenness' (x_{14}); and factor 4 on 'academic' and, perhaps, 'honesty'. A varimax rotation clarifies the interpretation; the rotated loadings are given on the right-hand side of Table 12.9. The fourth factor now seems to be one of academic ability and is not loaded heavily on any of the other variables,

Table 12.8 48 applicants rated on each of 15 characteristics. From Kendall *et al* (1983)

	Characteristic														
	1	2	3	4	5	6	7	8	9	10	11	12	13	14	15
1	6	7	2	5	8	7	8	8	3	8	9	7	5	7	10
2	9	10	5	8	10	9	9	10	5	9	9	8	8	8	10
3	7	8	3	6	9	8	9	7	4	9	9	8	6	8	10
4	5	6	8	5	6	5	9	2	8	4	5	8	7	6	5
5	6	8	8	8	4	4	9	2	8	5	5	8	8	7	7
6	7	7	7	6	8	7	10	5	9	6	5	8	6	6	6
7	9	9	8	8	8	8	8	8	10	8	10	8	9	8	10
8	9	9	9	8	9	9	8	8	10	9	10	9	9	9	10
9	9	9	7	8	8	8	8	5	9	8	9	8	8	8	10
10	4	7	10	2	10	10	7	10	3	10	10	10	9	3	10
11	4	7	10	0	10	8	3	9	5	9	10	8	10	2	5
12	4	7	10	4	10	10	7	8	2	8	8	10	10	3	7
13	6	9	8	10	5	4	9	4	4	4	5	4	7	6	8
14	8	9	8	9	6	3	8	2	5	2	6	6	7	5	6
15	4	8	8	7	5	4	10	2	7	5	3	6	6	4	6
16	6	9	6	7	8	9	8	9	8	8	7	6	8	6	10
17	8	7	7	7	9	5	8	6	6	7	8	6	6	7	8
18	6	8	8	4	8	8	6	4	3	3	6	7	2	6	4
19	6	7	8	4	7	8	5	4	4	2	6	8	3	5	4
20	4	8	7	8	8	9	10	5	2	6	7	9	8	8	9
21	3	8	6	8	8	8	10	5	3	6	7	8	8	5	8
22	9	8	7	8	9	10	10	10	3	10	8	10	8	10	8
23	7	10	7	9	9	9	10	10	3	9	9	10	9	10	8
24	9	8	7	10	8	10	10	10	2	9	7	9	9	10	8
25	6	9	7	7	4	5	9	3	2	4	4	4	4	5	4
26	7	8	7	8	5	4	8	2	3	4	5	6	5	5	6
27	2	10	7	9	8	9	10	5	3	5	6	7	6	4	5
28	6	3	5	3	5	3	5	0	0	3	3	0	0	5	0
29	4	3	4	3	3	0	0	0	0	4	4	0	0	5	0
30	4	6	5	6	9	4	10	3	1	3	3	2	2	7	3
31	5	5	4	7	8	4	10	3	2	5	5	3	4	8	3
32	3	3	5	7	7	9	10	3	2	5	3	7	5	5	2
33	2	3	5	7	7	9	10	3	2	2	3	6	4	5	2
34	3	4	6	4	3	3	8	1	1	3	3	3	2	5	2
35	6	7	4	3	3	0	9	0	1	0	2	3	1	5	3
36	9	8	5	5	6	6	8	2	2	2	4	5	6	6	3
37	4	9	6	4	10	8	8	9	1	3	9	7	5	3	2
38	4	9	6	6	9	9	7	9	1	2	10	8	5	5	2
39	10	6	9	10	9	10	10	10	10	10	8	10	10	10	10
40	10	6	9	10	9	10	10	10	10	10	10	10	10	10	10
41	10	7	8	0	2	1	2	0	10	2	0	3	0	0	10
42	10	3	8	0	1	1	0	0	10	0	0	0	0	0	10
43	3	4	9	8	2	4	5	3	6	2	1	3	3	3	8
44	7	7	7	6	9	8	8	6	8	8	10	8	8	6	5
45	9	6	10	9	7	7	10	2	1	5	5	7	8	4	5
46	9	8	10	10	7	9	10	3	1	5	7	9	9	4	4
47	0	7	10	3	5	0	10	0	0	2	2	0	0	0	0
48	0	6	10	1	5	0	10	0	0	2	2	0	0	0	0

Table 12.9 Principal component loadings for data of Table 12.8. From Kendall *et al* (1983)

	Raw loadings				Rotated loadings			
Variable	1	2	3	4	1	2	3	4
Letter	0.45	0.62	0.37	−0.12	0.12	0.83	0.11	−0.14
Appearance	0.58	−0.05	−0.02	0.29	0.44	0.15	0.40	0.23
Academic	0.11	0.34	−0.50	0.71	0.06	0.13	0.01	0.93
Likeability	0.62	−0.18	0.58	0.36	0.22	0.25	0.87	−0.08
Self-confidence	0.80	−0.36	−0.30	−0.18	0.92	−0.10	0.17	−0.06
Lucidity	0.87	−0.19	−0.18	−0.07	0.86	0.10	0.26	0.00
Honesty	0.43	−0.58	0.36	0.45	0.22	−0.24	0.86	0.00
Salesmanship	0.88	−0.06	−0.25	−0.23	0.92	0.21	0.09	−0.05
Experience	0.37	0.80	0.10	0.07	0.08	0.85	−0.05	0.21
Drive	0.86	0.07	−0.10	−0.17	0.80	0.35	0.16	−0.05
Ambition	0.87	−0.10	−0.26	−0.21	0.92	0.16	0.11	−0.04
Grasp	0.91	−0.03	−0.13	0.09	0.81	0.26	0.34	0.15
Potential	0.91	0.04	−0.08	0.21	0.74	0.33	0.42	0.23
Keenness	0.71	−0.11	0.56	−0.23	0.44	0.36	0.54	−0.52
Suitability	0.65	0.61	0.10	−0.03	0.38	0.80	0.08	0.08

except negatively on ‘keenness’. Factor 3 is loaded most heavily on ‘likeability’ and ‘honesty’, so could be interpreted as a general likeability of a person apart from particular qualifications. Factor 2 seems to be one of experience, while factor 1 is the usual blend of required qualities such as self-confidence, salesmanship, ambition and lucidity.

Turning now to maximum likelihood, the raw and varimax rotated factor loadings obtained by this method are shown in Table 12.10.

Table 12.10 Maximum likelihood loadings for data of Table 12.8. From Kendall *et al* (1983)

	Raw loadings				Rotated loadings			
Variable	1	2	3	4	1	2	3	4
Letter	0.46	−0.14	0.53	−0.23	0.13	−0.10	0.72	0.13
Appearance	0.45	0.30	0.06	0.13	0.45	0.17	0.14	0.24
Academic	−0.11	0.52	0.39	0.26	0.07	0.69	0.12	−0.01
Likeability	0.70	−0.20	0.14	0.51	0.23	−0.05	0.24	0.83
Self-confidence	0.72	0.46	−0.40	−0.07	0.92	−0.09	−0.10	0.15
Lucidity	0.77	0.44	−0.12	0.05	0.84	0.05	0.12	0.29
Honesty	0.48	−0.08	−0.25	0.62	0.25	−0.02	−0.22	0.75
Salesmanship	0.79	0.44	−0.10	−0.23	0.90	−0.07	0.24	0.07
Experience	0.28	0.10	0.70	−0.23	0.09	0.19	0.76	−0.06
Drive	0.80	0.31	0.09	−0.16	0.76	−0.05	0.39	0.18
Ambition	0.78	0.44	−0.14	−0.18	0.90	−0.06	0.19	0.11
Grasp	0.80	0.44	0.08	0.12	0.78	0.17	0.28	0.36
Potential	0.80	0.44	0.21	0.22	0.73	0.27	0.35	0.45
Keenness	0.91	−0.39	−0.02	−0.04	0.42	−0.57	0.41	0.56
Suitability	0.56	−0.16	0.60	−0.24	0.36	0.10	0.78	0.06

Table 12.11 Maximum likelihood loadings for seven-factor fit to data of Table 12.8, after varimax rotation. From Kendall *et al* (1983)

	Factor loadings						
Variable	1	2	3	4	5	6	7
Letter	0.13	0.07	0.67	−0.10	0.02	−0.04	0.27
Appearance	0.33	0.24	0.18	0.10	0.61	−0.01	−0.01
Academic	0.05	−0.02	0.10	0.69	0.04	0.01	0.01
Likeability	0.25	0.76	0.25	−0.06	0.09	−0.10	0.20
Self-confidence	0.88	0.18	−0.08	−0.07	0.19	0.06	−0.05
Lucidity	0.91	0.27	0.14	0.05	−0.04	−0.29	−0.02
Honesty	0.20	0.91	−0.22	−0.01	0.17	−0.09	−0.20
Salesmanship	0.88	0.08	0.26	−0.08	0.14	0.04	−0.06
Experience	0.07	−0.03	0.72	0.16	0.07	0.04	0.01
Drive	0.78	0.20	0.39	0.03	−0.05	0.40	−0.02
Ambition	0.87	0.04	0.16	−0.05	0.38	0.14	0.21
Grasp	0.78	0.35	0.29	0.17	0.14	−0.16	0.11
Potential	0.70	0.41	0.35	0.33	0.14	0.07	0.19
Keenness	0.43	0.54	0.38	−0.54	−0.01	0.10	0.28
Suitability	0.31	0.08	0.91	0.05	0.14	0.03	−0.21

Comparing these loadings with the previous ones, we see that qualitatively there is a good similarity between the two rotated solutions (although the ordering of factors has changed and the precise numerical values show some variability between methods). Thus, the first factor is almost identical in the two cases; maximum likelihood factor 3 is clearly principal factor 2, but with the high loadings slightly less pronounced; maximum likelihood factor 2 is similarly a slightly less definite version of principal factor 4; while maximum likelihood factor 4 and principal factor 3 are almost coincident. It turns out, however, that four factors do not provide an adequate fit, the test of **12.22** yielding a chi-square value of 103.9 on 51 degrees of freedom. Adopting the model selection strategy of **12.25**, we need to go on to seven factors before we obtain an adequate fit. The varimax rotated loadings for this seven-factor solution are shown in Table 12.11. The previous factors 1, 2, 3 and 4 can be identified as factors 1, 4, 3 and 2 here (albeit again with slightly different emphases in each case). In addition, we have an 'appearance' factor (5) and, perhaps, a 'drive' factor (6). The last factor does not seem to be easily interpretable.

In studies of this kind, where the scores have been allotted by human beings on the basis of subjective judgement, and the kind of variables to be included have also been subjectively determined, the interpretation of the results is especially complicated. It may be that we have successfully delimited dimensions of variation in the candidates, but it may also be that we have only explored the interpretation put by the judges on expressions such as 'drive', and it may also be that we should consider whether we are not overweighting some dimensions at the expense of others.

Factor analysis: some extensions and generalisations

12.28 The most obvious generalisation of the linear factor model (12.32) is to allow terms that are nonlinear in the common factors $\boldsymbol{y}$. This is the *nonlinear* factor model that has been developed by McDonald (1962, 1965, 1967a,b,c, 1979, 1986) and Etezadi-Amoli and McDonald (1983). Assuming once more that there are m common factors, but now allowing polynomial terms up to degree q in each of them together with interaction terms in the form of pairwise products between them, the model for the manifest variables $\boldsymbol{x}$ becomes

$$x_i = \mu_i + \sum_{r=1}^{m}\sum_{s=1}^{q} \alpha_{irs} y_r^s + \sum_{r=1}^{m-1}\sum_{s=1}^{m} B_{irs} y_r y_s + e_i, \tag{12.58}$$

for $i = 1, \ldots, p$. Since this model is only nonlinear in the common factors (but still linear in the loadings α_{irs} and B_{irs}), it can be reparametrised into the form (12.32) if the vector of common factors is expanded into

$$(y_1, \ldots, y_m, y_1^2, \ldots, y_m^q, y_1 y_2, \ldots, y_{m-1} y_m)'.$$

Etezadi-Amoli and McDonald (1983) discuss the estimation of parameters in this model using either maximum likelihood or least squares, and describe a computer program for carrying out the computations when $q \leqslant 3$. They also give an extensive numerical illustration of its application. Note that such nonlinear factor models have some affinity with nonlinear principal components and other nonlinear ordination techniques, already discussed in Chapter 8 of Part 1.

Identifying the basic factor model with that in regression analysis, but with unobservable explanatory variables, it is evident that many of the well-known variants of regression modelling could, in principle, be applied to factor analysis. The above model is the equivalent of polynomial regression; the case of isotonic (or monotone) regression has been implemented by Kruskal and Shepard (1974), who term the resulting method 'nonmetric' factor analysis, while factor models in growth studies have been considered by Browne (1993) and Muthen (1993). See also Amemiya (1993).

12.29 A second possible direction of generalisation of the linear factor model is to its simultaneous imposition on several data matrices. This situation arises in the analysis of sets of individuals: either from repeated observation of the same individuals, or for different individuals taken from several identifiable groups. For example, a class of children might sit a set of IQ tests on several different occasions (e.g. over a period of years), and we wish to determine what relationships exist among the underlying factors and their loadings across occasions. Alternatively, several different classes of children (e.g. from different parts of the country) might sit the same set of IQ tests and we wish to investigate the corresponding relationships across classes. A single such class of children (on a single occasion) will yield an $(n \times p)$ matrix $\boldsymbol{X}$ of IQ test scores that can be factor analysed using model (12.32). Such a matrix is sometimes referred to as *two-mode*, because it connects the two 'modes', individuals and tests: each observation x_{ij} is labelled by one item from each mode, here individual i and test j. If we now have repeated testing of the same children, or testing of several different groups of children, then we will obtain a *set* of such score matrices, $\boldsymbol{X}_1, \ldots, \boldsymbol{X}_g$, one for each repeat or group. The matrices can be thought of as stacked into a three-dimensional structure, so the full set of data is *three-mode* with x_{ijk} giving the score obtained by child i on test j in group (or repeat) k.

Generalisation of model (12.32) to such data structures has been considered by Levin (1965) and Tucker (1966) under the name 'three-mode factor analysis', and by Harshman and Lundy (1984) using 'parallel factor analysis' (PARAFAC). The basic idea of all these generalisations is to represent each data value by a sum of products of three (or more) unknown sets of loadings/scores plus a residual, in place of the sum of products of two unknowns plus residual in the two-mode model (12.32). For example, Tucker's main model is of the form

$$x_{ijk} = \sum_r \sum_s \sum_t a_{ir} b_{js} c_{kt} g_{rst} + e_{ijk} \tag{12.59}$$

but many variations on this theme have been proposed. We thus do not give details here, but refer the reader to the cited references where estimation methods and illustrative examples may be found in addition to model discussion.

It is evident that such models have close connections with many other multivariate techniques that have already been discussed in these volumes. Most notably, the premise underlying (12.59) is closely matched by that for individual differences scaling (Tucker and Messick, 1963; Carroll and Chang, 1970; Part 1, **5.13**), the models in each case being based on generalisations of the Eckart and Young (1936) decomposition of a matrix. In fact, a generalisation to the n-way case of the three-way PARAFAC model was proposed by Carroll and Chang (1970) as the CANDECOMP (for CANonical DECOMPosition) model underlying the scalar product form of the individual differences model INDSCAL. When limited to the three-way case, CANDECOMP results in a model equivalent to PARAFAC; when this general model is further restricted to a data array symmetric in two of its three ways then both CANDECOMP and PARAFAC become equivalent to the scalar product form of INDSCAL. Tucker (1972) explores the connections between three-mode factor analysis and individual differences scaling in some detail, while Harshman and Lundy (1984) discuss the relations between INDSCAL, PARAFAC and CANDECOMP. All these methods are placed in the wider context of three-way scaling and clustering by Arabie *et al* (1987), while Leurgans and Ross (1992) give a thorough discussion of various multi-way models in the context of analysis of spectroscopic data.

A further connection of three-mode factor analysis is with the various constrained forms of principal component analysis, such as common principal components and partial common principal components (Flury, 1988; Part 1, **4.5**); the forms of covariance matrix derived from models such as (12.59) have close resemblance to those in common principal component analysis. Moreover, the connections between factor analysis and principal component analysis extend directly to the three- and higher-mode situations; see Kroonenberg and ten Berge (1989) for a very full algebraic exploration.

12.30 The final extension of the basic factor model (12.32) that we consider is to the case where the manifest variables are categorical rather than continuous. Given the prevalence of categorical variables in the social sciences, where factor analysis is also heavily used, it is evident that there has been considerable interest in such extension over the years. We concentrate on the case of binary variables first, and briefly indicate at the end how the methods can be applied to more general polytomous variables.

Historically, the first approaches to the problem attempted to transpose the continuous variable model directly on to the binary variable situation, so that the existing methods could be used for the analysis. One way of achieving this is to assume a *threshold* model for the manifest variables: if the observed binary variables are $z_1, \ldots, z_p$, then we assume the existence of p underlying continuous variables $x_1, \ldots, x_p$ and p associated constants (thresholds) $d_1, \ldots, d_p$ such that $z_i = 1$ or 0 according to whether $x_i \geqslant$ or $< d_i$ for each i. The underlying variables $\boldsymbol{x}$ are then assumed to follow the normal linear factor model (12.32), with its implication (12.33) for the dispersion matrix of $\boldsymbol{x}$. Since this dispersion matrix is estimated by the matrix of *tetrachoric correlations* (Stuart and Ord, 1991; **26.32**) between the binary variables, early approaches to the factor analysis of binary variables simply applied the usual methodology to this matrix. While this procedure may indeed give satisfactory results, problems arise when the tetrachoric correlation matrix is not positive definite and a more theoretically based approach is necessary.

Following through the consequences of the threshold model, we find that the likelihood of the sample can be written

$$L = \prod_{i=1}^{n} \int_{a_{i1}}^{b_{i1}} \int_{a_{i2}}^{b_{i2}} \cdots \int_{a_{ip}}^{b_{ip}} \phi(\boldsymbol{y}_i) d\boldsymbol{y}_i \qquad (12.60)$$

where $\phi(\boldsymbol{y}_i)$ is the multivariate normal density function with dispersion matrix (12.33) and the limits of integration depend on the values of the binary manifest variables: $a_{ij} = d_j, b_{ij} = \infty$ if $z_{ij} = 1$; $a_{ij} = -\infty, b_{ij} = d_j$ if $z_{ij} = 0$ for $i = 1, \ldots, n; j = 1, \ldots, p$. Starting from first principles, therefore, we need to maximise L with respect to the model parameters. However, these parameters include the unknown threshold constants $\boldsymbol{d}' = (d_1, \ldots, d_p)$, so iterative numerical evaluation of the multiple integral in L is required and this involves very heavy computation. Christofferson (1975) gives a computational scheme based on fitting the expected marginal proportions of single variables ($z_i = 1$) and pairs of variables ($z_i = 1, z_j = 1$) to their observed values, while Muthen (1978) improves the computational procedure by basing the fit on the comparison of observed and expected thresholds and tetrachoric correlations.

12.31 An alternative approach is to model the *probabilities* of obtaining the two values 0 or 1 of each binary variable z_i by means of a *response function* of the latent variables $\boldsymbol{y}$. This approach has also been used in item analysis within educational testing (Rasch, 1960; Solomon, 1961; Lord and Novick, 1968) and in signal detection theory (Egan, 1975). In both of these applications $\boldsymbol{y}$ is usually one-dimensional, the response function being known as the *item characteristic curve* in the former case and as the *receiver operating character-*

istic in the latter. The shape of these curves thus shows how the probability of a positive response increases with the underlying y.

Bartholomew (1980) has studied the general m-factor case, and for convenience assumed that the factors were independently uniformly distributed on $(0, 1)$. He started by considering a desirable set of properties that should be satisfied by a response function, and the following form for such a function:

$$G\{\pi_i(\boldsymbol{y})\} = \alpha_{i0} + \sum_{j=1}^{m} \alpha_{ij} H(y_j) \tag{12.61}$$

where $\pi_i(\boldsymbol{y}) = \Pr(z_i = 1|\boldsymbol{y})$ and $G(.)$, $H(.)$ are arbitrary functions. He then went on to show that, in order to satisfy these desirable properties, the choices for G and H are effectively limited either to $\text{logit}(v) = \log\{v/(1-v)\}$ or to $\text{probit}(v) = \Phi^{-1}(v)$, where $\Phi(.)$ is the standard normal distribution function. The case $m = 1$ had previously been treated by Lord and Novick (1968) using the logit function for G and the probit for H, and by Bock and Lieberman (1970) using the probit for both. A maximum likelihood computing scheme (not based on the E-M algorithm) for the logit/probit hybrid had also been given by Sanathanan and Blumenthal (1978). Bartholomew (1980) went on to the general case with the logit for both, while Bartholomew (1987, Chapters 5 and 6) treated the general case with logit for G and probit for H. The benefit of probit for H with uniformly distributed latent variables is that this choice yields a linear normal factor model, consistent with our previous development. Thus, assuming normal latent variables, the model is

$$\text{logit}\{\pi_i(\boldsymbol{y})\} = \alpha_{i0} + \sum_{j=1}^{m} \alpha_{ij} y_j \tag{12.62}$$

the joint probability function is

$$f(\boldsymbol{x}) = \int_{-\infty}^{+\infty} \cdots \int_{-\infty}^{+\infty} \prod_{i=1}^{p} \{\pi_i(\boldsymbol{y})\}^{x_i} \{1 - \pi_i(\boldsymbol{y})\}^{1-x_i} \phi(y_i) dy_1 \ldots dy_m \tag{12.63}$$

where $\phi(.)$ is the standard normal density function; and the log-likelihood for a sample $\boldsymbol{x}_1, \ldots, \boldsymbol{x}_n$ is

$$L = \sum_{s=1}^{n} \log f(\boldsymbol{x}_s) \tag{12.64}$$

Maximisation of this log-likelihood with respect to the model parameters can be effected by the E-M algorithm (Bock and Aitkin, 1981); for details see Bartholomew (1987, pp. 108–110). Factor scores can also be obtained readily (Bartholomew, 1987, pp. 102–103).

12.32 Although two ostensibly different methods have been outlined above for factor analysis of binary data, Bartholomew (1987, Section 5.4) demonstrates that they are in fact equivalent. He shows that, for suitable parametrisations, $\Pr\{\bigcap_{i \in S}(z_i = 1)\}$ can be written in the same form under either model, where S is any non-empty subset of $\{1, 2, \ldots, p\}$. Thus, for any response function model, there exists a threshold model that is empirically indistinguishable from it. If we now move on to polytomous (i.e. ordered categorical) manifest variables, however, then each of these two approaches can be generalised in a fairly obvious way but they are no longer equivalent to each other and distinct analyses result.

Assume that the data have the form of **12.13**: the ith manifest variable has c_i possible categories ($i = 1, \ldots, p$) and is denoted by the *vector* $\mathbf{z}_i' = (z_{i1}, \ldots, z_{ic_i})$, where $z_{ij} = 1$ and $z_{is} = 0$ for $s \neq j$ if the response on variable i is in the jth category (so that $\sum_{u=1}^{c_i} z_{iu} = 1$, $i = 1, \ldots, p$), and the complete response vector is $\mathbf{z}' = (\mathbf{z}_1', \ldots, \mathbf{z}_p')$. Considering first the response function approach, the single function $\pi_i(\mathbf{y}) = \Pr(z_i = 1|\mathbf{y})$ must now be replaced by the set of functions $\pi_{ij}(\mathbf{y}) = \Pr(z_{ij} = 1|\mathbf{y})$ for $j = 1, \ldots, c_i$ and $i = 1, \ldots, p$, with the added requirement that $\sum_j \pi_{ij}(\mathbf{y}) = 1$. With this requirement in mind, a natural generalisation of (12.62) is then

$$\pi_{ij}(\mathbf{y}) = \pi_{ij} \exp \sum_{s=1}^{m} \alpha_{ijs} y_s \Big/ \sum_{r=1}^{c_i} \pi_{ir} \exp \sum_{s=1}^{m} \alpha_{irs} y_s \tag{12.65}$$

where $\pi_{ij} = \exp \alpha_{ij0} / \{1 + \exp \sum_{r=1}^{c_i} \alpha_{ir0}\}$ is the baseline response probability (in fact, for the 'median' individual) ($j = 1, \ldots, c_i$; $i = 1, \ldots, p$). Maximisation of the likelihood arising from this model can be done as usual by using the E-M algorithm; details of the calculations are given by Bartholomew (1987, Chapter 8).

The threshold model is generalised from the binary case in the obvious way, by assuming that underlying each categorical manifest variable there is a continuous variable whose range is divided into intervals, and the category recorded for an individual indicates which interval that individual's underlying variable occupies. Thus, we assume that corresponding to $\mathbf{z}_i$ there is a continuous variable x_i and a sequence of thresholds $d_{i1}, \ldots, d_{ic_i}$ (where $d_{i1} = -\infty$ and $d_{ic_i} = \infty$) such that $z_{ij} = 1$ if $d_{ij} \leqslant x_i < d_{i,j+1}$ and $z_{ij} = 0$ otherwise, for $j = 1, \ldots, c_i$ and $i = 1, \ldots, p$. The x_i then again follow model (12.32) and its implication (12.33) for their dispersion matrix. This matrix is estimated by the set of *polychoric correlations* (Martinson and Hamdan, 1975; Olsson, 1979), which are the generalisations of tetrachoric correlations to the polytomous case. Maximum likelihood estimation of the parameters in the threshold model is possible in principle, but becomes computationally difficult if there are many manifest variable categories. Satisfactory results are generally obtained by factor analysing the matrix of polychoric correlations, and Bartholomew (1987) also gives two approximate methods.

Covariance structure modelling

12.33 So far in this chapter we have been concerned with the use of latent variables to explain associations between manifest variables, and the behavioural sciences have been highlighted as a main area of application of the various techniques. For further discussion of such latent variable methodology within the specific disciplines of sociology, psychology and economics see Bielby and Hauser (1977), Bentler (1980, 1986) and Aigner *et al* (1984) respectively. However, we now turn to a somewhat wider issue. Situations often arise in which the manifest variables can be partitioned into two logical groups, with those in one group being considered as explanatory for the values observed in those of the other group. Researchers then want to model the relationships that exist between the variables of the two sets, so as to gain deeper understanding of the system under study and, perhaps, to determine the causal mechanism or structure underlying it. A general statistical technique much employed in this respect is regression analysis, and this technique has been extended over the years as a consequence of particular requirements arising within specific disciplines. Thus, for example, the study of regression models with errors in variables was motivated mainly by the type of data gathered in social investigations, while complex simultaneous sets of regression relationships in which variables can appear as dependent (y; endogenous) in some equations but as independent (x; exogenous) in others arise frequently in econometrics. Path analysis, as described in the previous chapter, provides a general framework for systematic investigation of relationships and causality, while inclusion of latent variables adds further possibilities to the modelling process. Each of these techniques has implications for the covariances between observed variables; putting them all together leads us to the modern topic of *covariance structure modelling*. Good introductions to this topic may be found in Everitt (1984a) or Everitt and Dunn (1991), while an exhaustive treatment has been given by Bollen (1989); the following brief overview draws on all these accounts.

Example 12.5 (Bollen, 1989)
To motivate the discussion, and to provide a suitable illustrative example, consider the following set of variables (described in more detail by Bollen, 1979, 1980): $y_1, \ldots, y_4$ are indicators of political democracy in developing countries as of 1960 (freedom of the press, freedom of group opposition, fairness of elections, and the elective nature of the legislative body), while $y_5, \ldots, y_8$ are the same indicators for 1965, both sets of indicators being scores assigned by a panel of political experts. We might wish to relate these indicators to measures of industrialisation, and three such measures obtained in 1960 were: x_1, the gross national product (GNP) per capita; x_2, the energy consumption per capita; and x_3, the percentage of the labour force in industrial occupations. Thus, we can envisage the existence of three latent variables, 'political democracy in 1960' (η_1), 'political democracy in 1965' (η_2) and 'industrialisation' (ξ_1), such that $y_1, \ldots, y_4$ are all surrogate measurements of η_1; $y_5, \ldots, y_8$ are all surrogate measurements of η_2; and x_1, x_2, x_3 are all surrogate measurements of ξ_1. Suitable models for these measurements might therefore be written

$$y_i = \lambda_i \eta_1 + \epsilon_i \quad i = 1, \ldots, 4;$$
$$y_i = \lambda_i \eta_2 + \epsilon_i \quad i = 5, \ldots, 8;$$

and

$$x_i = \lambda_{8+i} \xi_1 + \delta_i \quad i = 1, \ldots, 3$$

where all variables are assumed measured about their means, and the ϵ_i, δ_i are zero-mean mutually uncorrelated errors of measurement of the y_i and x_i respectively. Furthermore, we might assume that for any country, political democracy in 1960 was some function of industrialisation, while

Table 12.12 Covariance matrix for political democracy and industrialisation variables for 75 developing countries. From Bollen (1989)

	y_1	y_2	y_3	y_4	y_5	y_6	y_7	y_8	x_1	x_2	x_3
y_1	6.89										
y_2	6.25	15.58									
y_3	5.84	5.84	10.76								
y_4	6.09	9.51	6.69	11.22							
y_5	5.06	5.60	4.94	5.70	6.83						
y_6	5.75	9.39	4.73	7.44	4.98	11.38					
y_7	5.81	7.54	7.01	7.49	5.82	6.75	10.80				
y_8	5.67	7.76	5.64	8.01	5.34	8.25	7.59	10.53			
x_1	0.73	0.62	0.79	1.15	1.08	0.85	0.94	1.10	0.54		
x_2	1.27	1.49	1.55	2.24	2.06	1.81	2.00	2.23	0.99	2.28	
x_3	0.91	1.17	1.04	1.84	1.58	1.57	1.63	1.69	0.82	1.81	1.98

political democracy in 1965 depended on that in 1960 as well as on industrialisation. This suggests the further relationships

$$\eta_1 = \gamma_{11}\xi_1 + \zeta_1$$

and

$$\eta_2 = \beta_{21}\eta_1 + \gamma_{21}\xi_1 + \zeta_2,$$

where the ζ_i are zero-mean random errors, uncorrelated with each other and with the previous error terms ϵ_i, δ_i.

Data collected on 75 developing countries produced the variances and covariances given in the lower triangular matrix of Table 12.12. An obvious question is thus: are these values consonant with the explanatory system of models given above? If they are, what are the best estimates of the model parameters? It is with such questions that the subject of covariance structure modelling is concerned.

12.34 Example 12.5 is a special case of the general model that links the $(p \times 1)$ vector of manifest variables $\boldsymbol{y}$ to the $(q \times 1)$ vector of manifest variables $\boldsymbol{x}$ via the $(m \times 1)$ and $(n \times 1)$ vectors $\boldsymbol{\eta}$, $\boldsymbol{\xi}$ through the system of equations

$$\boldsymbol{y} = \boldsymbol{\Lambda}_y\boldsymbol{\eta} + \boldsymbol{\epsilon}, \tag{12.66}$$

$$\boldsymbol{x} = \boldsymbol{\Lambda}_x\boldsymbol{\xi} + \boldsymbol{\delta}, \tag{12.67}$$

and

$$\boldsymbol{\eta} = \boldsymbol{B}\boldsymbol{\eta} + \boldsymbol{\Gamma}\boldsymbol{\xi} + \boldsymbol{\zeta}. \tag{12.68}$$

Here, $\boldsymbol{\eta}$ and $\boldsymbol{\xi}$ are latent endogenous and exogenous random variables respectively, $\boldsymbol{B}$ is the $(m \times m)$ coefficient matrix showing the influence of the endogenous variables on each other, $\boldsymbol{\Gamma}$ is the $(m \times n)$ coefficient matrix for the effects of $\boldsymbol{\xi}$ on $\boldsymbol{\eta}$, while $\boldsymbol{\Lambda}_y$ and $\boldsymbol{\Lambda}_x$ are the $(p \times m)$ and $(q \times n)$ coefficient matrices showing the relations of $\boldsymbol{y}$ to $\boldsymbol{\eta}$ and $\boldsymbol{x}$ to $\boldsymbol{\xi}$ respectively. The disturbance vectors $\boldsymbol{\epsilon}$, $\boldsymbol{\delta}$ and $\boldsymbol{\zeta}$ are all assumed to have mean vectors zero and to be uncorrelated with each other and with all other random variables in the system, the matrix $(\boldsymbol{I}-\boldsymbol{B})$ is assumed to be non-singular and, to simplify matters, $\boldsymbol{\eta}$, $\boldsymbol{\xi}$, $\boldsymbol{x}$ and $\boldsymbol{y}$ are all assumed to be measured about their means.

This formulation is known as the *linear structural relationships*, or LISREL, model. Equations (12.66) and (12.67) are together referred to as the *measurement* part of the model, while (12.68) constitutes the *latent variable* part. This model was first studied by Jöreskog (1970, 1973a,b), and has since been much used in the social sciences. Since this model predominates in such applications we focus entirely on it in what follows, but it should be noted that other formulations of covariance structure models are possible. Two that have received some attention are the EQS (Bentler, 1989) and the MIMIC (Robinson, 1974) models.

12.35 If we write $\boldsymbol{\Psi}$, $\boldsymbol{\Phi}$, $\boldsymbol{\Theta}_\epsilon$ and $\boldsymbol{\Theta}_\delta$ for the dispersion matrices of $\boldsymbol{\zeta}$, $\boldsymbol{\xi}$, ϵ and $\boldsymbol{\delta}$ respectively, and if we set $\boldsymbol{u}' = (\boldsymbol{y}', \boldsymbol{x}')$, then the covariance matrix of $\boldsymbol{u}$ can readily be shown (Exercise 12.6) to be

$$\boldsymbol{\Sigma} = \begin{pmatrix} \boldsymbol{\Sigma}_{11} & \boldsymbol{\Sigma}_{12} \\ \boldsymbol{\Sigma}_{21} & \boldsymbol{\Sigma}_{22} \end{pmatrix} \tag{12.69}$$

where

$$\boldsymbol{\Sigma}_{11} = \boldsymbol{\Lambda}_y(\boldsymbol{I}-\boldsymbol{B})^{-1}(\boldsymbol{\Gamma\Phi\Gamma}' + \boldsymbol{\Psi})(\boldsymbol{I}-\boldsymbol{B})'^{-1}\boldsymbol{\Lambda}_y' + \boldsymbol{\Theta}_\epsilon,$$

$$\boldsymbol{\Sigma}_{12} = \boldsymbol{\Lambda}_y(\boldsymbol{I}-\boldsymbol{B})^{-1}\boldsymbol{\Gamma\Phi\Lambda}_x',$$

$$\boldsymbol{\Sigma}_{21} = \boldsymbol{\Sigma}_{12}',$$

and

$$\boldsymbol{\Sigma}_{22} = \boldsymbol{\Lambda}_x\boldsymbol{\Phi\Lambda}_x' + \boldsymbol{\Theta}_\delta.$$

To use the LISREL model, we need to specify the pattern of elements in each of the eight matrices $\boldsymbol{B}$, $\boldsymbol{\Gamma}$, $\boldsymbol{\Lambda}_y$, $\boldsymbol{\Lambda}_x$, $\boldsymbol{\Phi}$, $\boldsymbol{\Psi}$, $\boldsymbol{\Theta}_\epsilon$, $\boldsymbol{\Theta}_\delta$. This will naturally involve substantive knowledge of the research area, to formulate as appropriate and parsimonious a model as possible, and will also be dictated to some extent by the hypotheses that are to be tested. In any particular application, we can envisage one of three possibilities for each parameter.

- The parameter is *fixed*, i.e. has been assigned a particular value.
- The parameter is *constrained*, i.e. its value is unknown but equal to one or more other parameters (or functions of them).
- The parameter is *free*, i.e. its value is unknown and unconstrained.

Fixed parameters usually have values 0 or 1. The former case arises when either knowledge of the area, or formulation of the hypothesis to be tested, excludes the corresponding variable from the model. The latter case is generally a reflection of scaling of the variables. Since $\boldsymbol{\eta}$ and $\boldsymbol{\xi}$ are unobserved, the origin and scale of each element are arbitrary. To specify the model completely, this arbitrariness must be removed. The origin has already been dealt with by assuming zero means for each variable. A simple way of specifying scale is then to fix one element in each column of $\boldsymbol{\Lambda}_y$ and $\boldsymbol{\Lambda}_x$ to be unity, thereby defining the unit of measurement of each latent variable to be the same as one of the observed variables. Constrained parameters generally reflect either symmetries in the data or some group structure among the variables, while free parameters are ones about which there is no external information.

Example 12.5 (continued)
Returning to the political democracy example, it might be reasonable to postulate that the same pattern holds between the 1965 latent democracy variable η_2 and manifest variables $y_5, \ldots, y_8$ as between the 1960 latent variable η_1 and corresponding manifest variables $y_1, \ldots, y_4$. Thus, the parameters λ_{4+i} are constrained to equal λ_i for $i = 1, \ldots, 4$, and we can therefore relabel λ_9, λ_{10} and λ_{11} as λ_5, λ_6 and λ_7. Moreover, to remove arbitrary scaling we set both λ_1 and λ_5 equal to 1. The measurement model is thus given by (12.66) and (12.67) with

$$\boldsymbol{\Lambda}_y = \begin{pmatrix} 1 & 0 \\ \lambda_2 & 0 \\ \lambda_3 & 0 \\ \lambda_4 & 0 \\ 0 & 1 \\ 0 & \lambda_2 \\ 0 & \lambda_3 \\ 0 & \lambda_4 \end{pmatrix}$$

and

$$\boldsymbol{\Lambda}_x = \begin{pmatrix} 1 \\ \lambda_6 \\ \lambda_7 \end{pmatrix}$$

while the latent variable model is given by (12.68) with

$$\boldsymbol{B} = \begin{pmatrix} 0 & 0 \\ \beta_{21} & 0 \end{pmatrix}$$

and

$$\boldsymbol{\Gamma} = \begin{pmatrix} \gamma_{11} \\ \gamma_{21} \end{pmatrix}$$

It seems reasonable to assume that ζ_1 and ζ_2 are uncorrelated, thus making $\boldsymbol{\Psi}$ diagonal, and that the only free and non-zero elements of $\boldsymbol{\Theta}_\epsilon$ are those on the main diagonal (i.e. all the variances) plus those corresponding to covariances between measurement errors for the same indicator at the two points of time [elements $(i, i+4)$ and $(i+4, i)$ for $i = 1, \ldots, 4$] and to covariances between indicators that come from the same data source [elements (4,2), (8,6) and their transposes]. Moreover, $\boldsymbol{\Phi}$ is a scalar (since there is only one industrialisation latent variable ξ_1), and there is no reason to suppose that $\boldsymbol{\Theta}_\delta$ is other than diagonal. This specifies all the parameters of the model.

12.36 Having specified all the parameters of a model, we next want to estimate them. However, before we can do so it is necessary to satisfy ourselves that they can be estimated uniquely, i.e. that the model is *identified.* Denoting the complete set of parameters by the vector $\boldsymbol{\theta}$, it follows from above that the dispersion matrix of the manifest variables is a function $\boldsymbol{\Sigma}(\boldsymbol{\theta})$ of these parameters. The model is identified if $\boldsymbol{\Sigma}(\boldsymbol{\theta}_1) = \boldsymbol{\Sigma}(\boldsymbol{\theta}_2)$ implies $\boldsymbol{\theta}_1 = \boldsymbol{\theta}_2$. In the linear factor analysis model, it was easy to find a universal set of constraints that ensured that the parameters were always identified. With covariance structure modelling, however, each application is unique in terms of parameter patterns and values and there is no universal means of establishing identifiability.

Providing the model is a simple one, identifiability can be established algebraically: if each element of $\boldsymbol{\theta}$ can be expressed as a function of one or more elements σ_{ij} of $\boldsymbol{\Sigma}$ (by solving the series of equations defined by the matrices of **12.35**), then the model is identified. However, if the model is too complicated for this treatment, then we may not be able to establish identifiability unequivocally as no conditions for identifiability that are both necessary and sufficient have yet been discovered. A necessary condition for identifiability is $t < \frac{1}{2}(p+q)(p+q+1)$, where t is the number of (free and unconstrained) parameters to be estimated; Bollen (1989) gives two sufficient conditions, the two-step rule and the MIMIC rule (for a particular special case of the general model, see also Stapleton, 1977); while Jöreskog and Sorbom (1986) suggest that the model is almost certainly identified if the information matrix is positive definite. These are currently the only available diagnostics.

12.37 Estimation of parameters in covariance structure models follows the same general principle as was adopted for factor analysis in **12.24**: we seek those values of the parameters, $\hat{\boldsymbol{\theta}}$, that make the 'fitted' covariance matrix $\boldsymbol{\Sigma}(\hat{\boldsymbol{\theta}})$ for the manifest variables as close to the observed sample covariance matrix $\boldsymbol{S}$ as possible. In order to achieve this aim, we need to define a suitable function measuring the discrepancy between $\boldsymbol{S}$ and $\boldsymbol{\Sigma}(\boldsymbol{\theta})$, and then find the value $\boldsymbol{\Sigma}(\hat{\boldsymbol{\theta}})$ that minimises this function. A number of discrepancy functions exist, paralleling those in factor analysis. If we can assume multivariate normality of manifest variables, then comparison of the likelihood when $\boldsymbol{\Sigma} = \boldsymbol{\Sigma}(\boldsymbol{\theta})$ with its maximum value yields the discrepancy function

$$V_1 = \log|\boldsymbol{\Sigma}(\boldsymbol{\theta})| - \log|\boldsymbol{S}| + \mathrm{tr}[\boldsymbol{S}\boldsymbol{\Sigma}(\boldsymbol{\theta})^{-1}] - (p+q)$$

If multivariate normality is not assumed then we can use one of the variants of least squares to define the function. Ordinary least squares gives the function

$$V_2 = \frac{1}{2}\mathrm{tr}\{[\boldsymbol{S} - \boldsymbol{\Sigma}(\boldsymbol{\theta})]^2\}$$

while generalised least squares provides

$$V_3 = \frac{1}{2}\mathrm{tr}\{[\boldsymbol{I} - \boldsymbol{\Sigma}\boldsymbol{S}^{-1}]^2\}$$

These functions are minimised numerically within specialised software packages, such as LISREL VII (Jöreskog and Sorbom, 1988) or EQS (Bentler, 1989). Standard errors of parameter estimates are obtained from the information matrix in the usual way.

Table 12.13 Parameter estimates and their standard errors (in parentheses) for the political democracy model, obtained by three different methods. From Bollen (1989) with permission

	Method		
Parameter	ML	OLS	GLS
λ_1	1.00 (—)	1.00	1.00 (—)
λ_2	1.19 (0.14)	1.21	1.18 (0.15)
λ_3	1.18 (0.12)	1.14	1.22 (0.13)
λ_4	1.25 (0.12)	1.29	1.23 (0.13)
λ_5	1.00 (—)	1.00	1.00 (—)
λ_6	2.18 (0.14)	2.06	2.24 (0.16)
λ_7	1.82 (0.15)	1.62	1.89 (0.18)
γ_{11}	1.47 (0.40)	1.32	1.81 (0.45)
γ_{21}	0.60 (0.23)	0.41	0.72 (0.28)
β_{21}	0.87 (0.08)	0.92	0.80 (0.08)

Example 12.5 (continued)
Returning to the LISREL formulation of the political democracy data, and the observed covariance matrix in Table 12.12, Bollen (1989) gives the estimates of parameters as obtained from each of the three fitting functions specified above. These values are reproduced in Table 12.13, where ML, OLS and GLS denote maximum likelihood, ordinary least squares and generalised least squares respectively. Bollen gives standard errors of free parameters using ML and GLS, and these are also reproduced in the table. We only report the estimates of linear coefficients of the model here; for estimates of the variance and covariance terms in the model, see Bollen (1989). We can note that the estimates of parameters in the measurement part of the model are similar for all three methods, but greater divergence occurs for the estimates in the latent variable part of the model. Also, for these data, the GLS and ML estimates are closer to one another than either is to the OLS estimates (but no inference can be drawn about such relationships among the methods in general).

12.38 Once a covariance structure model has been specified and its parameters have been estimated, we need some method for testing its goodness-of-fit. More generally, we might wish to find the best fitting model from among a set of candidate models. If we can assume normality of data, and if we have used the fitting criterion V_1 (based on the likelihood of the sample), then a standard likelihood ratio test of the null hypothesis H_0: $\boldsymbol{\Sigma} = \boldsymbol{\Sigma}(\boldsymbol{\theta})$ against the alternative H_a: $\boldsymbol{\Sigma}$ arbitrary can be applied here as it was for maximum likelihood factor analysis (**12.22**). The test statistic is the minimum value of V_1 (i.e. the value $V_{1,min}$ at $\boldsymbol{\theta} = \hat{\boldsymbol{\theta}}$), and standard theory (Part 1, Appendix **A.6**) shows that $(n-1)V_{1,min}$ has an asymptotic χ^2 distribution on $\frac{1}{2}(p+q)(p+q+1)-t$ degrees of freedom under H_0, where t is the number of estimated parameters. This result enables us either to test the fit of a particular model, or to compare a series of nested models. In the former case a significant value of $(n-1)V_{1,min}$ indicates lack of fit, while in the latter case a significant value of the difference between $(n-1)V_{1',min}$ for two nested models, judged against the chi-squared distribution on the difference in their degrees of freedom, indicates that the additional parameters in the 'larger' model have made a genuine improvement to model fit.

The above results all pertain to the ML criterion V_1, assuming multivariate normality of data. Browne (1982, 1984) has explored their extension to other criteria and to non-normal distributions; results in Shapiro (1987) are also relevant here. A summary of the position is as follows. The estimator $\hat{\boldsymbol{\theta}}$ is consistent whichever method of estimation (i.e. ML, GLS or OLS) is employed, for arbitrary multivariate distributions (subject to standard regularity conditions); it is asymptotically efficient under ML and GLS for multivariate normal, elliptic or non-normal distributions that have no excess kurtosis; while the usual asymptotic covariance matrix and chi-squared tests apply for ML or GLS under multivariate normality or non-normal distributions having no excess kurtosis. Thus, a test using Mardia's (1970, 1974) measure of multivariate kurtosis (see also Part 1, Chapter 3) may need to be done before applying any of the above procedures.

12.39 Despite the above theoretical basis, however, the chi-squared approach has limited practical use as a measure of goodness-of-fit of a model, and has been a cause of dissatisfaction among researchers in the social sciences. This is because it is a function of sample size as well as of the differences between observed and fitted covariances. Typical practical experience has been that the chi-squared test is significant in a particular instance, implying that the model should be rejected, but the differences between observed and fitted covariances appear to be small in absolute terms, suggesting that the model is not such a bad one after all. This state of affairs has led to the search for a suitable goodness-of-fit *index*, which is a standardised measure between 0 (lack of fit) and 1 (perfect fit) that assesses the fit of a model to data and that can be used to choose between models.

The first goodness-of-fit index was defined by Tucker and Lewis (1973) as

$$TLI = \frac{\chi_0^2/f_0 - \chi_k^2/f_k}{\chi_0^2/f_0 - 1}$$

where χ_0^2 is the chi-squared value for the null model (uncorrelated variables) on f_0 degrees of freedom, and χ_k^2 is the chi-squared value for the structural model under consideration on f_k degrees of freedom. Tucker and Lewis based this index on a reliability index obtained from mean squares in mixed-model ANOVA, and defined it for factor analysis; Bentler and Bonett (1980) subsequently extended it to more general covariance structures and popularised it as the Non-Normed Fit Index. Bentler and Bonett also defined the Normed Fit Index as

$$NFI = \frac{\chi_0^2 - \chi_k^2}{\chi_0^2}$$

while Bentler (1990) proposed the Comparative Fit Index

$$CFI = 1 - \frac{\chi_k^2 - f_k}{\chi_0^2 - f_0}$$

Here, $\chi^2 - f$ is an estimate of the non-centrality parameter of the corresponding model's lack of fit; the same index was called a Relative Fit Index by McDonald and Marsh (1990). More generally, Mulaik *et al* (1989) suggested the General Normed Fit Index

$$GNFI = (F_0 - F_k)/(F_0 - F_h)$$

where F_0 denotes *any* lack of fit index (e.g. chi-squared measure, sum of squared residuals, unbiased estimate of model non-centrality parameter, etc) for the most restricted model in a nested sequence of models, F_h is the corresponding index for an intermediate model in the sequence, and F_k is the same index for the least restricted model in the sequence.

Any of these indices can be multiplied by the *parsimony ratio* $P = f_k/f_0$, to make it reflect the disconfirmability of the model relative to the data (Mulaik *et al*, 1989; Carlson and Mulaik, 1993). This factor adjusts the index to take account of degrees of freedom, in the way that AIC adjusts the likelihood ratio statistic. For a comprehensive survey and evaluation of these, and other, indices see Mulaik *et al* (1989). Further discussion and surveys are also given by Bentler and Bonett (1980), Sobel and Bohrnstedt (1985), and Bollen (1989, Chapter 7).

12.40 An alternative approach to model assessment and choice has been explored by Cudeck and Browne (1983) and Browne and Cudeck (1989,1992), who describe various cross-validation schemes for achieving these objectives. In their simplest form these schemes require the available sample to be split into two independent halves, a calibration sample and a validation sample. Let $\boldsymbol{S}_c$ be the sample covariance matrix obtained from the calibration sample. Each competing model is fitted to just the calibration sample using one of the criteria above, so that $\hat{\boldsymbol{\Sigma}}$ is obtained for each model from $\boldsymbol{S}_c$. The validation sample then provides a separate estimate $\boldsymbol{S}_v$ of the true dispersion matrix $\boldsymbol{\Sigma}$, and this estimate is substituted into the chosen criterion function with each $\hat{\boldsymbol{\Sigma}}$ in turn to yield a set of cross-validation criterion values on which the model selection is based. In this way, the matrices $\hat{\boldsymbol{\Sigma}}$ and $\boldsymbol{S}$ that are being compared in the model selection process have not come from the same set of data. As a simplification of this technique, Browne and Cudeck (1989) propose a way of avoiding the split into two samples by using an asymptotic approximation for the expected value of the difference between the cross-validation index and the calibration sample criterion value. Cudeck and Henly (1991) discuss sample size issues that arise in such cross-validation assessment.

Example 12.5 (continued)

Bollen (1989) presents a set of measures of fit of the models specified in Table 12.13. The chi-squared statistic is 39.6 for the ML estimates and 38.6 for the GLS estimates, both on 38 degrees of freedom. In neither case is the null hypothesis rejected (p values of 0.40 and 0.44 respectively), indicating adequate fit of the model. Moreover, all the goodness-of-fit indices are very close to 1.00, strongly supporting this conclusion.

12.41 The above account has focused entirely on the basic formulation of the linear structural relationships model, but there has also been much effort placed on extending various aspects of this model. McDonald (1978, 1980), McArdle and McDonald (1984) and Bentler and Weeks (1980) have all explored different model representations; McDonald (1980), Lee (1980) and Bentler and Lee (1983) have discussed incorporation of more complex parameter constraints into the model; while Kenny and Judd (1984), Busemeyer and Jones (1983) and Heise (1986) have considered inclusion of quadratic and interaction terms in the model. It is clear that all possibilities for extension have not yet been exhausted.

Confirmatory factor analysis

12.42 Direct comparison of the relevant equations shows that (12.66) and (12.67) are both just restatements of the mean-centred version of factor model (12.32) but in different symbols. Thus, factor analysis as discussed earlier in this chapter can be viewed as a special case of the linear structural relationships model in which there is only one set of manifest variables and all parameters are free. Such a factor analysis is generally termed *exploratory* factor analysis, because we are seeking a simple structure with which to explain the observed covariances among a set of variables. Such an analysis will involve determining the number of factors for adequate explanation, and rotation of factors to facilitate their interpretation.

In some situations, however, the investigator may have (from previous research, perhaps) some preconceived idea about the factor structure. He or she may thus be willing to specify in advance the number of factors and the pattern of zero and non-zero loadings on them, and will then want to test this hypothesis on the data collected in the experiment or survey. Such an analysis is termed a *confirmatory* factor analysis, as the data are being used to confirm or refute the prior hypothesis. The only difference between this situation and the earlier exploratory one is that some of the parameters are now constrained, so we need to obtain maximum likelihood estimates of the parameters in a *restricted* model. There has been much interest in such situations, and there is quite a large literature on the topic. For references to the early work in this area, see Chapter 7 in Lawley and Maxwell (1971). However, much of this work has now been superseded because confirmatory factor analysis constitutes a special case of the general covariance structure model, so that any required hypotheses can be tested through a standard application of LISREL.

Example 12.6
Everitt (1984a) provides a confirmatory factor analysis of some data from Child (1970), relating to correlations between eight tests of mental ability. Tests 1 and 2 are conventional intelligence tests, believed to measure 'convergent thinking', tests 3–6 require verbal responses and are believed to measure 'verbal divergent thinking', while tests 7 and 8 require non-

Table 12.14 Correlation matrix for eight tests of mental ability. From Everitt (1984a) with permission

	x_1	x_2	x_3	x_4	x_5	x_6	x_7	x_8
x_1	1.00							
x_2	0.54	1.00						
x_3	0.08	0.10	1.00					
x_4	0.18	0.05	0.58	1.00				
x_5	0.20	0.07	0.51	0.46	1.00			
x_6	0.13	−0.01	0.26	0.40	0.46	1.00		
x_7	0.10	0.08	0.24	0.27	0.40	0.11	1.00	
x_8	0.05	0.00	0.22	0.22	0.21	0.18	0.51	1.00

verbal responses and are believed to measure 'non-verbal divergent thinking'. The observed correlation matrix for a sample of subjects is shown in Table 12.14. The background theory and beliefs thus suggest that these correlations should be explainable by three common factors, the first one loading only on tests x_1 and x_2, the second one loading only on tests x_3 to x_6, and the third loading only on tests x_7 and x_8. Moreover, allowance should be made for a correlation between factors 2 and 3 (since both measure 'divergent thinking'), but both of these factors should be uncorrelated with factor 1 ('convergent thinking'). We are therefore interested in testing the fit of a factor analysis model (12.32) in which the loadings take the form

$$\boldsymbol{\Gamma} = \begin{pmatrix} \gamma_{11} & 0 & 0 \\ \gamma_{21} & 0 & 0 \\ 0 & \gamma_{32} & 0 \\ 0 & \gamma_{42} & 0 \\ 0 & \gamma_{52} & 0 \\ 0 & \gamma_{62} & 0 \\ 0 & 0 & \gamma_{73} \\ 0 & 0 & \gamma_{83} \end{pmatrix}$$

but which allows a covariance ϕ between factors 2 and 3. It can be seen readily that such a model is equivalent to the portion (12.67) of the general covariance structure model with $\boldsymbol{\Lambda}_x$ given by $\boldsymbol{\Gamma}$ above, with

$$\boldsymbol{\Phi} = \begin{pmatrix} 1.0 & & \\ 0.0 & 1.0 & \\ 0.0 & \phi & 1.0 \end{pmatrix}$$

and with diagonal matrix of specific variate covariances $\boldsymbol{\Theta}_\delta = \text{diag}(\theta_1, \ldots, \theta_8)$.

This model needs to be examined first for identifiability. Using the algebraic approach outlined in **12.36**, Everitt (1984a) shows that we need the additional constraints $\gamma_{11} = \gamma_{21}$ and $\theta_1 = \theta_2$ for all parameters to be identifiable. Having thus specified the model, LISREL gave the following

Table 12.15 Predicted correlation matrix for eight tests of mental ability. From Everitt (1984a)

x_1	1.00							
x_2	0.54	1.00						
x_3	0.00	0.00	1.00					
x_4	0.00	0.00	0.52	1.00				
x_5	0.00	0.00	0.51	0.37	1.00			
x_6	0.00	0.00	0.36	0.37	0.37	1.00		
x_7	0.00	0.00	0.29	0.29	0.29	0.20	1.00	
x_8	0.00	0.00	0.22	0.22	0.22	0.16	0.51	1.00
	x_1	x_2	x_3	x_4	x_5	x_6	x_7	x_8

parameter estimates and standard errors (in parentheses):

$$\hat{\boldsymbol{\Lambda}}_x = \begin{pmatrix} 0.73(0.10) & 0 & 0 \\ 0.73(0.10) & 0 & 0 \\ 0 & 0.71(0.10) & 0 \\ 0 & 0.73(0.10) & 0 \\ 0 & 0.72(0.10) & 0 \\ 0 & 0.51(0.11) & 0 \\ 0 & 0 & 0.81(0.15) \\ 0 & 0 & 0.62(0.13) \end{pmatrix}$$

$$\hat{\boldsymbol{\Phi}} = \begin{pmatrix} 1.0 & & \\ 0.0 & 1.0 & \\ 0.0 & 0.49(0.12) & 1.0 \end{pmatrix}$$

and

$$\hat{\boldsymbol{\Theta}}_\delta = \text{diag}(0.46, 0.46, 0.49, 0.47, 0.48, 0.74, 0.33, 0.61).$$

Standard errors of the $\hat{\theta}_i$ are 0.06, 0.06, 0.10, 0.10, 0.10, 0.12, 0.21 and 0.15 respectively.

Table 12.15 shows the predicted correlation matrix using this model. The chi-squared goodness-of-fit test for the model yielded a value of 26.11 on 20 degrees of freedom, indicating a fairly good fit of the model to the data.

12.43 We conclude this chapter with a brief comparison between factor analysis and principal component analysis. Both techniques can be used to reduce the dimensionality of large data sets, and we have seen that some computational algorithms for factor analysis rely heavily on principal component calculation, so the two techniques have been much confused over the years. It is thus worth clearly establishing the distinctions between them.

First and foremost, factor analysis is a model-based technique that provides opportunities for testing hypotheses about, or exploring the structure of, the *associations* among a set of manifest variables. By contrast, principal component analysis aims to explain the *variances* in the system *without* assuming any underlying model. Hence, there is no hypothesis-testing implication, and it is generally used in a descriptive fashion. From the computational schemes discussed earlier, if the factor model holds and the specific variances are small then we would expect to obtain similar results from both analyses; but if the specific variances are large then they will be absorbed into the principal components but separated from the factors, and the two analyses will give different interpretations. In factor analysis there is a simple relationship between the results obtained from the correlation matrix (i.e. standardised data) and those from the covariance matrix (i.e. raw data), but not in principal component analysis. Finally, factor score estimation requires a subsidiary operation, whereas principal component scores come directly out of the standard analysis. Before choosing which analysis to use, therefore, the researcher must be clear about the objectives of analysis and the use to be made of the results.

Exercises

12.1 Establish equations (12.12), (12.13) and (12.14) for the second derivatives of the log-likelihood with respect to the parameters of the latent class model with binary manifest variables.

12.2 Derive the estimating equations (12.18), (12.19) and (12.20) for the parameters of the latent class model with multicategorical manifest variables.

(Bartholomew, 1987, pp. 28–30)

12.3 Using (12.41), show that $\sigma^{ii} \leqslant \psi_i^{-1}$, where σ^{ii} and ψ_i are the ith diagonal elements of $\boldsymbol{\Sigma}^{-1}$ and $\boldsymbol{\Psi}$ respectively. Hence, using the sample covariance matrix for the manifest variables $\boldsymbol{S}$ as an estimate of $\hat{\boldsymbol{\Sigma}}$, find an upper bound that can be used as a starting value in the iterations for $\hat{\psi}_i$.

From standard regression theory, and using $\psi_i = \text{var}(x_i|y)$, show that two further possible starting values are $s_{ii}(1 - R_i^2)$ and $s_{ii}(1 - \max r_{ij}^2)$, where s_{ij}, r_{ij} are the (i, j)th elements of the sample covariance and correlation matrices among the manifest variables and R_i^2 is the multiple correlation coefficient when x_i is regressed on the other manifest variables.

(Bartholomew, 1987, p54)

12.4 Consider the factor analysis model (12.32) for the manifest variables $\boldsymbol{x}$, and let $\tilde{\boldsymbol{S}} = \boldsymbol{\Psi}^{-1/2}\boldsymbol{S}\boldsymbol{\Psi}^{-1/2}$ where $\boldsymbol{S}$ is the sample covariance of $\boldsymbol{x}$. Let $\boldsymbol{z} = \boldsymbol{D}\boldsymbol{x}$ be a scale transformation (i.e. $\boldsymbol{D}$ is a diagonal matrix whose diagonal elements all have the same sign). Show that $\tilde{\boldsymbol{S}}$ is unchanged by this transformation. Deduce that maximum likelihood factor analysis is independent of the scales of measurement of the manifest variables.

(Seber, 1984, p218)

12.5 Obtain equation (12.56) for the expected value of the latent variables $\boldsymbol{y}$ given that the manifest variables $\boldsymbol{x}$ have value $\boldsymbol{x}_i$ for the normal linear factor model.

12.6 Show that the covariance matrix $\boldsymbol{\Sigma}$ of the manifest variables in the linear structural relationships model has the form (12.69).

(Bollen, 1989, p323)

13

Repeated Measures and Growth Curves

Introduction

13.1 The term 'repeated measures' is usually taken to mean that the same measurements are applied to a number of individuals or subjects at several fixed occasions. For example:

- a response (say blood pressure or ventilation) to a drug is recorded on patients at fixed intervals from administration until the effect is exhausted;
- reading tests are given to children at different stages of their education, and their scores are recorded;
- measurements, such as height and weight, are taken on children or animals at different ages;
- measurements of soil moisture are taken at different depths, over the area of an agricultural experiment (Roberts and Raison, 1983).

Usually, the individuals belong to different groups or receive different treatments, and one of the objectives is to assess the effect of treatment on the set of responses. When the response is a single measurement at different times, the analysis may either treat each observation as a univariate response with (correlated) replicates at the different times for each individual, or treat the set of measurements at different times as a vector response for each individual.

Growth curves are special cases of repeated measures. Usually they span a substantial period, in terms of the individual's lifetime, and any assumption that the observations at different times are identically distributed is likely to be false. The measurements are likely to change monotonically, typically increasing towards an asymptote.

Two approaches to the analysis of data of this sort are definitely wrong, although occasionally published. First, an assumption that errors are independent and identically distributed, leading to a simple two-way analysis of variance, is clearly false. Secondly, a separate analysis at each time period is misleading, even as a preliminary to another form of analysis. Apart from raising problems of multiple testing, and the obvious loss of efficiency, the questions asked are not of any real interest.

There remain several possible approaches to the problem. For the moment, consider only the case of a single measurement repeated in time. The types of analysis commonly employed are as follows.

(1) Univariate repeated measures analysis, which is the name given, in many computer packages, to an analysis of variance approach treating the observations at different times as having identically distributed errors. The correlations between measurements on the same individual are assumed equal, so that there is one variance estimate between individuals and another within individuals. The analysis is formally identical to the 'split plot design' (Yates, 1937), which is well known in agricultural field experiments. The assumptions are seldom strictly fulfilled, but modifications to the significance tests ensure that the analysis is conservative, and reasonably efficient if the departures from the model are slight.
(2) Full multivariate analysis of variance (see Part 1, **7.5**), in which the observations at different times are treated as a vector of random variables, with the covariance matrix estimated from the data. This method is valid, but fails to take account of prior knowledge that the observations form a sequence; and usually, for example, the correlation will be greater between observations close together in time than between those further apart. In very large data sets, this feature may be unimportant; otherwise it may lead to considerable loss of efficiency.
(3) A modified multivariate analysis involving a covariance matrix that is constrained to follow a simple time-series model—for example, an $AR(1)$ scheme. If the right model is chosen, this is almost certainly the best approach. The computation, however, is considerably heavier than in the two methods described above.
(4) A preliminary regression in which a single curve, of fixed type dependent on certain parameters, is fitted to each individual response, followed by an analysis carried out on the fitted parameters. The great advantage of this analysis is its flexibility. Missing values, or observations taken at irregular times, do not bias the parameter estimates (though they have some effect on their variance), and provided the curves fitted are appropriate, the analysis is informative and reasonably efficient. This idea was introduced by Wishart (1938), and has since been developed by Rao (1965, 1966b) and others with adjustments for covariates.

Inference about these models should be treated with caution. The selection of the model fitted usually depends on preliminary analysis of the data, and subsequent analysis assumes that the model is correct. Resulting confidence intervals and significance tests are conditional on that assumption.

Each of these methods can be extended to the situation where several different variables are recorded at each time point. This is a very common situation and, while the notation gets more and more complicated, the generalisation is not difficult.

Univariate analysis of repeated measures

13.2 Suppose y_{ij} is the response of subject i, $i = 1 \ldots n$, at time j, $j = 1 \ldots q$. The individuals are divided into g groups, and a univariate analysis of variance takes the form given in Table 13.1.

The sums of squares between individuals, between groups, between occasions, and the total are calculated in the usual way. The sum of squares between individuals within groups is obtained from the first two by subtraction. The groups × occasions interaction is obtained by calculating the sum of squares from a table of group totals on each occasion, and subtracting the group and occasion sums of squares, and finally the error sum of squares (the interaction between occasions and individuals within groups) is found by subtraction.

The table has the structure of a split-plot design, with individuals corresponding to main plots and occasions to sub-plots. Differences among group means are tested against the 'between individuals within groups' mean square, and this mean square is used to calculate confidence limits for group means or differences between pairs of group means. Occasions, and the groups × occasions interaction, are both tested against the error mean square, which is the interaction between occasions and individuals within groups, a measure of random variation from time to time within individuals.

The expected mean squares in the table are taken from Crowder and Hand (1990). The model assumes random variation between individuals, a random individual × occasion interaction, and purely random variation within individuals. The corresponding variance components are σ_I^2, σ_{IO}^2 and σ^2. Notice that, in this simple design, the three components are not identifiable; in the split-plot design, the random interaction component is usually assumed to be zero. The fixed effects D_O, D_G and D_{GO} represent differences between occasion means, differences between group means, and the interaction. Each is an estimate of the mean squared deviation of the corresponding means from the overall mean, or, in the case of the interaction term, from the additive model.

The direct calculation of these expected mean squares is straightforward, but extremely tedious. Hartley (1967) first gave a general method for finding these

Table 13.1 Analysis of variance for repeated measures design

Source of variation	*d.f.*	*Expected mean square*
Between individuals	$n-1$	
Between groups	$g-1$	$\frac{D_G}{g-1} + q\sigma_I^2 + \sigma^2$
Between individuals within groups	$n-g$	$q\sigma_I^2 + \sigma^2$
Between occasions	$q-1$	$\frac{D_O}{q-1} + \frac{q}{q-1}\sigma_{IO}^2 + \sigma^2$
Groups × occasions interaction	$(g-1)(q-1)$	$\frac{D_{GO}}{(g-1)(q-1)} + \frac{q}{q-1}\sigma_{IO}^2 + \sigma^2$
Error	$(n-g)(q-1)$	$\frac{q}{q-1}\sigma_{IO}^2 + \sigma^2$
Total	$nq-1$	

expressions by 'synthesis'; his method was generalised by Rao (1968). Crowder and Hand (1990) describe the method, and give examples of the calculations, both by the direct method and by 'synthesis'.

As a rule, interest centres on the interaction term. Differences among occasions are usually large and obvious. The group mean differences are often masked by individual variation—although this may be reduced by covariance on measures taken before treatments start, if these are available. The interaction term represents differences in the shape of the response curve in the different groups. It may be divided into appropriate components. In particular, orthogonal polynomial fitting can isolate linear, quadratic and cubic components if these can reasonably reflect the differences in the curves.

The analysis is based on two linear models, one between and one within individuals. The model implies intra-class correlation; the error structure is accounted for entirely by the differences among the individuals and by the random variation from time to time, including possible random interactions between individuals and occasions.

Writing $y_{ij(k)}$ for the response of subject i, in group k, at time j,

$$y_{ij(k)} = \mu + \alpha_i + \beta_j + \gamma_k + \delta_{jk} + \epsilon_{ij}.$$

The assumption underlying the analysis of variance is now

$$E(\epsilon_{ij}^2) = \sigma^2 \tag{13.1}$$

$$E(\epsilon_{ij}\epsilon_{il}) = \rho\sigma^2. \tag{13.2}$$

This assumption is likely to be a reasonable approximation for 'short' experiments, in which the subjects do not change much. It is unrealistic for models of growth curve type, in which in which observations may differ greatly in magnitude and in which observations close in time may be more closely correlated than those far apart.

Even when the approximation is reasonable, deviations from the error structure mean that confidence intervals may be too short, and significance tests may have a larger type I error than the nominal one. This can give misleading results, and these can be avoided by modifying the significance tests to allow for departure from the assumptions.

13.3 The assumptions underlying the univariate analysis of variance imply, first, that the error covariance matrix is the same for all groups, and secondly that it is 'spherical'; that is, that it has the form $\sigma^2\boldsymbol{I}$, where $\boldsymbol{I}$ is the unit matrix. These two assumptions can be tested. Box (1949) modified the likelihood ratio test for homogeneity of covariance matrices first given by Wilks (1932); see Part 1, **A.8**. Mauchly (1940) gave the first test of sphericity. A modified test was given by Bartlett (1950); see Part 1, **4.6**. These tests may be applied to the matrix of residuals, after adjusting for individual means, and for occasion means within each group. The covariance matrix of these residuals, treating the observations on each occasion as separate variables, is singular, of rank $q-1$. It must first be transformed into a set of $q-1$ orthonormal contrasts, using, for example, the Helmert transformation or orthogonal polynomials with normalized coefficients. The tests are then carried out on the covariance matrix of these contrasts, say $\hat{\boldsymbol{\Sigma}}_c$. If the original observations were uncorrelated, the matrix would be an estimate of a matrix $\boldsymbol{\Sigma}_c = \sigma^2 \boldsymbol{I}$.

Separate matrices of this sort may be calculated for each group, and tested for homogeneity. If the result is non-significant, they may be pooled, and the pooled matrix tested for sphericity. If this test gives a non-significant result, the analysis is justified.

These tests all depend on the assumption of a multivariate normal distribution, and in fact are notoriously sensitive to departures from normality. Since the analysis of variance is quite robust against non-normality, a less formal approach is probably appropriate. The assumptions are likely to break down in one of two ways. First, there may be differences in the variability among the treatments, usually associated with large differences in the means. This is a familiar problem in all linear models, and can often be remedied by a suitable transformation. Secondly, the pooled covariance matrix may show autocorrelation, with substantial positive correlation close to the diagonal. If the tests show significant, but not large, differences, and these differences do not suggest either of the effects discussed above, the analysis can be regarded as valid.

13.4 If $\boldsymbol{\Sigma}_c$ is not spherical, significance tests based on the variance ratio can be seriously misleading. Greenhouse and Geisser (1959) first investigated the possibility of modifying the simple univariate tests to allow for non-sphericity. If

$$\begin{aligned}\epsilon &= \frac{(\text{tr}\,\boldsymbol{\Sigma}_c)^2}{(q-1)\text{tr}\,(\boldsymbol{\Sigma}_c)^2} \\ &= \frac{(\sum_{i=1}^{q-1}\lambda_i)^2}{(q-1)\sum_{i=1}^{q-1}\lambda_i^2}\end{aligned}$$

where λ_i is the ith eigenvalue of $\boldsymbol{\Sigma}_c$, it may be shown that the variance ratio test statistic F has approximately the standard F distribution with *both* degrees of freedom multiplied by ϵ. Thus the test statistic for occasions in Table 13.1, distributed as $F_{q-1,(q-1)(n-g)}$ under sphericity, has the distribution corresponding to $F_{\epsilon(q-1),\epsilon(q-1)(n-g)}$ and the degrees of freedom for the groups × occasions interaction is similarly modified. In general, of course, the degrees of freedom are non-integral.

There remains the problem of estimating ϵ. The value lies between 1 and $1/(q-1)$, and if the minimum value gives a significant result no further test is needed. Greenhouse and Geisser (1959) suggested first testing with $\epsilon = 1/(q-1)$ and, if the result is not significant, testing with $\hat{\epsilon}$ estimated by replacing $\boldsymbol{\Sigma}_c$ in the equations above by $\hat{\boldsymbol{\Sigma}}_c$. This natural estimate is obviously biased, and in particular is always less than 1 even when sphericity holds. As a result, the Greenhouse–Geisser test is conservative, particularly when n is small and sphericity is not obviously violated. Huynh and Feldt (1976) suggested a modified estimate. Calculate

$$\begin{aligned}a &= n(q-1)\hat{\epsilon} - 2 \\ b &= (q-1)[n-g-(q-1)\hat{\epsilon}] \\ \tilde{\epsilon} &= \min(1, a/b)\end{aligned}$$

This estimate is based on unbiased estimates of the numerator and denominator of ϵ, and usually gives a less conservative test than the Greenhouse–Geisser test.

Many standard computer packages have analysis of variance programs suitable for univariate repeated measures analysis. BMDP program 2V gives p-values for a sphericity test, Greenhouse–Geisser and Huynh–Feldt estimates of ϵ, and p-values corresponding to the standard variance ratio tests as well as the Greenhouse–Geisser and Huynh–Feldt tests. SPSS program MANOVA gives a battery of tests of homogeneity and sphericity, and either univariate or multivariate analysis of variance of repeated measures designs. SAS program GLM is a general linear model program, and includes the Greenhouse–Geisser and Huynh–Feldt tests. For a full description of these and other programs for analysis of repeated measures designs, see Crowder and Hand (1990); examples of the BMDP and SPSS output are also given.

In conclusion, the univariate analysis is appropriate if individuals change little during the course of the experiment, and observations are spaced so that individual differences account for the correlations observed. When these assumptions break down, the modified tests described are valid as significance tests, but it becomes difficult to model and interpret the differences among response curves under different conditions. When the simple univariate analysis is clearly invalid, it is probably better to consider other approaches.

Multivariate analysis of repeated measures

13.5 The standard multivariate linear model, or MANOVA model, has the form (Part 1, Chapter 7)

$$E(\boldsymbol{Y}) = \boldsymbol{A}\boldsymbol{\xi} \tag{13.3}$$

$$\text{var}(\text{vec}\,\boldsymbol{Y}) = \boldsymbol{\Sigma} \otimes \boldsymbol{I}, \tag{13.4}$$

where $\boldsymbol{Y}$ is an $n \times q$ data matrix, $\boldsymbol{A}$ is an $n \times m$ matrix of known constants—a design matrix—and $\boldsymbol{\xi}$ is an $m \times q$ matrix of unknown coefficients. The second equation means that the n rows of $\boldsymbol{Y}$, representing the individuals, are mutually independent, with the q observations on each individual having an unknown $q \times q$ covariance matrix $\boldsymbol{\Sigma}$. Assuming multivariate normality, hypotheses of the form

$$\boldsymbol{C}_{s\times m}\boldsymbol{\xi}_{m\times q}\boldsymbol{V}_{q\times v} = \boldsymbol{O}_{s\times v} \tag{13.5}$$

can be tested by any of the four tests discussed in Part 1, **A.8**.

In this model, the matrix $\boldsymbol{A}$ can represent a grouping of the data, or a regression model such as a polynomial fit to the occasions. In general, however, it is required to investigate hypotheses about the form of the curves, and about the parameters of the curves fitted to the different groups, and to attach suitable confidence bands to the curves. Potthoff and Roy (1964) suggested a modification to (13.3), of the form

$$E(\boldsymbol{Y}_{n\times q}) = \boldsymbol{A}_{n\times m}\boldsymbol{\xi}_{m\times p}\boldsymbol{P}_{p\times q}. \tag{13.6}$$

This is often called a generalised MANOVA model.

Potthoff and Roy (1964) give examples of the use of these equations; see also Roy *et al* (1971). For fitting a polynomial growth curve to a single group of individuals, the matrix $\boldsymbol{A}$ may be an $n \times 1$ vector of 1s, and $\boldsymbol{P}$ has elements $p_{ij} = t_j^{i-1}$, $j = 1, \ldots, q$, $i = 1, \ldots, p$, where the t_j values represent the times at which the observations are taken. The model then fits a polynomial of degree $p - 1$ to all the data.

Suppose now there are m groups of individuals. The matrix $\boldsymbol{A}$ has m columns, the ith column consisting of 1s for those individuals in group i and 0s elsewhere. This model fits a separate polynomial of degree $p - 1$ to each group.

Potthoff and Roy (1964) describe three tests that can be based on this model.

(1) Suppose the null hypothesis is that all growth curves are the same, regardless of group. In (13.5), set $v = q$, $\boldsymbol{V} = \boldsymbol{I}$, $s = m - 1$, and the matrix $\boldsymbol{C}$, $(m - 1) \times m$, is $\boldsymbol{I}_{m-1}$ with a final column in which each element is -1.
(2) If the null hypothesis is that all curves are equal except for the constant term, then $\boldsymbol{C}$ is the same matrix, and $\boldsymbol{V}$ is a $q \times (q - 1)$ matrix with the first row filled with zeros and the remaining rows corresponding to $\boldsymbol{I}_{q-1}$.
(3) To test whether all curves are of degree $q - 2$ or less, set $s = m$, $\boldsymbol{C} = \boldsymbol{I}$, and $\boldsymbol{V}$ to a $q \times 1$ vector with 1 as the last element and zero elsewhere.

The model of (13.6) can be reduced to a standard MANOVA model by a transformation of the form

$$\boldsymbol{Z} = \boldsymbol{Y}\boldsymbol{G}^{-1}\boldsymbol{P}'(\boldsymbol{P}\boldsymbol{G}^{-1}\boldsymbol{P}')^{-1} \tag{13.7}$$

Here, $\boldsymbol{G}^{-1}$ is a weight matrix. Any symmetric positive definite matrix can legitimately be used as $\boldsymbol{G}$ in the subsequent multivariate analysis of variance, but the best choice is $\boldsymbol{G} = \boldsymbol{\Sigma}$, or an arbitrary scalar multiple of $\boldsymbol{\Sigma}$. $\boldsymbol{\Sigma}$, however, is unknown, and the subsequent analysis is not valid if the weighting matrix is based on the data matrix. Potthoff and Roy (1964) discuss the choice of $\boldsymbol{G}$ in some detail. One possibility is simply to choose $\boldsymbol{G} = \boldsymbol{I}$, but usually experience of related data will make a better choice possible. In an example on the growth of a head measurement in children aged 8–14, Potthoff and Roy (1964) take $g_{ij} = \rho^{|i-j|}$, assuming that the variance at the different ages is approximately constant, and that the data follow an $AR(1)$ scheme; the value of ρ was taken as 0.824, based on an earlier study of a different measurement in children.

Example 13.1

The data in Table 13.2, from Potthoff and Roy (1964), show the change in a measurement on the heads of 11 girls and 16 boys. These data constitute a fairly typical set of growth measurements. The observations—with a few exceptions—increase with time, and there is also a slight suggestion that the variance increases with time. The following analysis is taken from the paper cited.

The generalised MANOVA model for this data set has $q = 4$, $n = 27$, $m = 2$. If the data are fitted by a second-order polynomial, $p = 3$

Table 13.2 Measurements of the distance from the centre of the pituitary to the pteryomaxillary fissure, on 11 girls and 16 boys at four ages. From Potthoff and Roy (1964), reproduced with the permission of Biometrika Trustees

Girls				
	Age in years			
Individual	8	10	12	14
1	21.0	20.0	21.5	23.0
2	21.0	21.5	24.0	25.5
3	20.5	24.0	24.5	26.0
4	23.5	24.5	25.0	26.5
5	21.5	23.0	22.5	23.5
6	20.0	21.0	21.0	22.5
7	21.5	22.5	23.0	25.0
8	23.0	23.0	23.5	24.0
9	20.0	21.0	22.0	21.5
10	16.5	19.0	19.0	19.5
11	24.5	25.0	28.0	28.0
Mean	21.18	22.23	23.09	24.09

Boys				
	Age in years			
Individual	8	10	12	14
1	26.0	25.0	29.0	31.0
2	21.5	22.5	23.0	26.5
3	23.0	22.5	24.0	27.5
4	25.5	27.5	26.5	27.0
5	20.0	23.5	22.5	26.0
6	24.5	25.5	27.0	28.5
7	22.0	22.0	24.5	26.5
8	24.0	21.5	24.5	25.5
9	23.0	20.5	31.0	26.0
10	27.5	28.0	31.0	31.5
11	23.0	23.0	23.5	25.0
12	21.5	23.5	24.0	28.0
13	17.0	24.5	26.0	30.0
14	22.5	25.5	25.5	25.0
15	23.0	24.5	26.0	30.0
16	22.0	21.5	23.5	25.0
Mean	22.87	23.81	25.72	27.47

(constant, linear and quadratic components), and $\boldsymbol{\xi}$ and $\boldsymbol{P}$ have the form

$$\boldsymbol{\xi} = \begin{pmatrix} \xi_{10} & \xi_{11} & \xi_{12} \\ \xi_{20} & \xi_{21} & \xi_{22} \end{pmatrix}$$

$$\boldsymbol{P} = \begin{pmatrix} 1 & 1 & 1 & 1 \\ -3 & -1 & 1 & 3 \\ 9 & 1 & 1 & 9 \end{pmatrix}$$

The matrix $\boldsymbol{G}$ was taken as proportional to the correlation matrix of an $AR(1)$ model with correlation $\rho = 0.824$, as described above.

The first step is to test whether the quadratic terms in the model are needed. This gives an exact variance ratio test, the same test for the four criteria, $F_{2,25} = 1.27$. This value is clearly not significant, so ξ_{12} and ξ_{22} are set to zero.

Next, the difference between the two curves is tested. The null hypothesis is $\xi_{10} = \xi_{20}$, $\xi_{11} = \xi_{21}$. Again, this gives an exact test, $F_{2,24} = 6.93$. This value is significant at the 5% level, so that there is some evidence of a difference between the curves for girls and boys. The authors show further how to construct confidence bands around the fitted lines for boys and girls.

It is interesting to compare this analysis with a univariate analysis. After fitting a full model, equivalent to fitting separate cubics to the two groups, Box's test for homogeneity of dispersion matrices is not quite significant at the 5% level ($P = 0.056$). The sphericity test on the pooled matrix is not significant ($P = 0.259$). The Greenhouse–Geisser estimate of ϵ is 0.876, but the Huynh–Feldt estimate is 1. The cubic and quadratic terms are not significant, but the linear term is highly significant, and the difference in the linear term between the groups is significant at the 5% level. In this example, the assumptions underlying the univariate model are not seriously violated, and conclusions from the two analyses are the same.

The covariance adjustment method

13.5 Rao (1965, 1966b) introduced a modification to the model of Potthoff and Roy (1964) based on a multivariate analysis of covariance. The underlying idea is that if the matrix $\boldsymbol{P}$ correctly represents the curve of time variation at the q observed points with $p < q$ parameters, other contrasts among the q variates may be used as covariates. This suggestion was first made by Leech and Healy (1959). In particular, if $\boldsymbol{P}$ represents the p orthogonal components of a polynomial of order $p-1$, the remaining $q-p$ orthogonal polynomials may be considered as potential covariates.

The theory is a straightforward application of multivariate analysis of covariance, assuming multivariate normal errors, but making no assumptions about the structure of the covariance matrix. The standard theory of inference for the model is used, but some caution is necessary. The order of the underlying model is based on analysis of the data, usually choosing p on the basis of significance tests and selecting covariates according to their effectiveness in reducing the random variability. The effect of these selections on the subsequent inference about the parameter estimates is uncertain.

The model of Potthoff and Roy (1964) may be written

$$Y_{n\times q} = A_{n\times m}\xi_{m\times p}P_{p\times q} + E_{n\times q} \tag{13.8}$$

where E is a matrix of residuals, with rows ϵ_i, $i = 1\ldots n$, independently, $N(\mathbf{0}, \Sigma)$. Now, following Rao (1965) consider a matrix

$$H_{q\times q} = (H_1 \vdots H_2)$$

where H_1 and H_2 are respectively $q \times p$ and $q \times (q-p)$ matrices such that

$$PH_1 = I_p, \quad PH_2 = O_{p\times(q-p)}.$$

Next consider the transformation

$$Z = (Z_1 \vdots Z_2) = YH.$$

It is easy to see that

$$E(Z_1, Z_2) = (A\xi, O)$$
$$\text{var}(\text{vec}(Z)) = H'\Sigma H \otimes I$$

and the elements of Z have a multivariate normal distribution. Now inference is based on $Z_1|Z_2$, where

$$E(Z_1|Z_2) = A\xi + Z_2B \tag{13.9}$$

$$\text{var}(\text{vec}(Z_1|Z_2)) = (P\Sigma^{-1}P')^{-1} \otimes I_n \tag{13.10}$$

$$B = (H_2'\Sigma H_2)^{-1}H_2'\Sigma H_1 \tag{13.11}$$

and, of course, $Y_1|Y_2$ follows a p-variate normal distribution.

The matrices H_1 and H_2 are not uniquely defined by the relationships above, but if P represents a polynomial of order $p-1$, it is natural to consider the transformation of Y to the p orthogonal polynomial vectors corresponding to Z_1, and the remaining $q-p$ orthogonal polynomials corresponding to Z_2. If the mean then has the form

$$E(Z) = A(\xi_1, \xi_2)\begin{pmatrix} P \\ P_0 \end{pmatrix} \tag{13.12}$$

where P_0 is a $(q-p) \times q$ matrix of additional coefficients, it is convenient to take

$$H_1 = P'(PP')^{-1} \tag{13.13}$$

and, omitting some calculations,

$$H_2 = (I - H_1P)P_0'\{P_0(I - H_1P)P_0'\}^{-1} \tag{13.14}$$

This leads to the estimator

$$\hat{\xi}_1 = (A'A)^{-1}A'\{YP'(PP')^{-1} - YV(V'\Sigma V)^{-1}V'\Sigma P'(PP')^{-1}\} \tag{13.15}$$

where $\boldsymbol{H}_2$ has been replaced by $\boldsymbol{V}$ and $\boldsymbol{\Sigma}$ is estimated by

$$\hat{\boldsymbol{\Sigma}} = \frac{1}{n-m}\boldsymbol{Y}'(\boldsymbol{I} - \boldsymbol{A}'(\boldsymbol{A}\boldsymbol{A})^{-1}\boldsymbol{A})\boldsymbol{Y} \tag{13.16}$$

The covariance matrix of this estimate is given by

$$\operatorname{var}(\hat{\boldsymbol{\xi}}_1) = k(\boldsymbol{A}'\boldsymbol{A})^{-1} \otimes (\boldsymbol{P}\boldsymbol{P}')^{-1}\boldsymbol{P}\{\boldsymbol{\Sigma} - \boldsymbol{\Sigma}\boldsymbol{V}(\boldsymbol{V}'\boldsymbol{\Sigma}\boldsymbol{V})^{-1}\boldsymbol{V}'\boldsymbol{\Sigma}\}\boldsymbol{P}'(\boldsymbol{P}\boldsymbol{P}')^{-1} \tag{13.17}$$

where

$$k = \frac{n-m-1}{n-m-q+p-1}$$

Further, these results generalise to the case when only a subset of the available covariates is used. Suppose $\boldsymbol{V}$ contains $c < q-p$ columns of $\boldsymbol{H}_2$. The expression for $\hat{\boldsymbol{\xi}}_1$ is then the same, and the covariance matrix has the same form, with $n-m-c-1$ replacing $n-m-q+p-1$ in the denominator of k.

It is thus possible to obtain estimates of $\boldsymbol{\xi}_1$ modified by adjustment for orthogonal polynomials not in the model, and these estimates may well have lower variance than unadjusted estimates. This will be the case if the estimates of $\boldsymbol{\xi}_1$ and $\boldsymbol{\xi}_2$ are strongly correlated, as a result of autocorrelation among the residuals of the original observations. In fact, the adjusted estimate using all available covariates is simply the maximum likelihood estimate of $\boldsymbol{\xi}_1$, assuming errors following a multivariate normal distribution with unknown covariance matrix (Grizzle and Allen, 1969; see also Baksalary *et al*, 1978).

The last point seems to cast doubt on the value of the formulation in terms of multivariate analysis of covariance. Maximum likelihood estimation can be applied more directly, leading to a generalised least squares solution; it is asymptotically efficient, but when q is large and n is not very large, estimation of the $q \times q$ covariance matrix may be inefficient or even impossible. The advantage of the covariance adjustment formulation of the problem is that certain rows of $\boldsymbol{P}_0$, those that have coefficients strongly correlated with elements of ξ_1, can be chosen as covariates. In practice, at most one or two such covariates are selected, and they are chosen to give more accurate estimates of the elements of $\boldsymbol{\xi}_1$. Kenward (1986) discusses some distributional results.

Selection of covariates

13.6 The problem of covariate selection is related to that of variable selection in multiple, or multivariate, regression. There are four possibilities (Kenward, 1985): (i) use none, (ii) use all, (iii) make a selection independently of the data under analysis, and (iv) make a selection using the data under analysis. Options (i)–(iii) involve no bias, but may be inefficient, unless prior information suggests a suitable choice under (iii). The fourth option is attractive, but makes exact inference impossible.

If covariates are to be selected using the data, various options are possible. Grizzle and Allen (1969) propose selecting covariates to minimise the generalised variance of the estimator $\hat{\boldsymbol{\xi}}$. Asymptotically, this leads to the inclusion of all covariates, and generally it will include a generous selection. Kenward

(1985) considers minimising the variance of each component of $\hat{\xi}$ separately, giving a different set of covariates for each element. He also, following Rao (1965) and Chinchilli and Carter (1984), discusses selecting covariates by a significance test, using a stepwise method. More recently, Verbyla (1986) discusses the relationship between autocorrelation structure in the covariance matrix and the redundancy of potential covariates, while Fujikoshi and Rao (1991) describe likelihood ratio tests and the use of Akaike's information criterion for the selection of covariates.

Example 13.2
An example from Kenward (1985) illustrates the method. A 2×2 experiment on rabbits, with factors sex and diet, gave weight measurements on seven occasions, at three or four day intervals. Fitting orthogonal components gave significant diet effects for the linear, quadratic and cubic components; sex was significant only for the linear component, and the interaction was not significant for any of the components.

Table 13.3 gives the correlations among the coefficients of the orthogonal polynomials. These correlations may guide the choice of suitable covariates; note that the cubic term is not considered as a potential covariate, since it showed a significant contribution to the diet effect.

Kenward (1985) compares different methods of covariate selection, for the linear and quadratic coefficients separately. For the linear coefficient, use of all covariates actually increases the unconditional standard error of the coefficient. No covariates are selected as significant in a stepwise approach, but the fourth- and fifth-order coefficients together do give a slight reduction in the unconditional standard error.

The quadratic coefficient is significantly correlated with the sixth-order coefficient, and the stepwise and variance-minimising approaches both select this term as a covariate. It gives a 10% reduction in the unconditional standard error, and is slightly better than using all covariates.

The example illustrates the properties of the covariance adjustment method, although in this case the benefits of adjustment, and the effects of different choices of covariates are slight.

13.7 The covariate selection method is attractive mainly because it is a simple analysis of variance approach to a complex modelling problem. It is flexible, and problems of missing values or observations taken at irregular intervals can largely be ignored; since the analysis is carried out on parameter estimates, there is only some loss of efficiency in estimating the parameters and carrying

Table 13.3 Correlations among coefficients of orthogonal polynomials

	Linear	Quadratic	4th-order	5th-order
Quadratic	0.247			
4th-order	−0.234	−0.188		
5th-order	−0.328	0.024	−0.231	
6th-order	−0.302	−0.477	0.122	0.210

out an unweighted analysis on them. The calculations are straightforward, and were well suited to the calculators in use when the methods were introduced.

On the other hand, the method is somewhat limited. It is restricted, in practice, to fitting polynomial growth curves, perhaps with seasonal effects incorporated. Covariance adjustment is an artificial device, often involving high-order polynomials that admit of no interpretation—and in any case there is no logical reason for picking covariates from the set of unused orthogonal polynomials, rather than any other orthogonal vector set. If it is possible to model the covariance structure satisfactorily, taking account of autocorrelated errors, that is surely a preferable approach.

Modelling covariance structure

13.8 Consider the model of Potthoff and Roy (1964) in the form

$$\boldsymbol{Y}_{n\times q} = \boldsymbol{A}_{n\times m}\boldsymbol{\xi}_{m\times p}\boldsymbol{P}_{p\times q} + \boldsymbol{E}_{n\times q} \tag{13.18}$$

Here $\boldsymbol{E}$ is a matrix of errors, with independent rows ϵ_i supposed to be normally distributed with zero mean and unknown covariance matrix $\boldsymbol{\Sigma}$. The problem is to choose a suitable structure for $\boldsymbol{\Sigma}$. The following treatment is based on Vlachonikolis and Vasdekis (1994); similar models are discussed by Diggle (1988) and Verbyla and Cullis (1990).

Suppose

$$\epsilon_{ij} = z_{ij} + u_i + q_{ij} \tag{13.19}$$

where z_{ij} represents independent and identically distributed variation within individuals, u_i are independent variations in average response between individuals, and q_{ij} are autocorrelated errors within individuals. The corresponding structure for $\boldsymbol{\Sigma}$ is

$$\boldsymbol{\Sigma} = \boldsymbol{T}^2 + \upsilon^2\boldsymbol{J} + \boldsymbol{V} \tag{13.20}$$

where

- $\boldsymbol{T}^2 = \text{diag}(\tau_1^2, \ldots, \tau_q^2)$; usually it will be reasonable to assume $\boldsymbol{T}^2 = \tau^2\boldsymbol{I}$. Changes in variance with time can be incorporated in the matrix $\boldsymbol{V}$, but the full model may be useful to allow for, say, scoring by different judges on different occasions.
- υ^2 is an unknown parameter, and $\boldsymbol{J}$ is the $q \times q$ matrix with all elements unity.
- $\boldsymbol{V}$ is usually parametrised to give a simple autoregressive scheme. Note that $\boldsymbol{T}^2 = \tau^2\boldsymbol{I}$ and $\boldsymbol{V} = \boldsymbol{0}$ give the equal covariance model of the 'univariate' analysis.

This very general formulation makes it possible to model many types of error structure corresponding to different (stationary or non-stationary) time series models, and, if the errors are assumed to be normally distributed, to estimate the parameters from the likelihood. It is often convenient to write

$$\boldsymbol{V} = \boldsymbol{MRM}' \tag{13.21}$$

where $\boldsymbol{R}$ is a symmetric $q \times q$ positive definite matrix (usually a diagonal matrix of variances) and $\boldsymbol{M}$ is a $q \times q$ matrix specifying the correlations. For example, the case where V corresponds to a stationary $AR(1)$ scheme, i.e.

$$v_{ij} = \frac{\sigma^2}{1-\phi^2}\phi^{|i-j|}$$

may be put into this form by taking

$$\begin{aligned} m_{ij} &= \phi^{i-j}, \quad i \geqslant j \\ &= 0, \quad i < j \\ r_{11} &= \sigma^2/(1-\phi^2) \\ r_{ii} &= \sigma^2, \quad i \neq 1 \\ r_{ij} &= 0, \quad i \neq j \end{aligned}$$

and a more general first-order autoregressive scheme may be obtained by replacing σ^2 by σ_i^2. The stationary process implies $\phi < 1$; non-stationary $AR(1)$ processes with $\phi > 1$ are also possible, and may be realistic in growth curves.

Now the log-likelihood function is

$$\ell(\boldsymbol{\xi}, \boldsymbol{\Sigma}) = \text{constant} - \frac{n}{2}\log|\boldsymbol{\Sigma}| - \frac{1}{2}\text{tr}\,\{(\boldsymbol{Y} - \boldsymbol{A\xi P})'(\boldsymbol{Y} - \boldsymbol{A\xi P})\boldsymbol{\Sigma}^{-1}\} \qquad (13.22)$$

(see Part 1, **3.20**).

Maximum likelihood estimation for this model when a specific parametric form is assumed for $\boldsymbol{\Sigma}$ is not straightforward. The maximisation involves some awkward matrix differential calculus (Magnus and Neudecker, 1988, give a good account of the subject), and, as with all problems involving composite variances or covariance matrices, anomalous estimates involving negative variance components may occur. This is a minor problem; it usually means the model adopted is implausible or over-parametrised.

Another difficulty is that maximum likelihood estimates are, in general, biased, and in this context, when q is large and n not very large, the bias may be serious. Patterson and Thompson (1971) introduced the method now known as REML (residual, or restricted, maximum likelihood) for estimation of variance components. In simple, balanced, problems REML gives unbiased estimates of variance components based on dividing sums of squares by their appropriate degrees of freedom rather than by n. In more complicated situations, it is generally believed to give nearly unbiased estimates, whereas ML estimates are typically negatively biased. The underlying idea of REML is to maximise the likelihood associated with contrasts, restricting consideration to that part of the sample space with mean value zero.

REML estimation in the growth-curve context has been developed in a number of recent papers; see, in particular, McGilchrist (1989), McGilchrist and Cullis (1991), Verbyla and Cullis, (1990), Vasdekis (1993), and Vlachonikolis and Vasdekis (1994).

13.9 When the equal covariance model is unsuitable, the $AR(1)$ model is a natural choice. It is relatively easy to fit, and its adequacy can be tested by a likelihood ratio test against the unrestricted model, provided only that there are enough observations to fit the full model and give a reasonable estimate of error. The extension to general $AR(k)$ models is straightforward, and some work has been done on fitting *ARMA* models; see McGilchrist (1989).

Time series models of this sort, however, are seldom realistic. Gabriel (1961, 1962) introduced 'antedependence' models, essentially low-order autoregressive models in which the autocorrelation is not necessarily the same for the whole period of the observations. Models of this sort can be used, for example, when the intervals between successive occasions are not the same. They are also appropriate for many biological situations in which the environment of the individuals changes; for instance, plants or trees in seedbeds may be transplanted to permanent positions, or animals may be weaned. Jöreskog (1970) also considered models of this type.

Kenward (1987) gives an example of the analysis using this model. He shows that when the covariance structure is of this type (not involving random effects for each individual) the analysis reduces to a multivariate analysis of covariance. Another example of Kenward's method is given by Crowder and Hand (1990) (Example 9.1, p138).

Vasdekis (1993) and Vlachonikolis and Vasdekis (1994) consider models of this sort. For example, writing as before

$$\boldsymbol{V} = \boldsymbol{MRM}',$$

an $AR(1)$ change point model with one change point at occasion r may be specified by setting $\boldsymbol{R}$ as a diagonal matrix with terms $r_{ii} = \sigma_1^2$, $i \leqslant r$, $r_{ii} = \sigma_2^2$, $i > r$, and the rows of $\boldsymbol{M}$ as

$$\boldsymbol{m}_j' = (\phi_1^{j-1}, \phi_1^{j-2}, \ldots, 1, 0, \ldots, 0), \quad \text{if } j = 1, \ldots, r \tag{13.23}$$

$$\boldsymbol{m}_j' = (\phi_2^{j-r}\phi_1^{r-1}, \phi_2^{j-r}\phi_1^{r-2}, \ldots, \phi_2^{j-r}\phi_1, \phi_2^{j-r-1}, \phi_2^{j-r-2}, \ldots, 1, 0, \ldots, 0), \quad \text{if } j = r+1, \ldots, q. \tag{13.24}$$

Vasdekis (1993) and Vlachonikolis and Vasdekis (1994) discuss such change-point models as an alternative to simple $AR(1)$ models. On the basis of fairly extensive simulations, they conclude the following:

(1) Detection of a change point by a likelihood ratio test is straightforward, and the tests are powerful, particularly when $\phi_2 > 1$, giving non-stationary models.
(2) Model misspecification, fitting a model with no change point or with the change point in the wrong place, has little effect on the estimation of the parameters $\boldsymbol{\xi}$ of the growth curve. This is not unexpected; misspecification leads to inefficient weighting, not to biased estimation.
(3) Misspecification can, however, considerably inflate the errors of the estimates.

Multivariate growth curves

13.10 Multivariate growth curve models are concerned with data recorded on several occasions, on individuals receiving different treatments or divided into different classes, in which each record consists of measurements on a number of, presumably correlated, random variables. This is a very common situation; when individuals are followed up over a period, the cost of collecting the data is almost unaffected by the number of measurements taken on each occasion. Analysis of such data sets, however, is usually confined to repeated measures analysis on each variable separately.

If it can reasonably be assumed that the variance–covariance matrix of the observations remains constant, and that the correlations between observations on the same individual are independent of the interval between them, the situation is analogous to the equal covariance model. It is then easy to extend the 'split plot' analysis of variance to a multivariate analysis of variance. This model, however, is seldom plausible.

Some work has been done on truly multivariate models for this problem. Published work includes Reinsel (1982, 1984, 1985) and Anderson *et al* (1986). Vasdekis (1993) examines maximum likelihood and REML estimation for multivariate growth curves; the following discussion is based on his (unpublished) formulation of the model.

Suppose n individuals are divided into m groups, and a set of r observations are recorded on each individual on q occasions. These observations may be written y_{ijkl}, where $i = 1, \ldots, n_j$, $j = 1, \ldots, m$, and $\sum_{j=1}^{m} n_j = n$, $k = 1, \ldots, q$ and $l = 1, \ldots, r$. Write

$$\boldsymbol{y}_{ijk} = (y_{ijk1}, \ldots, y_{ijkr})' \tag{13.25}$$

for the vector of r measurements on individual i of the jth group at time k. Suppose

$$\boldsymbol{y}_{ijk} = \boldsymbol{\mu}_{ijk} + \boldsymbol{u}_{ijk} \tag{13.26}$$

where $\boldsymbol{\mu}_{ijk}$ and $\boldsymbol{u}_{ijk}$ represent the r-variate mean and error respectively, with $E(\boldsymbol{u}_{ijk}) = \boldsymbol{0}$ and $E(\boldsymbol{u}_{ijk}\boldsymbol{u}'_{ijk}) = \boldsymbol{\Sigma}_1$. Now define the $r \times q$ matrices

$$\boldsymbol{Y}_{ij} = \boldsymbol{M}_{ij} + \boldsymbol{U}_{ij} \tag{13.27}$$

with the q columns corresponding to the vectors in (13.26). Then

$$E(\boldsymbol{U}_{ij}) = \boldsymbol{0} \tag{13.28}$$

$$\text{var}\{\text{vec}(\boldsymbol{U}_{ij})\} = \boldsymbol{I}_q \otimes \boldsymbol{\Sigma}_1 \tag{13.29}$$

Next, the mean is modelled by a linear regression, typically on orthogonal polynomials. Suppose

$$\boldsymbol{Y}_{ij} = \boldsymbol{\Xi}_{ij}\boldsymbol{P} + \boldsymbol{U}_{ij} \tag{13.30}$$

where $\boldsymbol{\Xi}_{ij}$ is an $r \times p$ matrix of unknown parameters and $\boldsymbol{P}$ is a $p \times q$ matrix of known coefficients, usually representing a polynomial of order $p - 1$.

The next step is to partition $\boldsymbol{\Xi}_{ij}$ and $\boldsymbol{P}$ to give

$$\boldsymbol{\Xi}_{ij}\boldsymbol{P} = (\boldsymbol{\Xi}_{ij}^{(1)}, \boldsymbol{\Xi}_{j}^{(2)})\begin{pmatrix} \boldsymbol{P}_1 \\ \boldsymbol{P}_2 \end{pmatrix} \tag{13.31}$$

In this equation, $\boldsymbol{\Xi}_{ij}^{(1)}$ is an $r \times t$ matrix, and $\boldsymbol{\Xi}_{j}^{(2)}$ is $r \times (p - t)$. $\boldsymbol{P}$ is partitioned conformally. $\boldsymbol{\Xi}_{j}^{(2)}$ is a matrix of coefficients dependent on the group only, while

$\Xi_{ij}^{(1)}$ depends on the individuals, and is a random variable, with

$$E(\Xi_{ij}^{(1)}) = \Xi_j^{(1)} \tag{13.32}$$

$$\mathrm{var}\{\mathrm{vec}(\Xi_{ij}^{(1)})\} = \boldsymbol{I}_t \otimes \boldsymbol{\Sigma}_2. \tag{13.33}$$

This defines a fairly general model, with $\boldsymbol{\Sigma}_1$ representing the time structure of each variable, and $\boldsymbol{\Sigma}_2$ the variations in the response curves between individuals. Vasdekis (1993) goes on to consider maximum likelihood and REML estimation of the parameters involved, but the algebra is formidable and will not be discussed here.

Nonlinear growth curves

13.11 Linear regression models of the type discussed so far are more or less restricted to polynomials and periodic functions. Transformation of the time scale can give some extra flexibility, but the models fitted are not really suited to the curves to be modelled. Most growth curves have a sigmoid form, rising slowly at first, with the growth rate increasing to a maximum and then falling to zero, so that the curve approaches an asymptote. Responses to drugs or other stimuli show a fairly rapid increase, followed by a slower decline, approaching an asymptote equal, or nearly equal, to the starting value. They behave, in fact, rather like the derivative of a growth curve.

The shortcomings of polynomials as models for growth curves are discussed by Sandland and McGilchrist (1979). The most important shortcoming is that polynomials have no asymptotes. Polynomials can give good approximations to growth curves over a restricted range, but when the curve approaches a limiting value, high-order polynomials are required to give a reasonable fit, and of course any attempt at extrapolation is hopeless.

Various curves have been suggested to represent growth of an organism or a population. The best known are the logistic curve

$$\mu_t = \frac{\alpha}{1 + \beta \exp(-kt)}, \tag{13.34}$$

and the Gompertz curve

$$\mu_t = \alpha e^{\{-\beta \exp(-kt)\}}. \tag{13.35}$$

Rather more general models have been discussed by Richards (1959), Nelder (1961, 1962) and Heitjan (1991a,b). The full model of Heitjan has the form

$$\begin{aligned} \mu_t &= e^{\kappa_2}[1 + (e^{(\kappa_2 - \kappa_1)\kappa_4} - 1)\exp\{-\kappa_3(t - t_0)e^{\kappa_2\kappa_4}\}]^{-1/\kappa_4} && (13.36) \\ &\quad \kappa_4 \neq 0 \\ &= \exp[\kappa_2 + (\kappa_1 - \kappa_2)e^{-\kappa_3(t - t_0)}] && (13.37) \\ &\quad \kappa_4 = 0 \end{aligned}$$

as given by Lindsey (1993).

This model includes the logistic ($\kappa_4 = 1$) and Gompertz ($\kappa_4 = 0$) forms, and given sufficient data and a suitable model for error covariance, fitting by maximum likelihood is possible. Even with modern computers, however, the calculations are formidable; for an example, see Heitjan (1991b). Lindsey (1993) also uses this model and, for the same data set as analysed by Heitjan, compares it with a quadratic model. Disappointingly, the latter model gives a better fit.

The interest in nonlinear, and non-parametric, models in general statistical applications will certainly spread to the analysis of repeated measures and growth curves. In spite of their theoretical drawbacks, however, traditional polynomial models have the advantage of simplicity and are easy to interpret.

Categorical and discrete data

13.12 The analysis of repeated categorical data is usually included in books on repeated measures, but is rather remote from other aspects of multivariate analysis. The data consist of the categories into which a number of individuals fall on a sequence of occasions. The problem is thus one of modelling a stochastic process. The process is usually assumed to be Markovian, so that it is specified by a transition matrix. This may well be occasion-dependent, and it will usually depend on covariates. These include treatment groups, and sometimes *time-dependent covariates*—recorded on each occasion. The matrix is, in any case, probably not the same for individuals with the same set of covariates, so that models will typically show overdispersion.

The number of parameters increases rapidly with the number of states. The simplest situation, with two states, can be modelled as a binomial response, usually with a logit link, using a generalised linear model program, such as GLIM, and including a binary covariate to represent the previous state. Survival data can be regarded as a two-state stochastic process with an absorbing state.

More complex chains include sequences of animal behaviour, and assessment of severity of a disability at different times—the latter an example of ordered categories. In theory, these situations can be modelled in the same way, treating each state separately as a binary, yes or no, response, but coding the preceding states as binary variables. In practice, this is possible only with very simple models, or very large data sets.

Data consisting of repeated counts can be analysed using log-linear models, again with previous counts often appropriate as covariates. These models are rather easier to handle, but again overdispersion is common.

Crowder and Hand (1990) and Lindsey (1993) include a number of examples. Lee *et al* (1977) give an account of methods of estimating transition matrices. Lindsey (1989, 1992) and Aitkin *et al* (1989) give examples specifically of the use of GLIM in repeated measures and survival data.

Example 13.3

Lindsey (1993) analyses a data set giving counts of epileptic fits in four two-week periods. A total of 59 patients were divided into two groups receiving different treatments (placebo and progabide). Age, and a 'baseline count', based on an eight-week period before treatments began, were available as covariates.

Fitting linear and quadratic age and baseline effects and an interaction gives a deviance of 810.76 on 228 degrees of freedom. First and second-order autoregressive terms gave little improvement, but including interaction between the first-order autoregression and treatment and baseline brings the deviance down to 733.88 on 225 degrees of freedom. Fitting a factor for patients gives a deviance of 377.84 on 174 degrees of freedom.

There is obviously still a marked overdispersion in the model, and the significance of the effects fitted cannot be judged by treating the changes in deviance as χ^2 variables. Usually when this happens it is reasonable to treat the analysis of deviance as an analysis of variance, with the final deviance providing a divisor for F tests. For a discussion of this and other possibilities, see McCullagh and Nelder (1989).

Example 13.4

Table 13.4 (Fingleton, 1984; Lindsey, 1993) shows the votes cast by 1651 electors in the Swedish elections of 1964, 1968 and 1970. The four political parties form the four categories, and the table shows the counts of electors falling into the 64 possible sequences.

Various models can be fitted to data of this sort using a generalised linear modelling program, and the fit can be judged by the residual deviance. The sparseness of the data, and particularly the zero entries, mean that significance tests should be regarded with some caution.

Table 13.4 Votes of 1651 electors in the Swedish elections of 1964, 1968 and 1970. The four parties are Social Democrat, Centre, People's and Conservative. Reproduced with the permission of Oxford University Press from Lindsey (1993) and Fingleton (1984)

		1970				
1964	1968	SD	C	P	Con	Total
SD	SD	812	27	16	5	860
	C	5	20	6	0	31
	P	2	3	4	0	9
	Con	3	3	4	2	12
C	SD	21	6	1	0	28
	C	3	216	6	2	227
	P	0	3	7	0	10
	Con	0	9	0	4	13
P	SD	15	2	8	0	25
	C	1	37	8	0	46
	P	1	17	157	4	179
	Con	0	2	12	6	20
Con	SD	2	0	0	1	3
	C	0	13	1	4	18
	P	0	3	17	1	21
	Con	0	12	11	126	149
Total		865	373	258	155	1651

Lindsey fits a Markov chain, with residual deviance 207.33 on 36 degrees of freedom. This model implies that the transitions in the second period are independent of the vote in the 1964 election; the model is clearly wrong. Lindsey goes on to test equilibrium of the margins, and reversibility—symmetry of the transition matrix—for each of the two periods. The only hypothesis that seems acceptable is that of reversibility in the period 1968–1970.

Computer programs

13.13 Most of the large statistical packages have facilities for at least some of the standard methods of analysing repeated measurements and growth curves. Crowder and Hand (1990) give details of the relevant programs in BMDP, SPSS and SAS. All have suitable univariate analysis of variance software, with Greenhouse–Geisser and Huynh–Feldt adjustments. All have general multivariate linear model programs, with corresponding tests. BMDP 5V allows the user to fit some types of structured covariance matrices, including autoregressive models, and estimation is by way of maximum likelihood or REML. SAS CATMOD is a program for analysis of repeated measures on categorical variables. SPSS MANOVA gives general multivariate analysis of variance, with covariates specifiable, and orthogonal polynomials calculated within the program. Crowder and Hand (1990) give examples of the use of BMDP 2V (for univariate analysis) and SPSS MANOVA.

Lindsey (1993) gives a number of examples of the use of GLIM in fitting repeated measures models of various types, and mentions some more specialised programs. In particular, GROWTH is a program developed by D.F.Heitjan, for fitting nonlinear regression models to repeated measures.

Conclusions

13.14 This chapter has only touched the fringes of a very large subject, and has concentrated on classical methods of analysing repeated measures data. There are a number of books on the subject, Crowder and Hand (1990) being an excellent introduction. The bibliography lists about 550 references. Lindsey (1993) is more advanced and includes sections on survival data and renewal theory; his bibliography lists 1382 items.

Missing data are a problem in all repeated measure data sets. When they are missing 'completely at random' (Little and Rubin, 1987) standard imputation methods are appropriate. Particularly, however, in medical trials, patients may drop out of the trial for reasons that are related to the treatment, or to their condition. Little and Rubin (1987) define 'informative drop-out' as a process dependent on the observations that *would have been observed* had the individual not dropped out.

This situation presents difficulties, and serious bias can arise if due account is not taken of the dependence of drop-out on other factors. Diggle and Kenward (1994) discuss the problem, and that paper, with the following discussion, constitutes a good account of approaches at present available.

Exercises

13.1 Show that fitting a polynomial by maximum likelihood, with no restrictions on the covariance matrix, is equivalent to fitting the same model by the covariance adjustment, using all available higher order polynomial terms as covariates (Grizzle and Allen, 1969; see **13.5**).

13.2 Suppose that the matrix $\boldsymbol{V} \equiv \boldsymbol{H}_2$ (equation 13.14) is partitioned so that $\boldsymbol{V} = (\boldsymbol{V}_1, \boldsymbol{V}_2)$, with $\boldsymbol{V}_1(q \times c)$ and $\boldsymbol{V}_2(q \times (p-c))$, and the columns of $\boldsymbol{V}_1$ are used to adjust the ordinary least squares estimator. Show that the form of the new estimator is a generalised least squares estimator weighted by

$$\boldsymbol{S}^{-1} - \boldsymbol{S}^{-1}\boldsymbol{V}_2(\boldsymbol{V}_2'\boldsymbol{S}^{-1}\boldsymbol{V}_2)^{-1}\boldsymbol{V}_2'\boldsymbol{S}^{-1}$$

where $\boldsymbol{S} = (n-m)\hat{\boldsymbol{\Sigma}}$, as given in (13.16) (Vasdekis, 1993; the case $p = c$ corresponds to the previous exercise).

14

Miscellaneous Topics

In this chapter we cover a number of more or less unrelated topics, which do not fit naturally into any of the preceding chapters but without which no treatise on modern multivariate analysis would be complete. They include the definition and utilisation of distances between populations, the handling of high-dimensional and functional data, the analysis of shape, and the treatment of angular data. Some of the modern developments, particularly in functional data analysis and in the handling of shape, rest on sophisticated ideas in pure mathematics that are relatively unfamiliar to statisticians. As it is not possible in this volume to provide the relevant mathematical theory in such areas as functional analysis or Riemannian geometry, our treatment of the statistical ideas is necessarily brief and superficial. The aim is merely to outline the general lines of development, and to indicate some of the key results; for a deeper understanding the reader will need to go to the source references cited.

Distances between populations

14.1 The concept of distance lies at the heart of many problems involving multivariate description, multivariate inference, or multivariate classification. Specific techniques that depend on this concept include scaling methods (Part 1, Chapter 5), two-sample hypothesis tests (Part 1, Chapter 6 and Appendix), distance-based discriminant analysis (Chapter 9), and cluster analysis (Chapter 10). However, most of the uses that we have made of distance so far have been through the distances between *individuals*, and it is with the definition of such distances that we have been concerned. On the other hand, many situations exist in practice in which we need to consider distances between *groups* of individuals and so we now turn to this case. We first review the available theoretical results (i.e. distances between populations), and then discuss their practical implementation (i.e. distances between samples).

14.2 The first systematic attack on the problem was by Mahalanobis (1936), who considered the definition of distance between p-variate normal populations. Two such normal distributions are completely specified by the sets of parameters

$$\boldsymbol{\theta}_1' = (\mu_{11}, \ldots, \mu_{1p};\ \sigma_{111}, \ldots, \sigma_{1pp};\ \sigma_{112}, \sigma_{113}, \ldots, \sigma_{1,p-1,p})$$

and

$$\boldsymbol{\theta}_2' = (\mu_{21}, \ldots, \mu_{2p};\ \sigma_{211}, \ldots, \sigma_{2pp};\ \sigma_{212}, \sigma_{213}, \ldots, \sigma_{2,p-1,p})$$

where $\boldsymbol{\mu}_1' = (\mu_{11}, \ldots, \mu_{1p})$ and $\boldsymbol{\mu}_2' = (\mu_{21}, \ldots, \mu_{2p})$ are the two population mean vectors while $\boldsymbol{\Sigma}_1 = (\sigma_{1ij})$ and $\boldsymbol{\Sigma}_2 = (\sigma_{2ij})$ are the population dispersion matrices. Mahalanobis restricted attention to the equal dispersion case, so that $\boldsymbol{\Sigma}_1 = \boldsymbol{\Sigma}_2 = \boldsymbol{\Sigma} = (\sigma_{ij})$. For independent variables we have $\sigma_{ij} = 0$ if $i \neq j$, so that $\boldsymbol{\Sigma} = \boldsymbol{\Sigma}_0 = \text{diag}(\sigma_{11}, \ldots, \sigma_{pp})$. In such a case, Mahalanobis (1930) had already shown that a reasonable measure of distance between the populations is the Euclidean distance calculated from the standardised variables:

$$\Delta_1 = \{(\boldsymbol{\mu}_1 - \boldsymbol{\mu}_2)' \boldsymbol{\Sigma}_0^{-1} (\boldsymbol{\mu}_1 - \boldsymbol{\mu}_2)\}^{1/2} = \left\{ \sum_{i=1}^{p} \frac{(\mu_{1i} - \mu_{2i})^2}{\sigma_{ii}} \right\}^{1/2} \tag{14.1}$$

To generalise this result to the correlated variables case Mahalanobis (1936) introduced a statistical field, such that at each point in this field there is a specified set of parameters $\boldsymbol{\theta}$ defining a particular population. In other words, each point in the statistical field is the centre of a density-cluster belonging to a particular normal population completely specified by the value of $\boldsymbol{\theta}$ at that point. By establishing correspondence between the statistical field having constant dispersion (i.e. values of σ_{ij} equal across all points of the field) and the physical field in the theory of relativity, Mahalanobis was then able to use geometrical results from the latter theory to derive the distance between two normal populations with different means, $\boldsymbol{\mu}_1$ and $\boldsymbol{\mu}_2$, but the same dispersion matrix $\boldsymbol{\Sigma}$ as

$$\Delta_2 = \{(\boldsymbol{\mu}_1 - \boldsymbol{\mu}_2)' \boldsymbol{\Sigma}^{-1} (\boldsymbol{\mu}_1 - \boldsymbol{\mu}_2)\}^{1/2} \tag{14.2}$$

Δ_2 is known as *Mahalanobis' generalised distance* between the populations; it is one of the most influential results to have been obtained in multivariate theory. (Actually, in the original definitions, Mahalanobis included $\frac{1}{p}$ inside the curly brackets of both (14.1) and (14.2), but this term is now usually omitted.)

14.3 Rao (1945, 1949) continued the development of the above ideas, to arrive at a definition of distance between any two members of an arbitrary family of distributions. Suppose that $f(\boldsymbol{x}, \boldsymbol{\theta})$ is the probability density function of the family in question, depending on the parameter vector $\boldsymbol{\theta}' = (\theta_1, \ldots, \theta_k)$. Any member of this family is therefore specified by its value of $\boldsymbol{\theta}$, and this in turn can be represented by a point in the k-dimensional parameter space (or statistical field in the terminology above).

Let $\boldsymbol{\theta}' = (\theta_1, \ldots, \theta_k)$ and $(\boldsymbol{\theta} + d\boldsymbol{\theta})' = (\theta_1 + d\theta_1, \ldots, \theta_k + d\theta_k)$ be two contiguous points in this space. At any assigned values of $\boldsymbol{x}$, the differences in probability densities corresponding to these contiguous points is $df(\theta_1, \ldots, \theta_k)$ to first order. The distribution of df/f over the xs summarises the consequences of replacing θ_i by $\theta_i + d\theta_i$, $(i = 1, \ldots, k)$. The expectation of the square of the relative discrepancy df/f is given by

$$ds^2 = \sum_{i=1}^{k}\sum_{j=1}^{k} g_{ij}(\boldsymbol{\theta})d\theta_i d\theta_j \tag{14.3}$$

where g_{ij} are the elements of the (positive definite, symmetric) *information matrix*, i.e. $g_{ij} = E\left(\frac{1}{f}\frac{\partial f}{\partial\theta_i}\cdot\frac{1}{f}\frac{\partial f}{\partial\theta_j}\right) = E\left(\frac{\partial \ln f}{\partial\theta_i}\cdot\frac{\partial \ln f}{\partial\theta_j}\right)$. It thus follows that the distance between the distributions identified by parameter values $\boldsymbol{\theta}_1$ and $\boldsymbol{\theta}_2$ can be taken as the geodesic distance between the corresponding points in the parameter space, with respect to the metric (14.3) of a Riemannian geometry.

In principle, this solves the problem of finding the distance between any two members of a specified family of distributions. In practice, however, the evaluation of such a distance can be extremely difficult, as it can involve the solution of complicated differential equations to find the geodesic followed by complicated integrations along this geodesic. Atkinson and Mitchell (1981) give details of these various steps, and obtain distances in a variety of simple cases (mainly univariate). Mitchell (1988) gives some further univariate results.

However, since (14.3) is invariant under transformations of $\boldsymbol{x}$ and $\boldsymbol{\theta}$, it is sometimes possible by appropriate transformations to reduce (14.3) to the simpler form

$$ds^2 = c\sum_{i=1}^{k}\sum_{j=1}^{k}(d\theta_i)^2 \tag{14.4}$$

for some constant c. This is now a metric of a Euclidean geometry referred to rectangular Cartesian coordinates. The geodesic distance $s(\boldsymbol{\theta}_1, \boldsymbol{\theta}_2)$ between distributions with parameters $\boldsymbol{\theta}_1$ and $\boldsymbol{\theta}_2$ is then given by $s(\boldsymbol{\theta}_1, \boldsymbol{\theta}_2) = \{c(\boldsymbol{\theta}_1 - \boldsymbol{\theta}_2)'(\boldsymbol{\theta}_1 - \boldsymbol{\theta}_2)\}^{1/2}$. Mitchell and Krzanowski (1985) exploited this approach to show that the distance between two members of any particular form of elliptic family distribution (Part 1, **2.11, 2.12**) is a constant multiple of Δ_2 provided that $\boldsymbol{\Sigma}$ exists.

Some further work on distance in the context of Riemannian geometry, including the investigation of metrics other than (14.3), is given by Burbea and Rao (1982, 1984) and Calvo and Oller (1990).

14.4 An alternative approach was initiated by Bhattacharyya (1943), who defined the *affinity* ρ between two probability density functions $f_1(\boldsymbol{x})$ and $f_2(\boldsymbol{x})$ as

$$\rho = \int \sqrt{f_1(\boldsymbol{x})f_2(\boldsymbol{x})}d\boldsymbol{x} \tag{14.5}$$

Affinity satisfies $0 \leqslant \rho \leqslant 1$. It is equivalent to the similarity between the two probability distributions, so that a measure of distance (also referred to as divergence or discriminatory information) between them can be obtained from any decreasing function of ρ. Three commonly used possibilities are

(1) $\Delta_3 = \cos^{-1}\rho$ (Bhattacharyya, 1946);
(2) $\Delta_4 = -\log\rho$ (Chernoff, 1952);
(3) $\Delta_5 = \sqrt{2(1-\rho)}$ (Matusita, 1956).

Δ_5 is also known as *Hellinger distance* (LeCam, 1970); it satisfies all the necessary metric conditions, and various properties are given by Matusita (1973) including transformations under which it is invariant.

More directly, measures of discriminatory information between two such probability density functions can be defined using ideas from information theory (Kullback and Leibler, 1951). These measures were originally devised to test hypotheses about the probability distribution $f(\boldsymbol{x})$ appropriate for a given situation, the mean discriminatory information for $H_1 : f = f_1$ versus $H_2 : f = f_2$ when H_1 is true being given by

$$I(1, 2) = \int f_1(\boldsymbol{x}) \ln\{f_1(\boldsymbol{x})/f_2(\boldsymbol{x})\} d\boldsymbol{x}.$$

However, this is not a symmetric measure in f_1, f_2 and is thus not ideal as a distance function. A more appropriate measure is the composite of $I(1, 2)$ and $I(2, 1)$ defined by

$$\Delta_6 = \int \{f_1(\boldsymbol{x}) - f_2(\boldsymbol{x})\} \ln\{f_1(\boldsymbol{x})/f_2(\boldsymbol{x})\} d\boldsymbol{x} \tag{14.6}$$

(Jeffreys, 1948). This is symmetric and invariant under non-singular transformation of $\boldsymbol{x}$ (Kullback, 1959, p21). However, it still does not fully qualify as a distance measure since it does not satisfy the triangle inequality. Nonetheless, it has proved useful as a 'pseudo-distance' in practical applications. A related measure was defined by Sibson (1969) as

$$\Delta_7 = \frac{1}{2} \int \{f_1(\boldsymbol{x}) \ln[f_1(\boldsymbol{x})/f(\boldsymbol{x})] + f_2(\boldsymbol{x}) \ln[f_2(\boldsymbol{x})/f(\boldsymbol{x})]\} d\boldsymbol{x}, \tag{14.7}$$

where $f(\boldsymbol{x}) = \frac{1}{2}[f_1(\boldsymbol{x}) + f_2(\boldsymbol{x})]$.

The above measures are relatively easy to handle with the more common probability density functions, and have thus been used to find the distance between two populations in a range of cases that bear on practical situations. In particular, Bhattacharyya (1946) has evaluated Δ_3 for two multinomial distributions (for use when analysing qualitative data), Matusita (1967) has found Δ_5 between two arbitrary multivariate normal distributions (for modelling quantitative data), while Krzanowski (1983b) and Bar-Hen and Daudin (1995) have evaluated Δ_5 and Δ_6 respectively in the case of two conditional Gaussian distributions (for mixed qualitative and quantitative data). The Bhattacharyya distance for multinomial distributions and some approximations to it have already been given in Part 1, Chapter 5 (equations 5.36, 5.37 and 5.38). Further discussion of these distance measures may be found in Adhikari and Joshi (1956), Ali and Silvey (1966), Edwards (1971), Matusita (1973) and Ben-Bassat (1982). In the case of multivariate normal distributions with equal dispersion matrices, all the measures yield a monotonic function of the Mahalanobis distance Δ_2.

14.5 Another general approach to the definition of distance between distributions is as a decreasing function of the misclassification errors associated with discriminant functions derived from these distributions. This line was initiated by Rao (1948), who termed such a distance the 'overlap distance'. It has been further investigated and developed by Glick (1973), Lissack and Fu (1976) and Ray (1989a,b). Finally, if we are able to define a distance $\delta(\boldsymbol{x}_1, \boldsymbol{x}_2)$ between any two single individuals $\boldsymbol{x}_1, \boldsymbol{x}_2$ using, say, the measures of Chapter 10, then we can define the distance between two populations from the dissimilarity coefficient (9.38) already discussed in Chapter 9.

14.6 Of course, all the above discussion has been in terms of theoretical populations while, in practice, we are faced with having to find the distance between two groups of individuals, $\boldsymbol{x}_{11}, \ldots, \boldsymbol{x}_{1n_1}$ and $\boldsymbol{x}_{21}, \ldots, \boldsymbol{x}_{2n_2}$ say. The standard approach is then to treat these groups as independent random samples from two separate populations, fit an appropriate density function to each sample, and use one of the above distance measures on the fitted densities. The fitting of the density functions can be achieved by parametric (either frequentist or Bayesian) or by non-parametric methods.

Parametric frequentist methods require specification of a functional form for the densities in the two groups, followed by the use of estimative methods to fit these densities to the data (done by replacing all unknown parameters in the densities by their estimates from the sample data). For quantitative data this process usually leads to two fitted normal distributions having means $\bar{\boldsymbol{x}}_1 = \frac{1}{n_1}\sum_{i=1}^{n_1} \boldsymbol{x}_{1i}$, $\bar{\boldsymbol{x}}_2 = \frac{1}{n_2}\sum_{i=1}^{n_2} \boldsymbol{x}_{2i}$ and common dispersion matrix

$$W = \frac{1}{n_1 + n_2 - 2}\left\{\sum_{i=1}^{n_1}(\boldsymbol{x}_{1i} - \bar{\boldsymbol{x}}_1)(\boldsymbol{x}_{1i} - \bar{\boldsymbol{x}}_1)' + \sum_{i=1}^{n_2}(\boldsymbol{x}_{2i} - \bar{\boldsymbol{x}}_2)(\boldsymbol{x}_{2i} - \bar{\boldsymbol{x}}_2)'\right\},$$

so the appropriate distance is the sample Mahalanobis distance D defined by

$$D^2 = (\bar{\boldsymbol{x}}_1 - \bar{\boldsymbol{x}}_2)' W^{-1} (\bar{\boldsymbol{x}}_1 - \bar{\boldsymbol{x}}_2). \tag{14.8}$$

As D is a biased estimate of Δ_2 (see **9.36**), the unbiased version (9.44) is sometimes preferred (but note the discussion in **9.36**).

For categorical data it is appropriate to assume multinomial distributions in the two populations: if each distribution has k multinomial cells, and if there are n_{1i}, n_{2i} observations in the ith cell of the two populations respectively $(i = 1, \ldots, k)$, then the estimated Bhattacharyya distance between the two populations is $\cos^{-1}\sum_{i=1}^{k}\sqrt{(n_{1i}n_{2i}/n_1 n_2)}$. For mixed categorical and continuous variables the conditional Gaussian distribution provides a suitable model; Krzanowski (1983b) discusses various estimative procedures for this model.

Parametric Bayesian methods also require specification of a functional form for the densities in the two groups, but these densities are then averaged over the posterior distributions of the unknown parameters to form the appropriate *predictive* distributions from which to calculate distances. For quantitative data, the multivariate normal model with equal dispersion matrices plus independent vague priors for the unknown parameters leads to multivariate Student's t predictive distributions (Aitchison and Dunsmore, 1975). The results of Mitchell and Krzanowski (1985) then show that the sample Mahalanobis distance is again the appropriate measure in this case. Some results relevant for categorical and mixed binary/continuous data are given by Vlachonikolis (1990).

Finally, a non-parametric approach requires estimation of the probability densities, say by kernel methods (Silverman, 1986), followed by numerical integration to obtain the distances.

14.7 In addition to the systematic approaches outlined above, various individual distance or dissimilarity coefficients between two samples have been suggested from time to time. The earliest of these was the coefficient of racial likeness

$$D_r = \left\{ \frac{1}{p} \sum_{i=1}^{p} \frac{n_1 n_2}{n_1 + n_2} \left(\frac{\bar{x}_{1i} - \bar{x}_{2i}}{s_i} \right)^2 \right\}^{1/2} \tag{14.9}$$

where $\bar{x}_{1i}, \bar{x}_{2i}$ and s_i are the two group means and the pooled within-group standard deviation for the ith variable. This coefficient is thus based on the sum of squared two-group t-statistics for each variable, and can be thought of as the sample analogue of Δ_1. It was introduced by Pearson (1926), and was used in early anthropometric studies until superseded by Mahalanobis' distance.

Alternatives to the Mahalanobis distance have also been suggested from time to time, to allow for different dispersion matrices in the two groups. Chaddha and Marcus (1968) investigated some possibilities empirically; the best one seemed to be the distance proposed by Anderson and Bahadur (1962)

$$D_a = \max_t \frac{2\boldsymbol{b}_t'\boldsymbol{d}}{(\boldsymbol{b}_t'\boldsymbol{S}_1\boldsymbol{b}_t)^{1/2} + (\boldsymbol{b}_t'\boldsymbol{S}_2\boldsymbol{b}_t)^{1/2}} \tag{14.10}$$

where $\boldsymbol{d} = \bar{\boldsymbol{x}}_1 - \bar{\boldsymbol{x}}_2$ and $\boldsymbol{b}_t = (t\boldsymbol{S}_1 + (1-t)\boldsymbol{S}_2)^{-1}\boldsymbol{d}$ for individual sample covariance matrices $\boldsymbol{S}_1$, $\boldsymbol{S}_2$. More recently, Krzanowski (1990) has evaluated Δ_5 for two multivariate normal populations whose dispersion matrices follow the common principal component model (Part 1, **4.5** and **6.28**), and the sample version of this distance has been used in the empirical study of Tyteca and Dufrêne (1993).

Finally, a number of Mahalanobis-like distances have been proposed for categorical data by Balakrishnan and Sanghvi (1968) and Kurczynski (1970), and some empirical comparisons of them have been made by Kurczynski (1970) and Krzanowski (1971).

High-dimensional and functional data

14.8 The development of automatic data-recording instruments in recent years has led to a dramatic increase in data sets that have very large numbers of variables observed on each of a relatively small number of sample individuals. Such data matrices appear structurally to be transposes of those encountered in classical multivariate analysis, where the number of individuals (n) might be large but the number of variables (p) is usually relatively small. Certainly, most of the techniques described so far in these two volumes have assumed that $n > p$, whereas in the modern data sets we often find that $p \gg n$. The types of problems requiring solution are generally the same as those for classical data, but given the relationship between n and p simple application of foregoing techniques is either very difficult (because of enormous covariance or correlation matrices) or impossible (because of singularity of matrices that require inversion).

Examples of such data sets arise in spectroscopy (e.g. infra-red reflectances obtained at many different wavelengths for a sample of substances of given chemical compositions); sound or speech analysis (e.g. intensity of sound at a particular frequency obtained at a large number of time points for individuals of a particular type); or analysis of molecular activity of various chemicals. It is not uncommon to find as many as 2000 variables measured in this way on each individual, whereas the difficulties of creating suitable chemical compounds, or of obtaining suitable individuals for participation in an experiment, might limit the sample size to as few as 40 or 50 individuals.

14.9 An associated source of data is from a continuous monitoring instrument, which provides a continuous trace or function for each individual. For example, hospital EEG and ECG traces record the amount of electrical activity in a patient's brain and heart continuously as a function of time; atmospheric pressures or surface temperatures are recorded continuously at different weather stations in a region; and Rice and Silverman (1991) report data in which angular rotations in the sagittal plane of the knee and hip were recorded continuously over a gait cycle of one double step for 39 normal 5-year-old children. Such data can always be reduced to the form above by discretisation, i.e. by taking values off the continuous trace at successive points of time; the closer together are these points of time, the more 'variables' are generated in this way. However, it may be preferable to treat each individual's observations as a continuous function and to attempt an analysis of the whole set of functions. Conversely, discrete data of the above form can be turned into continuous curves by fitting, for example, polynomial splines through each individual's values and then analysing these splines as a set of functions.

There has been some development in recent years of the analysis of both forms of data; some authors have addressed the problems arising when standard methodology is applied to discrete high-dimensional data and have suggested ways of overcoming them, while other authors have worked on the extension of standard methodology to continuous curves via functional analysis. Both sets of developments provide useful practical methods of analysis, and we now survey each one briefly.

14.10 We first consider the case of discrete high-dimensional data. It is worth mentioning that such data may either have undergone, or may need to undergo, some form of preprocessing before being subjected to more formal analysis. Simple preprocessing includes addition or subtraction of constants to ensure that all individuals have the same origin or 'level', first or second differencing to remove 'instrument drift' effects, and simple transformations of values to ensure satisfactory properties of the data (e.g. in spectroscopic analysis it is customary to analyse log(1/reflectance) values rather than the original reflectances). More sophisticated preprocessing might involve subjecting the data to a fast Fourier transform to express them in the frequency domain, or to an alternative orthogonal decomposition such as is provided for example by a wavelet transformation (Daubechies, 1988). However, details of these various techniques take us outside the area of direct interest, so we shall simply assume that any necessary preprocessing has already been conducted and we are faced with the data to be analysed. The main areas of interest with high-dimensional discrete data have been regression and calibration, and discriminant analysis. They have been briefly mentioned already, in **7.3** (Part 1) and **9.56** respectively, but we now provide some further details.

14.11 The regression and, more specifically, calibration area has been developed by Sundberg and Brown (1989) and Denham and Brown (1993). We refer the reader to Brown (1982) and **7.3** (Part 1) for some general observations on multivariate calibration, while concentrating here specifically on the high-dimensional case. Denham and Brown (1993) motivate the development with reference to applications in the food and chemical industries on establishing the amounts of component ingredients present in a sample mixture of chemicals. Suppose that a set of n chemical samples is available for analysis. Accurate assessment of the p component amounts can be obtained by wet chemistry: let the proportion of component k for sample i be x_{ik}, forming the (i, k)th element of the $n \times p$ matrix $\boldsymbol{X}$. This, however, is a slow and expensive process. An alternative quick approach is to examine the infra-red spectrum of the sample and obtain its absorbances at q infra-red frequencies: let y_{ij} be the absorbance at the jth frequency of the ith sample, forming the (i, j)th element of the $n \times q$ matrix $\boldsymbol{Y}$. It is proposed in future to measure only the infra-red absorbances of any sample, from which it is hoped to predict its component proportions.

A traditional classical approach requires first the regression of absorbances on components. The model for this regression is

$$\boldsymbol{Y} = \boldsymbol{XB} + \boldsymbol{E} \tag{14.11}$$

where $\boldsymbol{B} = (\beta_{ij})$ is a $p \times q$ matrix of unknown regression coefficients and $\boldsymbol{E} = (\epsilon_1, \ldots, \epsilon_n)'$ is the matrix of error terms having $E(\epsilon_i) = \boldsymbol{0}$, $E(\epsilon_i\epsilon_i') = \boldsymbol{\Gamma}$ and $E(\epsilon_i\epsilon_l') = \boldsymbol{0}$ for $i \neq l$. Here, $\boldsymbol{\Gamma}$ is the $q \times q$ covariance matrix of the errors within an observation and, for notational simplicity, $\boldsymbol{X}$ and $\boldsymbol{Y}$ have been assumed centred *post hoc*. Standard regression analysis yields estimates $\hat{\boldsymbol{B}}$ and $\hat{\boldsymbol{\Gamma}}$ of parameters and error covariances respectively.

Now suppose that n_f future samples have been subjected to infra-red spectroscopy, yielding the absorbances $\boldsymbol{Y}_f$, and assume that the previous centring of $\boldsymbol{X}$ and $\boldsymbol{Y}$ has fixed the location of the absorbances and compositional variables in these n_f samples. The analogous regression model for the prediction set is thus

$$\boldsymbol{Y}_f = \boldsymbol{X}_f\boldsymbol{B} + \boldsymbol{E}_f \tag{14.12}$$

with error structure as before. Now, however, $\boldsymbol{X}_f$ is unknown and to be predicted/estimated. If we insert the estimates $\hat{\boldsymbol{B}}$ and $\hat{\boldsymbol{\Gamma}}$ in place of the unknown parameters into (14.12) and treat $\boldsymbol{X}_f$ as the 'parameter' matrix, then application of generalised least squares yields the estimator

$$\hat{\boldsymbol{X}}_f = \boldsymbol{Y}_f\hat{\boldsymbol{\Gamma}}^{-1}\hat{\boldsymbol{B}}'(\hat{\boldsymbol{B}}\hat{\boldsymbol{\Gamma}}^{-1}\hat{\boldsymbol{B}}')^{-1} \tag{14.13}$$

14.12 The difficulty with high-dimensional data is that q might be very large (e.g. of the order of 1000) while n is quite small (e.g. about 50), but $\hat{\boldsymbol{\Gamma}}$ is singular whenever $n < q + 1$ so that (14.13) will not yield a unique estimator with such data. A way round this problem is to model the error structure in such a way that when this additional model is fitted then $\hat{\boldsymbol{\Gamma}}$ has a unique inverse. Denham and Brown (1993) propose treating the ϵ_i as uncorrelated random vectors arising from the same autoregressive process of order m, and then go on under this model to give an explicit form for the estimator $\hat{\boldsymbol{X}}_f$ that does not require the storage or inversion of matrices of order $q \times q$.

A further aspect of such spectroscopic data addressed by Denham and Brown is the underlying continuity, given that the absorbances have been measured at discrete points of a continuous wavelength scale. They introduce this continuity into the model by assuming that that the coefficient matrix $\boldsymbol{B}$ represents the values of p continuous functions at q frequencies $\omega_1, \ldots, \omega_q$. In particular, they assume that these functions can be represented as cubic spline functions. Applying spline function theory to the resulting model, they then show that the optimum estimator $\tilde{\boldsymbol{B}}$ of $\boldsymbol{B}$ is the same as the estimator obtained by applying regression splines to $\hat{\boldsymbol{B}}$. Standard algorithms (e.g. de Boor, 1978) can be applied to obtain this estimator, whence $\hat{\boldsymbol{X}}_f$ is given on replacing $\hat{\boldsymbol{B}}$ by $\tilde{\boldsymbol{B}}$ in (14.13).

14.13 An alternative approach to those above is based on inverse regression, i.e. regressing $\boldsymbol{X}$ on $\boldsymbol{Y}$ instead of $\boldsymbol{Y}$ on $\boldsymbol{X}$. The regression model in this case is

$$\boldsymbol{X} = \boldsymbol{Y}\boldsymbol{B}_* + \boldsymbol{E}_* \tag{14.14}$$

where $\boldsymbol{B}_* = (\beta_{*ij})$ is now a $q \times p$ matrix of unknown regression coefficients while $\boldsymbol{E}_*$ is the matrix of error terms with means, variances and covariances equivalent to those following equation (14.11) except that the covariance matrix $\boldsymbol{\Gamma}_*$ is a $p \times p$ matrix. The analogous regression model for the prediction set is then

$$\boldsymbol{X}_f = \boldsymbol{Y}_f\boldsymbol{B}_* + \boldsymbol{E}_{f*}, \tag{14.15}$$

with corresponding error structure. Sundberg and Brown (1989) show the best linear predictor of $\boldsymbol{X}_f$ to be $\hat{\boldsymbol{X}}_f = \boldsymbol{Y}_f\hat{\boldsymbol{B}}_*$ where $\hat{\boldsymbol{B}}_*$ is given by

$$(\boldsymbol{Y}'\boldsymbol{Y})\hat{\boldsymbol{B}}_* = \boldsymbol{Y}'\boldsymbol{X}. \tag{14.16}$$

When $n < q + 1$, however, $\boldsymbol{Y}'\boldsymbol{Y}$ is not of full rank and $\hat{\boldsymbol{B}}_*$ is not uniquely defined: any estimator of the form $\hat{\boldsymbol{B}}_* = (\boldsymbol{Y}'\boldsymbol{Y})^-\boldsymbol{Y}'\boldsymbol{X}$, where $(\boldsymbol{Y}'\boldsymbol{Y})^-$ is a generalised inverse of $\boldsymbol{Y}'\boldsymbol{Y}$, satisfies (14.16). To obtain a unique estimator we must choose a particular generalised inverse, and Denham and Brown (1993) suggest using the Moore–Penrose generalised inverse; this choice yields the *minimum length* estimator (as measured by the trace of $\hat{\boldsymbol{B}}_*'\hat{\boldsymbol{B}}_*$).

For several practical examples of the use of all these methods, see Denham and Brown (1993).

14.14 The other main area that can cause problems with high-dimensional data is discriminant analysis. Here we have a number (g) of *a priori* groups of samples for each of which a spectroscopic analysis (say) has been conducted, and the objective is either to produce a discriminant function for distinguishing between the groups on the basis of the resulting measurements or to provide an allocation rule for assigning a new sample to one of the groups. Given the continuous nature of most such data sets, a reasonable assumption to make is that the vector $\boldsymbol{x}' = (x_1, \ldots, x_p)$ of measurements for a particular sample has a multivariate normal distribution with mean $\boldsymbol{\mu}_i$ depending on the group π_i, $(i = 1, \ldots, g)$ from which it is taken but with a common dispersion matrix $\boldsymbol{\Sigma}$ in all groups. Standard theory (**9.11**) then shows that the optimal procedure for classifying a new individual $\boldsymbol{x}$ to one of the groups requires calculation of the functions

$$(\boldsymbol{x} - \boldsymbol{\mu}_i)'\boldsymbol{\Sigma}^{-1}(\boldsymbol{x} - \boldsymbol{\mu}_i) \tag{14.17}$$

for $i = 1, \ldots, g$. In the two-group case, this procedure reduces to classification by Fisher's linear discriminant function

$$(\boldsymbol{\mu}_1 - \boldsymbol{\mu}_2)'\boldsymbol{\Sigma}^{-1}\boldsymbol{x}_i \tag{14.18}$$

Given training data for each group, comprising n_i samples in group π_i $(i = 1, \ldots, g)$ with $n = \sum_i n_i$, the standard procedure (**9.15**) is to estimate the $\boldsymbol{\mu}_i$ by the corresponding training set means $\bar{\boldsymbol{x}}_i$, and $\boldsymbol{\Sigma}$ by the pooled within-set covariance matrix $\boldsymbol{W}$, and to use these estimates in place of the unknown parameters in (14.17) and (14.18). If the objective is simply discrimination between the given training samples, then the appropriate technique is canonical variate analysis (Part 1, **4.11**, and **9.6**). In the two-group case this again reduces to calculation of Fisher's linear discriminant function.

14.15 The problem with direct application of this standard methodology to high-dimensional data is once more caused by singularity of matrices; p is usually much greater than $n - g$ so that the $p \times p$ matrix $\boldsymbol{W}$ is of less than full rank and therefore singular. In this case, let rank $(\boldsymbol{W}) = r < p$ and consider its spectral decomposition:

$$\boldsymbol{W} = \boldsymbol{LDL}' \tag{14.19}$$

where $\boldsymbol{D}$ is the diagonal $(p \times p)$ matrix of ranked eigenvalues $d_1 \geqslant d_2 \geqslant \ldots \geqslant d_r > d_{r+1} = \ldots. = d_p = 0$ of $\boldsymbol{W}$, and $\boldsymbol{L}$ is the orthonormal $(p \times p)$ matrix whose columns $\boldsymbol{\ell}_1, \ldots, \boldsymbol{\ell}_p$ are the corresponding eigenvectors of $\boldsymbol{W}$. We can therefore write

$$\boldsymbol{W} = [\boldsymbol{L}_1 \ \boldsymbol{L}_2]\begin{bmatrix} \boldsymbol{D}_1 & \boldsymbol{O} \\ \boldsymbol{O} & \boldsymbol{O} \end{bmatrix}\begin{bmatrix} \boldsymbol{L}_1' \\ \boldsymbol{L}_2' \end{bmatrix} \tag{14.20}$$

where $\boldsymbol{L}_1$ contains the first r columns of $\boldsymbol{L}$, $\boldsymbol{L}_2$ contains $p - r$ columns that are orthogonal to each other and to those of $\boldsymbol{L}_1$ but otherwise arbitrary, and $\boldsymbol{D}_1$ is the $(r \times r)$ diagonal matrix containing the non-zero d_i only.

Krzanowski *et al* (1995) discuss a number of ways in which the singularity problem might be overcome. Some of these methods are well established, while others are new and as yet relatively untried. The following is a brief summary of the various options.

(1) *Preliminary transformations*
The simplest way to circumvent the problem of singular $\boldsymbol{W}$ is by conducting a preliminary transformation from $\boldsymbol{x}$ to $\boldsymbol{y} = (y_1, \ldots, y_m)'$, where $m \leqslant n - g$, and then applying the standard theory to $\boldsymbol{y}$. Possible approaches for selection of $\boldsymbol{y}$ are to choose the first m components in

either a preliminary principal component analysis (Part 1, Chapter 4) or a partial least squares analysis (Chapter 11) of the data. The main problems with such an approach are that an arbitrary decision must be made regarding how many components to retain and how many to discard, and that valuable between-group information may be lost in the discarded components if principal component analysis is used for the selection (see Chang, 1983).

(2) *Modified canonical analysis*
Campbell and Atchley (1981) demonstrated that canonical analysis in the non-singular case can be achieved by a two-stage spectral decomposition process. The first stage is the spectral decomposition $\boldsymbol{LDL}'$ of $\boldsymbol{W}$ as given above. This is followed by the transformation from $\boldsymbol{x}$ to $\boldsymbol{w} = \boldsymbol{D}^{-\frac{1}{2}}\boldsymbol{L}'\boldsymbol{x}$ and calculation of the between-group covariance matrix with respect to $\boldsymbol{w}$, namely $\boldsymbol{C} = \frac{1}{g-1}\sum_{i=1}^{g} n_i(\bar{\boldsymbol{w}}_i - \bar{\boldsymbol{w}})(\bar{\boldsymbol{w}}_i - \bar{\boldsymbol{w}})'$ in obvious notation. Finally the spectral decomposition of $\boldsymbol{C}$ yields eigenvalues $\lambda_1, ..., \lambda_s$ and eigenvectors $\boldsymbol{c}_1, ..., \boldsymbol{c}_s$ (where $s = \min[p, g-1]$), from which the required canonical variate coefficients are recovered as $\boldsymbol{a}_i = \boldsymbol{LD}^{-\frac{1}{2}}\boldsymbol{c}_i$.
If $\boldsymbol{W}$ is singular then only r of the eigenvalues d_i are non-zero. In modified canonical analysis we set $\boldsymbol{D}^{-\frac{1}{2}} = \text{diag}(d_1^{-\frac{1}{2}}, ..., d_r^{-\frac{1}{2}}, 0, ..., 0)$ in the standard canonical procedure, which is equivalent to the use of the Moore–Penrose generalised inverse of $\boldsymbol{W}$.

(3) *Zero variance discrimination*
If $\mathcal{A}$ denotes the p-dimensional space of all vectors $\boldsymbol{a}$, then when $\boldsymbol{W}$ has rank r we can write $\mathcal{A} = \mathcal{R} + \mathcal{N}$ where $\mathcal{R}$ and $\mathcal{N}$ denote the range and null spaces of $\boldsymbol{W}$ respectively. One way of viewing the modified canonical analysis above is as the maximisation of $V = (\boldsymbol{a}'\boldsymbol{Ba})/(\mathrm{a}'\boldsymbol{Wa})$ for $\boldsymbol{a}$ in $\mathcal{R}$, where $\boldsymbol{B} = \frac{1}{n-g}\sum_{i=1}^{g} n_i(\bar{\boldsymbol{x}}_i - \bar{\boldsymbol{x}})(\bar{\boldsymbol{x}}_i - \bar{\boldsymbol{x}})'$ is the between-group covariance matrix (see 9.6). Krzanowski *et al* (1995) show that a corresponding maximum of $\boldsymbol{a}'\boldsymbol{Ba}$ in $\mathcal{N}$, i.e. over those vectors $\boldsymbol{a}$ such that $\boldsymbol{Wa} = \boldsymbol{0}$, is given by the system of canonical variates $y_i = \boldsymbol{a}_i'\boldsymbol{x}$ where $\boldsymbol{a}_i = \boldsymbol{L}_2\boldsymbol{\alpha}_i$ and $\boldsymbol{\alpha}_i$ are eigenvectors corresponding to non-zero eigenvalues of $\boldsymbol{L}_2'\boldsymbol{BL}_2$. These canonical variates have zero variance within groups, but might carry useful between-group information.

(4) *Generalised ridge discrimination*
Instead of reducing dimensionality, as in (i), to overcome singularity, we might try to 'augment' $\boldsymbol{W}$ to make it non-singular. In achieving this aim, one clearly wants to minimise the perturbation to the original matrix in order to disturb the original data as little as possible; Krzanowski *et al* (1995) list a number of criteria that such a perturbation should satisfy. These criteria are all upheld by the matrix $\boldsymbol{V}$ defined by

$$\boldsymbol{V} = \frac{1}{c}[\boldsymbol{L}_1\ \boldsymbol{L}_2]\begin{bmatrix}\boldsymbol{D}_1 + \alpha\boldsymbol{I} & \boldsymbol{O} \\ \boldsymbol{O} & (\alpha+\beta)\boldsymbol{I}\end{bmatrix}\begin{bmatrix}\boldsymbol{L}_1' \\ \boldsymbol{L}_2'\end{bmatrix} \tag{14.21}$$

where α and β are parameters satisfying $\alpha \geqslant 0$, $\beta < d_r$ and $\alpha + \beta > 0$, and $c = \left\{\alpha p + \beta(p-r) + \sum_{i=1}^{r} d_i\right\} / \sum_{i=1}^{r} d_i$ is a normalising constant. When $\beta = 0$ and α is a small positive finite constant, the two-group version of this procedure reduces to the usual expression for ridge discriminant analysis (Friedman, 1989). Estimation of the parameters α and β can be achieved empirically by optimising leave-one-out classification error rates.

(5) *Antedependence modelling*
All the preceding approaches are either empirical or purely data-based. An alternative general approach is to postulate and fit to the data a suitable stochastic model that has a relatively small number of parameters, taking care that the dispersion matrix implied by the model is non-singular. Having estimated the model parameters we can then obtain the inverse of the fitted matrix and use it in the standard theory.

Most spectroscopic data are obtained by a 'moving window' process, which aggregates points within a pre-specified width while scanning across the range of wavelengths involved. Since a typical point will appear in three or four successive windows, the resulting data will be serially correlated and will exhibit the general features of a non-stationary time series. A nested series of models suitable for such data structures are the antedependence models introduced by Gabriel (1962) and used for the analysis of repeated measurements (Kenward, 1987; **13.9**). A set of p ordered variables is said to have an antedependence structure of order r if the ith variable ($i > r$), given the preceding r, is independent of all further preceding variables. Under the antedependence structure of order r, the inverse of the covariance matrix has non-zero elements only on the leading diagonal and on the r diagonals immediately above and immediately below it. Complete independence ($r = 0$) and general dependence ($r = p - 1$) are special cases of this structure.

Suppose therefore that $X_i = \mu_i + e_i$ where the error terms e_i follow an antedependence structure of order r, i.e. $e_i = \sum_{j=1}^{r} e_{i-j}\lambda_{ji} + f_i$ and the f_i are independent $(0, \sigma_i^2)$ variables for $i = 1, \ldots, n$. Let

$$\boldsymbol{F}_r = \begin{pmatrix} 1 & 0 & 0 & 0 & & \cdots & 0 \\ -\lambda_{11} & 1 & 0 & 0 & & \cdots & 0 \\ -\lambda_{22} & -\lambda_{12} & 1 & 0 & & \cdots & 0 \\ \vdots & \vdots & \vdots & \vdots & & & \vdots \\ -\lambda_{rr} & -\lambda_{r-1,r} & -\lambda_{r-2,r} & & & \cdots & 0 \\ 0 & -\lambda_{r,r+1} & -\lambda_{r-1,r+1} & & & \cdots & 0 \\ \vdots & & & & & & \vdots \\ 0 & 0 & 0 & \cdots & -\lambda_{2,p-2} & -\lambda_{1,p-1} & 1 \end{pmatrix}$$

i.e. a lower triangular matrix with ones on the leading diagonal and r non-zero sub-diagonals. Estimates of λ_{ij} are obtained by setting the values in the lower r bands (ignoring the diagonal) of $\boldsymbol{F}_r\boldsymbol{W}$ to zero, and solving the resulting series of linear equations recursively. This gives $\hat{\boldsymbol{F}}_r$, from which we derive the inverse of the fitted covariance matrix as $\hat{\boldsymbol{F}}_r'\boldsymbol{G}_r\hat{\boldsymbol{F}}_r$, where $\boldsymbol{G}_r = \text{diag}(\hat{\boldsymbol{F}}_r\boldsymbol{W})$.

Krzanowski *et al* (1995) describe the application of all the above methods, including antedependence models of orders 1, 2 and 3, to a number of spectroscopic data sets. No single method comes out uniformly best, but all perform well on at least one of the data sets and all seem worthy of consideration in practice. There is clearly scope for further work in this general area.

14.16 We now turn to data in which each 'individual' is a continuous record, i.e. function, plotted over some fixed interval of an argument variable. In principle, this argument can be any variable, but in practice it is frequently time, and so will be denoted in the following by t. Some specific examples have already been mentioned in **14.9**; other practical applications include records of speech utterances in which the position of the tongue dorsum height is measured by ultrasound sensing (Besse and Ramsay, 1986), and continuous traces of the force exerted by subjects pinching a force meter between thumb and forefinger (Ramsay *et al* 1995). Typically, most of the variation in a single record will appear to follow a smooth curve of some sort but there will be some local noise in each observation. Also, while the smooth variation might vary noticeably from record to record, all records will exhibit similar general features. Rapid development of data collection hardware has led to increasing prevalence of such data and, consequently, to the need for well-founded methods of their analysis. In particular, such analysis should make use of the regularity in the data and should be able to take into account any information contained either in the argument values t_j or in the derivatives or integrals of the consequent functions. We now survey briefly the available techniques for displaying important features of such curves, for summarising the information they contain, and for conducting more formal analysis on them.

14.17 Jones and Rice (1992) demonstrate the problems encountered when trying to display the important features of a large collection of curves, by superimposing 100 functions $f_i(x)$ $(i = 1, \ldots, 100)$ each of which is a kernel density estimate (Silverman, 1986) based on a different random sample of size 50 from a standard normal distribution. The smoothing parameter (i.e. bandwidth—see **9.27**) in each of these kernel density estimates was chosen by the method of Hall *et al* (1991). The resulting plot is shown in Fig. 14.1, and it is evident that the behaviour of each individual curve is almost totally obscured because of the presence of so many coincident and interfering lines. Jones and Rice suggest that only a few of these curves should actually be displayed, and the ones selected for display should be those curves that best reflect the important modes of variation present in the whole collection. Such curves can be identified by treating each of the n functions as a vector of p function values on some equispaced grid, conducting a principal component analysis on the resulting (unstandardised) $n \times p$ data matrix, labelling each curve by its score on each principal component, deciding how many components are 'important' (Part 1, **4.6**), and using the scores to pick a few representative curves on each important component. Jones and Rice suggest plotting just three curves on each chosen component: the ones corresponding to the median and to two extreme quantile scores, one at each end. They applied this procedure to the kernel densities by choosing a grid of $p = 101$ points for each curve. The first two components accounted for 40.6% and 36.4% respectively of the overall variance, with a steep drop in succeeding components, so that two components seem to be adequate for representing the important modes of variation in the data. They then identified the curves corresponding to the median, minimum

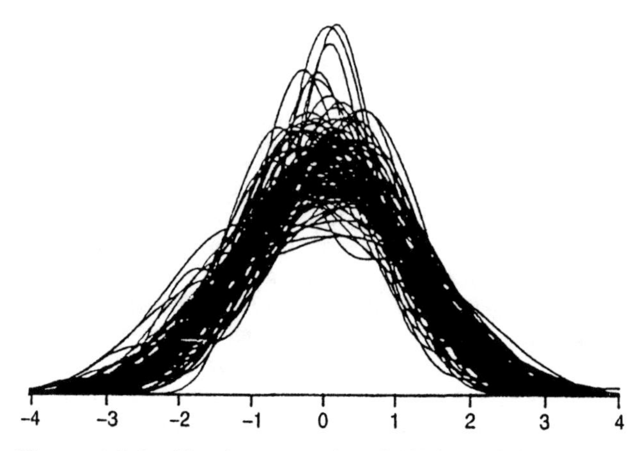

Figure 14.1 Simultaneous plot of 100 kernel density estimates, each based on a random sample of size 50 from a standard normal distribution. Reproduced from Jones and Rice (1992) with permission.

and maximum score on each component and plotted the two sets of three curves on two diagrams. The three curves for the first component are shown in Fig. 14.2, and the three for the second component in Fig. 14.3.

It is evident that the first component has identified the major source of variation among the curves as being due to their dispersion: the two extreme scores correspond to curves with very low and high spreads respectively, while the median score belongs to a curve with intermediate spread. The second component, on the other hand, identifies variation due to location of the mode and skewness of the curves: the curves corresponding to the extreme scores are markedly skew in opposite directions, while the median curve is much less skew (if not entirely symmetric). Subsequent investigation revealed that the

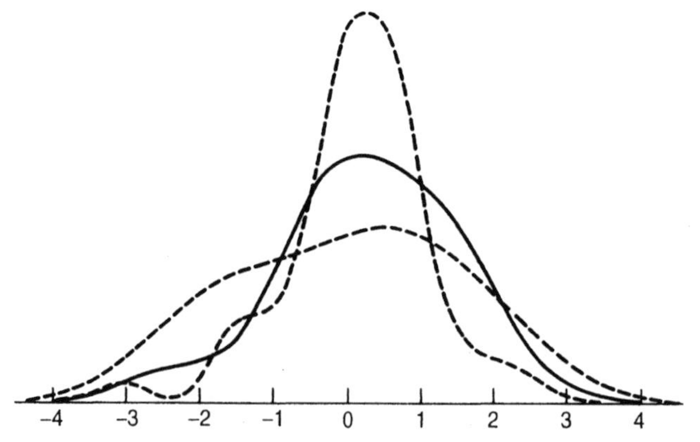

Figure 14.2 Curves from Fig. 14.1 corresponding to the median (solid curve), minimum and maximum (dashed curves) scores on the first principal component. Reproduced from Jones and Rice (1992) with permission.

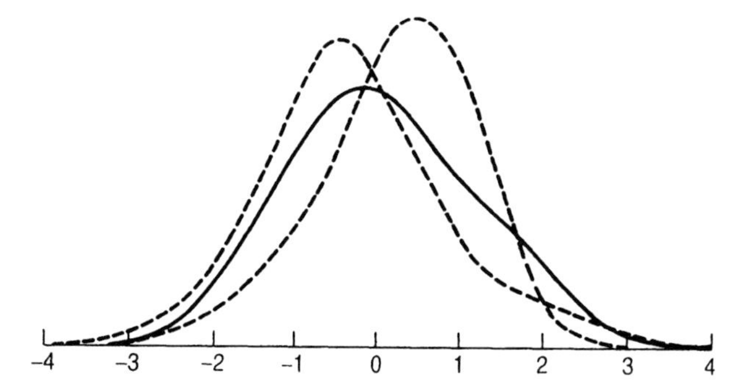

Figure 14.3 Curves from Fig. 14.1 corresponding to the median (solid curve), minimum and maximum (dashed curves) scores on the second principal component. Reproduced from Jones and Rice (1992) with permission.

scores on the first principal component exhibited a clear linear relationship with the bandwidths of the corresponding fitted kernel density estimates. Thus, the spread of the fitted density depended to some extent on the chosen bandwidth, and hence the latter quantity was a major determinant in the differences between curves. When the scores on the second component were investigated, however, no such relationship was discerned between bandwidth and skewness. Thus, variation in skewness of densities can be attributed solely to the sampling variation of skewness.

The above approach is very effective at displaying extreme curves of various types. By contrast, Flury and Tarpey (1993) propose an alternative technique, based on principal points, for plotting a small number of curves that are *representative* of the whole collection. The k principal points of a p–dimensional random vector $\boldsymbol{x}$ are the k p–element vectors $\boldsymbol{y}_1, \ldots, \boldsymbol{y}_k$ such that the expected squared distance of $\boldsymbol{x}$ to the nearest $\boldsymbol{y}_i$ is a minimum. Full details of the theory and computation of principal points are given by Flury (1990, 1993); for application to the kernel density curves we use the same $n \times p$ data grid as before. Flury and Tarpey (1993) used the normal theory maximum likelihood estimators of $k = 4$ principal points from this grid, then transformed these estimated points back into the coordinate system of the original variables and connected the vector coordinates by lines to obtain the four 'representative' curves of the data shown in Fig. 14.4. Note that these curves are not actual curves from the sample, and they represent much less variability than do those of Jones and Rice (1992). However, they do seem to capture the same important modes of variation as before, the four curves representing skew left and right and high and low spread respectively.

14.18 Moving on from display to summarisation, Rice and Silverman (1991) consider non-parametric estimation of the mean and covariance structure of a collection of curves. As a motivating example, they use records of the angular rotations of the hip in 39 normal 5-year-old children taken over a gait cycle consisting of one double step for each child. The resulting 39 curves are displayed in Fig. 1.2 (Part 1, **1.3**). Rice and Silverman model the sample curves

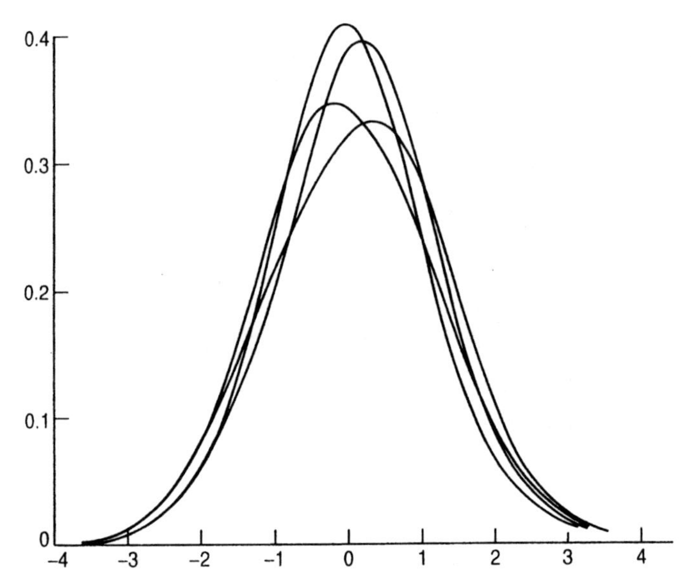

Figure 14.4 Four representative curves for the data of Fig. 14.1, based on principal points. Reproduced from Flury and Tarpey (1993) with permission.

as independent realisations of a stochastic process $x(t)$ that has mean $E\{x(t)\} = \mu(t)$ and covariance function $\text{cov}\{x(s), x(t)\} = \gamma(s, t)$. They assume that there is an expansion in terms of orthogonal eigenfunctions $\phi_i(t)$ (i.e. $\int \phi_i(t)\phi_j(t)dt = 0$ for $i \neq j$), so that

$$\gamma(s, t) = \sum_i \gamma_i \phi_i(s)\phi_i(t) \tag{14.22}$$

and a random curve from the population may be expressed as

$$x(t) = \mu(t) + \sum_i \xi_i \phi_i(t) \tag{14.23}$$

where the ξ_i are uncorrelated random variables with zero means and variances $E(\xi_i^2) = \gamma_i$. Furthermore, they assume that the mean curve and the first few eigenfunctions are smooth, and that the eigenvalues γ_i tend to zero rather rapidly so that the variability of a curve is predominantly of large scale and is well described as a sum of a small number of smooth curves with random amplitudes. The sample data then consist of n curves, each observed at times $t_1, \ldots, t_p$; the jth point on the ith curve is denoted by $x_i(t_j) = x_{ij}$ and we can write $\boldsymbol{x}_i' = (x_{i1}, \ldots, x_{ip})$ for $i = 1, \ldots, n$. The covariance matrix $\boldsymbol{\Gamma}$ of the observations on each curve then has elements $\gamma(t_j, t_k)$.

Standard Gauss–Markov theory for the general linear model leads to an estimate of the mean function obtained by averaging the data values separately at each time point t_j. However, since the mean function is assumed to be a smooth curve, Rice and Silverman (1991) suggest that the accuracy of estimation may be improved further by smoothing. This can be done with the use of penalised least squares, which reduces to seeking the curve $\hat{\mu}$ that minimises

$$\sum_j \{\bar{x}_j - \mu(t_j)\}^2 + \alpha \int \mu''(t)^2 dt \tag{14.24}$$

where α is a positive smoothing parameter and the integrated squared second derivative term quantifies the 'roughness' of μ (Green and Silverman, 1993). This is just spline smoothing (Silverman, 1986) applied to the pointwise averages, yielding as estimator $\hat{\mu}$ a particular cubic spline with knots at the data points t_j. Rice and Silverman suggest choosing the smoothing parameter α by cross-validation *leaving out an entire curve at a time.* As well as justifying this procedure they give full details of the most economical computational strategy.

To analyse variability, Rice and Silverman propose smoothed estimators of the eigenfunctions of the covariance kernel. These eigenfunctions are estimated sequentially, starting with the one corresponding to the largest eigenvalue. Let $\boldsymbol{G}$ denote the sample covariance matrix. A class of estimates is obtained by bounding the roughness of the eigenfunction, maximising $\boldsymbol{u}'\boldsymbol{G}\boldsymbol{u}$ subject to $\|\boldsymbol{u}\| = 1$ and $\boldsymbol{u}'\boldsymbol{D}\boldsymbol{u} \leqslant \beta$, where β is a smoothing parameter and $\boldsymbol{D}$ is a roughening matrix for example of the form $\boldsymbol{F}'\boldsymbol{F}$ where $\boldsymbol{F}$ is a second-differencing operator. The solution turns out to be the eigenvector of $\boldsymbol{G} - \lambda\boldsymbol{D}$ corresponding to the largest eigenvalue λ. Succeeding eigenfunctions are found iteratively. Suppose that smooth orthogonal eigenfunctions $\boldsymbol{u}_1, \ldots, \boldsymbol{u}_k$ have already been found. Obtain any $n \times (n-k)$ matrix $\boldsymbol{Q}$, all of whose columns are mutually orthogonal and orthogonal to these eigenfunctions, and find the leading eigenvector $\boldsymbol{v}$ of $\boldsymbol{Q}'(\boldsymbol{G} - \lambda\boldsymbol{D})\boldsymbol{Q}$. Then the next eigenfunction is given by $\boldsymbol{u}_{k+1} = \boldsymbol{Q}\boldsymbol{v}$. Once again, Rice and Silverman discuss a cross-validation method of choosing the smoothing parameter that involves leaving a curve out at a time, and describe ways of optimising the computations involved. As well as giving some examples, they conclude with a brief discussion on extensions of their methodology to the continuous time case. This leads us naturally on to the detailed analysis of such continuous record data.

14.19 In order to provide a firm foundation for the statistical analysis of continuous functions, we need to extend to such situations the stochastic modelling ideas familiar in the analysis of finite-dimensional random vectors. The first step in this direction was taken by Ramsay (1982), while a more systematic edifice was constructed by Besse and Ramsay (1986), Besse (1988) and Ramsay and Dalzell (1991). Facility with these techniques requires some familiarity with the details of functional analysis, for which the interested reader must consult an appropriate text (e.g. Aubin, 1979); here we can do no more than give a brief sketch of the resulting methods of analysis.

The basic assumption is that the sampled function $f(t)$ lies within the vector space $H^m(T)$ of functions defined on an interval $T = [a, b]$, possessing $m-1$ absolutely continuous derivatives, and for which the square of the mth derivative has a finite Lebesgue integral over T. Any function in this space can be decomposed into 'model' and 'residual' components

$$f(t) = u(t) + e(t), \tag{14.25}$$

where u satisfies the homogeneous linear differential equation

$$Lu = \sum_{j=0}^{m} a_j D^j u = 0 \tag{14.26}$$

in which $D^j \equiv \frac{d^j}{dt^j}$, $D^0 u = u$, and the coefficients a_j can be either continuous functions or constants. The class of functions $u(t)$ constitutes an m–dimensional subspace H_1 of H^m, and this class is specified by the differential operator L. For example, if $L = D^m$, then $Lt^j = 0$ for $j = 1, \ldots, m-1$ and H_1 is the space of polynomials of degree $m-1$; while if $L = I + D$, then H_1 is of dimension 1 and is spanned by the function $\exp(-t)$. Thus, by appropriate specification of L we can define the form of the model component. The residual function $e(t)$ is also within space H^m, but for decomposition (14.25) to be unique the subspace H_2 in which it lies must not contain elements of H_1. Besse and Ramsay (1986) specify appropriate boundary constraints to ensure that this condition is met, in which case the original space has the direct sum decomposition $H^m = H_1 \oplus H_2$.

Techniques of statistical analysis presume the existence of an inner product defined on the vector space, and when this is done for H^m and its subspaces they become *Hilbert* spaces. The natural extension to pairs of functions f, g in H^m of the familiar inner product between finite-dimensional vectors is

$$< f, g > \; = \int f(t) g(t) dt, \tag{14.27}$$

with corresponding definitions for functions in H_1 and H_2.

14.20 With this framework in place, it is now possible to extend many of the statistical techniques available in ordinary multivariate analysis to functional data. Ramsay and Dalzell (1991) give a full survey. As an example we will provide a sketch of one such technique, functional principal component analysis, following Besse and Ramsay (1986).

Consider n functions $x_i(t)$, and assume for simplicity that they have been mean-centred, i.e. that $\sum_i x_i(t) = 0$ for all $t \in T$. Then the *covariance function* is given by

$$v(s, t) = \frac{1}{n} \sum_{i=1}^{n} x_i(s) x_i(t), \quad s, t \in T, \tag{14.28}$$

and corresponding to this function is the *covariance operator* V defined by

$$Vw(s) = \int v(s, t) w(t) dt \; = \; < v(s, \cdot), w > \tag{14.29}$$

for any function $w(\cdot)$. Then, just as in conventional multivariate analysis principal components are defined as the solutions of the eigenproblem $\boldsymbol{V\xi} = \lambda \boldsymbol{\xi}$ where $\boldsymbol{V}$ is a covariance matrix and $\boldsymbol{\xi}$ is an eigenvector, so in functional data analysis the principal components are defined to be solutions of

$$V\xi(s) = \lambda \xi(s), \tag{14.30}$$

where V is now the covariance operator defined by (14.29) while $\xi(s)$ is an eigen*function*. Such an analysis may be conducted in whichever of the spaces H^m, H_1, H_2 is deemed appropriate for the problem in hand.

In practice, of course, we will only have each function value at a finite number $t_1, \ldots, t_p$ of sampled values. Thus, to apply the above ideas we need some representative function h in H^m whose values at $t_1, \ldots, t_p$ are 'close' to these sampled values. Besse and Ramsay (1986) discuss various considerations necessary to obtain a reasonable function, and show that they lead to an *interpolating spline* based on the *reproducing kernel* associated with the spaces H^m, H_1 and H_2. Essentially, a reproducing kernel for a Hilbert space of functions defined on an interval T is a bivariate function $k(\cdot, \cdot)$ defined on $T \times T$, which plays the same role as $\boldsymbol{M}^{-1}$ in a finite-dimensional vector space with metric $\boldsymbol{M}$. Besse and Ramsay (1986) summarise the more important properties of reproducing kernels and list some reproducing kernels for a variety of Hilbert spaces. They go on to show how to obtain interpolating splines from reproducing kernels, and establish a number of connections between the principal component analysis of interpolated functions and the classical PCA of sampled values in a particular metric. We refer the interested reader to their account for full technical details. Besse (1988) explores these ideas further to obtain an optimal metric for graphical displays.

Example 14.1

Besse and Ramsay (1986) illustrate their methods on a set of data consisting of the records of tongue dorsum height obtained by ultrasound sensing during the utterance of the sound 'kah' by 42 subjects. Cubic spline smoothing of each record produced the 42 curves shown in Fig. 14.5, where the interval of observation has been arbitrarily normalised to $[0, \pi]$. Inspection of the data and consideration of the underlying physiology suggested the model

$$f_i(t) = c_{i1} + c_{i2}\sin(t) + c_{i3}\cos(t) + e_i(t) \tag{14.31}$$

for each individual curve $f_i(t)$, $i = 1, \ldots, 42$. In terms of (14.25), the model component $u_i(t)$ is thus given by $c_{i1} + c_{i2}\sin(t) + c_{i3}\cos(t)$, and $e_i(t)$ is the residual component, for each curve. The coefficients c_{ij} are easily estimated by ordinary least squares regression, but they were of little direct interest in the analysis. The main question was to investigate to what extent there was interesting variation in tongue dorsum behaviour after removal of this overall trend. As classical principal component analysis of the residuals shed little light on the question, functional analysis methods were then considered.

In terms of the general theory above, the observations lie in the interval $T = [0, \pi]$ and the model components $u_i(t)$ are linear combinations of 1, sin and cos, which satisfy the differential equation $Lu = Du + D^3u = 0$. Thus, the overall space is H^3 and H_1 is the three-dimensional subspace spanned by these components. Besse and Ramsay went on to discuss various sets of constraints for partitioning the overall space of functions; each set of constraints led to a different principal component analysis, and these analyses were also described and compared. The most informative

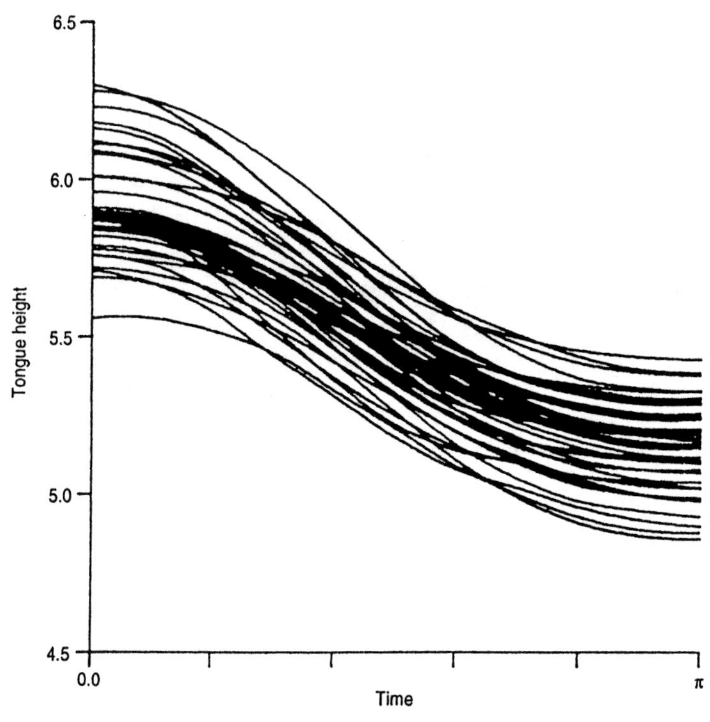

Figure 14.5 The movement of the back of the tongue in arbitrary units during 42 utterances of the sound 'kah'. Reproduced, with permission, from Besse and Ramsay (1986).

one was the analysis in the residual subspace H_2, which yielded two dominant eigenvalues accounting for 98.5% of the residual variation. The two corresponding eigenfunctions are shown in Fig. 14.6. The first eigenfunction accounts for a departure from simple harmonic motion, which is a simultaneous deviation in the same direction at the two end points, while the second harmonic describes the extent to which the tongue is too low initially and too high finally, or vice versa. These insights led to better description of the physiological process than had been possible with a classical principal component analysis of the data.

14.21 The above ideas have also been applied by Ramsay and Dalzell (1991) to functional regression analysis, while Kiiveri (1992) has extended them to the canonical variate situation of between- and within-group variation. Moreover, Kiiveri has explicitly allowed for the case of a large number of sampling points on each curve, and hence of singular sample matrices, by replacing observed matrices by fitted ones that are positive definite. Finally, Ramsay (1995) retreads the same general ground, but in such a way as to avoid the necessity of invoking any of the specialised functional analysis implicit in the above development.

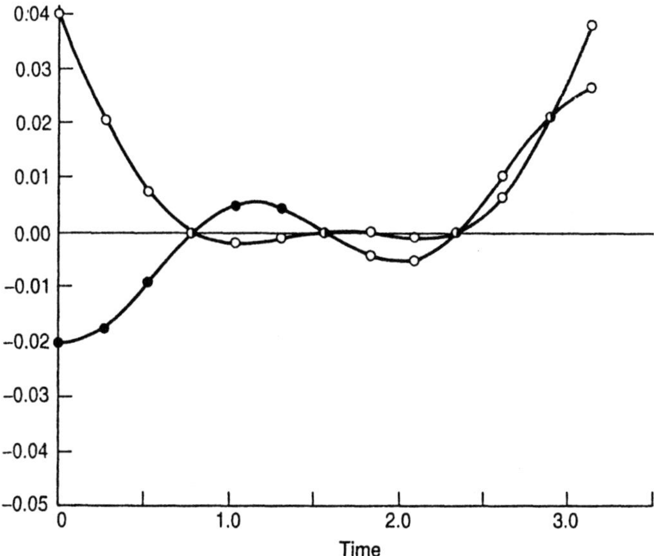

Figure 14.6 The first two components of variation in the residual subspace of the tongue data. Reproduced, with permission, from Besse and Ramsay (1986).

Analysis of shape

14.22 Size and shape play a central role in many branches of science, and have long been the focus of study. For example, Galileo was interested in questions of geometrical similarity, and considered the consequences of the decreasing ratio of surface area to volume in solids if geometrical similarity is maintained with size increases. His work had impact on forestry and biology (Sprent, 1972). Classical texts expounding on relationships between the growth and form of organisms are those of Thompson (1917) and Bower (1930). These books are concerned predominantly with applications in zoology and botany respectively, but more recent studies have involved such diverse areas as palaeontology, archaeology and astronomy, to name but three.

Given a set of p variables observed on each of a number of n organisms, one common problem is that of isolating measures of 'size' from those of 'shape' of the organisms. This general area is known as *morphometrics*; a comprehensive text is that by Reyment *et al* (1984). A common approach to the problem is via principal component analysis, typically of the logarithms of the variables (Part 1, **4.3**; Sprent, 1972). There has been considerable discussion concerning which variant of principal component analysis best extracts a 'size' component (see, for example, Jolicoeur, 1963; Mosimann, 1970; Mosimann and James, 1979; Bookstein *et al*, 1985; Somers, 1986, 1989; Rohlf and Bookstein, 1987), but in all cases the relevant component emerges as a linear combination of the original (transformed) variables. Sometimes, there may be more than one plausible component measuring 'size'; factor analysis (Chapter 12) is another technique that can be used for extracting such components (Hopkins, 1966). Procedures exist for testing hypotheses about, or constructing confidence intervals for, the coefficients in these components (Morrison, 1976, p250; Jolicoeur, 1984).

In general, however, 'size' is a nuisance factor in many studies and needs to be removed, the interest in the analysis featuring on what is then left over. If the 'size' factor(s) can be specified as linear combination(s) of the variables, then one way of achieving this removal is to compute the residual covariance matrix after subtracting the terms involving these components. For example, if there is a single size component $y = \boldsymbol{c}'\boldsymbol{x}$ where $\boldsymbol{c}'\boldsymbol{c} = 1$ and $\text{var}(y) = \lambda$, and if $\boldsymbol{S}$ is the covariance matrix of $\boldsymbol{x}$, then the residual covariance matrix is given by $\boldsymbol{S} - \lambda\boldsymbol{c}\boldsymbol{c}'$. A principal component analysis of this residual matrix ('size-constrained principal component analysis') will then extract the main 'shape' features of the data (Somers, 1986). The problem here, however, is that the resulting 'shape components' need not be orthogonal to the 'size' components, a distinct drawback when attempting any interpretations of the results. A better procedure is first to project the data on to the subspace orthogonal to the 'size' subspace, and then to analyse the scores in this subspace (Somers, 1989). For the example above, this procedure requires extraction of the principal components of $\boldsymbol{QSQ}$, where $\boldsymbol{Q} = \boldsymbol{I} - \boldsymbol{c}(\boldsymbol{c}'\boldsymbol{c})^{-1}\boldsymbol{c}' = \boldsymbol{I} - \boldsymbol{c}\boldsymbol{c}'$. This is the approach adopted by Burnaby (1966) and Gower (1976) when seeking canonical variates and discriminant functions that are free of 'size' effects (Exercise 14.2).

14.23 More recently, however, interest has focused on the analysis of shape and, much more directly, on the geometrical features of the measured entities. On any object whose shape is to be analysed there typically exists a small set of points of particular importance, which collectively help to determine the essential geometry of the object. These points are known as *landmarks*, and modern methods of shape analysis have been concerned with the analysis of landmarks. There have been several independent lines of development leading to the present body of knowledge in the subject, and several distinct approaches (albeit with considerable overlap in the foundations) can be identified.

Bookstein (1978) commenced one line of development, essentially drawing on the ideas of Thompson (1917), who was concerned with the study of different entities having a common form (e.g. different types of apes, or different types of a particular plant form). Since the form is common to all entities, they have the same landmarks measured on them. Thompson sought to quantify the difference between two entities of a single form by the amount of deformation in the coordinate axes that was needed to transform their two sets of landmarks to coincidence. This approach thus views the objects to be analysed as two-dimensional entities, so that sets of k landmarks observed upon a single form can be denoted by lists of k complex numbers $z_1, \ldots, z_k$ where each z_i specifies the coordinates of one landmark. The size of the object can be quantified by the distances between pairs of landmarks, and between arbitrarily weighted centroids of sets of landmarks; this set of distances generates one linear vector space. The shape of a single triangle ABC of landmarks may then be reduced to a single pair of shape coordinates that specify the position of the vertex C when landmarks A and B have been normalised to lie at $(0, 0)$ and $(1, 0)$ respectively; the span of all such shape coordinates provides a second linear space. Bookstein (1986) surveys the theory he developed from this basis. A suitable 'null' model is that of independent identical circular normal perturbations at each landmark, and this model leads to the identification of one particular size variable as having zero covariance with every shape variable, as well as providing an F test for associations between size and shape, and Hotelling's T^2 tests for existence of mean shape differences between groups of objects or for mean changes of shape. Bookstein and Sampson (1990) extend these ideas to allow a hierarchy of geometric components of shape difference or shape change, using a bivariate polynomial in the shape coordinates to model each component. A synthesis of this methodology is provided by Bookstein (1991).

14.24 An alternative approach to the analysis of shape was initiated independently by Kendall (1977), prompted by the observation made by some archaeologists that 'too many' of the 22 100 triples among the 52 standing stones near Land's End in Cornwall are 'too nearly' collinear for this feature to be a chance event and that it must have been deliberately planned. Kendall thus set out to characterise the mathematical description of the shape of n points in p dimensions (the 'landmarks' of Bookstein's theory). He defined 'shape' to mean 'what is left when the effects of translation, scaling and rotation have been removed'. If the $n \times p$ matrix $\boldsymbol{X}$ specifies the points, then we see that removal of these effects is closely related to the methods of Procrustes analysis (Part 1, Chapter 5).

First, translation is dealt with by mean centring the columns of $\boldsymbol{X}$ (i.e. moving the origin of axes to the centroid of the points). This can be done by multiplying $\boldsymbol{X}$ on the left by an $(n-1) \times n$ Helmert matrix $\boldsymbol{L}$ which has orthonormal rows, each orthogonal to $(1/\sqrt{p}, \ldots, 1/\sqrt{p})$. Let $\boldsymbol{Y} = \boldsymbol{LX}$. The simplest way of handling the scaling is then to divide each element of $\boldsymbol{Y}$ by the square root of the sum of squares of all its elements (i.e. standardising configurations so that $\text{tr}\boldsymbol{Y}'\boldsymbol{Y}$ is always unity). Hence the next step is to form $\boldsymbol{Z} = \boldsymbol{Y}/r$, where $r^2 = \text{tr}\boldsymbol{Y}'\boldsymbol{Y}$. Finally, the rotation can be taken care of by subjecting $\boldsymbol{Z}$ to the QR decomposition $\boldsymbol{Z} = \boldsymbol{T\Gamma}$ (Golub and van Loan, 1983). However, a complication here is that there are two alternative decompositions, one permitting reflections among the computational operations and the other restricting them to rotations only. In the former case, $\boldsymbol{\Gamma} \in O(p)$, the orthogonal group with $|\boldsymbol{\Gamma}| = \pm 1$ and $t_{pp} \geqslant 0$, while in the latter case $\boldsymbol{\Gamma} \in SO(p)$, where $|\boldsymbol{\Gamma}| = 1$ and t_{pp} is unrestricted. To distinguish the two cases it is customary to denote the former decomposition with superscript R: $\boldsymbol{Z} = \boldsymbol{T}^R\boldsymbol{\Gamma}^R$.

The matrices $\boldsymbol{X}$, $\boldsymbol{Y}$, $\boldsymbol{T}$ contain, respectively, the original landmark coordinates, the so-called *preshape* coordinates and the *shape* coordinates of the configuration. Furthermore, $r\boldsymbol{T}$ contains the *size-and-shape* coordinates, and we can add the prefix *reflection* if we wish to distinguish explicitly the case $\boldsymbol{T}^R$ from $\boldsymbol{T}$. Two figures $\boldsymbol{X}_1$, $\boldsymbol{X}_2$ thus have the same shape if they share the same matrix of shape coordinates, and the preceding steps show that this will be the case if they are related by the similarity transformation

$$\boldsymbol{X}_2 = r\boldsymbol{X}_1\boldsymbol{\Gamma} + 1_n\gamma' \tag{14.32}$$

where $\boldsymbol{\Gamma}$ is a $p \times p$ matrix with $|\boldsymbol{\Gamma}| = 1$, $\boldsymbol{\gamma}$ is a p-element vector, and $r > 0$ is a scalar. The triple $(\boldsymbol{\gamma}, r, \boldsymbol{\Gamma})$ specifies the translation, scale and rotation components of the similarity transformation from $\boldsymbol{X}_1$ to $\boldsymbol{X}_2$.

14.25 The foundations of any statistical theory of shape require a specification and study of the spaces of all such matrices as those defined above. This has been done in a series of papers by Kendall, of which Kendall (1984) and Kendall (1989) are the most central. A complete description of the geometrical structure of shape spaces is given by Le and Kendall (1993). This involves a number of modern concepts in Riemannian geometry as well as those of complex projective spaces, so much of the development is expressed in terms unfamiliar to applied statisticians. A full account would thus require the filling in of much background technical detail, which is beyond the scope of this particular volume. Hence, we do not attempt to summarise any results, but leave the interested reader to consult the cited references. Other relevant papers are those by Small (1988), who discusses connections between the Kendall and Bookstein geometries, and Ambartzumian (1982) and Carne (1990), who develop geometries for weaker concepts of shape such as affine and combinatorial shape respectively.

14.26 Given all the above basis, statistical methods for analysing collections of landmark matrices have been developed by Goodall (1991) and Goodall and Mardia (1993). Suppose that $\boldsymbol{X}_1, \ldots, \boldsymbol{X}_m$ contain the coordinates of n landmarks in p dimensions for each of m figures. These matrices may constitute a sample from a single population, with inference about the population mean shape the focus of interest, or they may represent samples from two or more populations where the interest lies in comparing the respective population mean shapes. For simplicity, the general ideas are presented in terms of the single sample case; extensions to the multi-sample case follow in an obvious way.

Any statistical analysis first requires specification of a model. The simplest model for landmark data is to assume that each figure $\boldsymbol{X}_i$ is the similarity transformation $(\boldsymbol{\gamma}_i, r_i, \boldsymbol{\Gamma}_i)$ of a zero-mean normal displacement of the $(n \times p)$ population mean $\boldsymbol{\mu}$, so that

$$\boldsymbol{X}_i = r_i(\boldsymbol{\mu} + \boldsymbol{E}_i)\boldsymbol{\Gamma}_i + 1_n\boldsymbol{\gamma}_i' \tag{14.33}$$

where $\text{vec}\boldsymbol{E}_i \sim N(\boldsymbol{0}, \boldsymbol{\Sigma})$. In the multi-sample case we parallel traditional multivariate methods by permitting different means $\boldsymbol{\mu}_j$ in each population, but requiring a common dispersion matrix $\boldsymbol{\Sigma}$ in all populations. We assume that the $(n \times p)\boldsymbol{E}_i$ are independent, and a common simplifying assumption is that $\boldsymbol{\Sigma} = \sigma^2\boldsymbol{I}_{np}$. In this case the shape coordinates of the m figures have independent isotropic matrix multivariate distributions with mean $\boldsymbol{\mu}$. A more general model sets $\boldsymbol{\Sigma} = \boldsymbol{\Omega} \otimes \boldsymbol{I}_p$, where $\boldsymbol{\Omega}$ is an $(n \times n)$ positive-definite variance–covariance matrix and $\otimes$ denotes the Kronecker product. With this model the errors at each landmark are isotropic but the errors between landmarks are correlated.

Statistical inference for shape is complicated by the nuisance parameters $\boldsymbol{\gamma}_i, r_i, \boldsymbol{\Gamma}_i$ $(i = 1, \ldots, m)$ for translation, scale and orientation. Goodall (1991) adopts a maximum likelihood approach using Procrustes methods, in which all parameters are estimated explicitly. He gives one-sample tests of location, two-sample tests of mean differences, and the k-sample one-way analysis of variance for a variety of covariance structures, and illustrates these methods on landmark data from radiographs of rats.

Goodall and Mardia (1993), on the other hand, integrate out the nuisance parameters and use marginal inference on the data. This approach requires the use of distributional results for shape coordinates arising from the normal model outlined above. Shape densities for planar figures comprising three or more landmarks had previously been given by Small (1981) and Kendall (1984) (for $\boldsymbol{\mu} = \mathbf{0}$) and by Mardia and Dryden (1989) (for the same situation but $\boldsymbol{\mu}$ unrestricted). Goodall and Mardia (1993) extend these results to higher dimensionalities and give some non-central results with Gaussian approximations that have implications for inference. Their results enable shape analysis to be carried out within the framework of familiar multivariate methodology. Kent (1994) discusses the complex Bingham distribution as a basis for the shape analysis of landmark data, and considers the use of principal component analysis on both Kendall and Bookstein coordinates. With all these methods, however, the technical details are such that it is not possible to give a brief summary of the main results without going into a great deal of involved mathematics. We go no further at this stage, therefore, but leave the interested reader to follow up the given references.

A related area is the analysis of directional and angular data, and a brief overview of relevant methodology provides the last of our miscellaneous topics in this chapter.

Angular distributions

14.27 Distributions of angles in two dimensions have many applications, including the directions of movement of animals, orientation of fibres in paper (lines, rather than vectors), and periodic data with angles representing a yearly or daily cycle. Batschelet (1981) is an excellent non-technical introduction to the subject, with many biological applications. Fisher (1993) is a comprehensive account of the analysis of circular data, with examples from many different fields.

Directional data in three dimensions are usually represented as a bivariate distribution of polar angles. Applications include directions of stars and the orientation of roots in soil, but the most important are concerned with geomagnetism, the direction of the magnetic field in rocks. A comprehensive account is given in Fisher *et al* (1987). Mardia (1972) discusses the distributional theory and the analysis of both types of data, and Mardia *et al* (1979), Chapter 15, give a shorter account of the subject.

It is not difficult to define angular distributions on hyperspheres of dimension higher than 3, but these have no obvious physical applications. Multivariate distributions in which the variables are angles may occur, but they are not usually suitable for representation as directions in multidimensional space. Recently, there has been some work on the relationship between angular distributions and shape analysis, in the sense of Kendall (1984) and Bookstein (1986); see Kent (1994).

Circular data

14.28 The univariate angular uniform distribution is

$$f(\theta) = \frac{1}{2\pi}, \quad 0 \leqslant \theta < 2\pi \tag{14.34}$$

It may also be useful to define uniform distributions on $(0, \pi)$ for undirected lines, or on $(0, \pi/2)$ for angles between lines.

Often, it is interesting to test whether a data set is consistent with a uniform angular distribution. The ordinary Pearson χ^2 test may be used, but tests based on the empirical distribution function are often preferred. The Kolmogorov–Smirnov test is unsuitable for angular data, because the value of the test statistic is dependent on the choice of the origin $\theta = 0$. A variation is Kuiper's test. Define x_j so that it is uniform on (0, 1), i.e. $x_j = \theta_j/2\pi$ for θ on $(0, 2\pi)$, and calculate

$$D_n^+ = \max\left(\frac{j}{n} - x_j\right), \quad j = 1, \ldots, n$$

$$D_n^- = \max\left(x_j - \frac{j-1}{n}\right), \quad j = 1, \ldots, n$$

$$V_n = D_n^+ + D_n^-$$

The test statistic is independent of the origin, and the test is valid for angular uniform distributions. $V_n\sqrt{n}$ is approximately $N(0, 1)$ for large samples from a uniform distribution. An alternative is Watson's test, for which the test statistic is given by

$$U^2 = \sum_{j=1}^{n} [x_j - (2j-1)/(2n)]^2 - n\left(\bar{x} - \frac{1}{2}\right) + 1/(12n).$$

The statistic U^2 is approximately $N(0, 1)$ for large samples from a uniform distribution.

For details and tables, see Pearson and Hartley (1972). Another possibility is to fit the von Mises distribution (see below) and test whether the concentration parameter $\kappa = 0$. This is known as Rayleigh's test (Rayleigh, 1919).

The most important distribution for directional data in the plane is the von Mises distribution (von Mises, 1918). It is given by

$$f(\theta) = \frac{1}{2\pi I_0(\kappa)} e^{\kappa\cos(\theta-\alpha)}, \quad (0 \leqslant \theta < 2\pi, 0 \leqslant \kappa < \infty) \tag{14.35}$$

The distribution has a single mode at $\theta = \alpha$. The constraint $\kappa \geqslant 0$ is necessary only to give a unique parametrisation; changing the sign of κ is equivalent to replacing α by $2\pi - \alpha$. The *concentration parameter* κ determines how closely the distribution clusters about the mode; $\kappa = 0$ gives the uniform distribution, while $\kappa \to \infty$ gives a single point distribution at $\theta = \alpha$.

In this equation, $I_0(\kappa)$ is an imaginary Bessel function of the first kind, defined as

$$I_0(\kappa) = \sum_{n=0}^{\infty} \frac{(\kappa^2/4)^n}{(n!)^2}$$

For large values of κ the distribution is nearly normal, with variance $1/\kappa$. This is the standard distribution for clustering about a fixed direction.

Fitting a von Mises distribution to a set of data involves the estimation of κ and, unless it is known, the angle α giving the preferred direction. If the direction is known, the projection of the data onto it is sufficient for κ. If it is not, the resultant vector, the sum of the data vectors, is sufficient for κ and α—and this property characterises the von Mises distribution.

The maximum likelihood estimate $\hat{\kappa}$ of κ is obtained by solving either

$$\frac{I_0'(\hat{\kappa})}{I_0(\hat{\kappa})} = \frac{|\boldsymbol{R}|}{n}, \quad \text{or} \quad \frac{I_0'(\hat{\kappa})}{I_0(\hat{\kappa})} = \frac{\sum_{i=1}^{n} \cos\theta_i}{n}$$

depending on whether α is unknown or known, where

$$\boldsymbol{R} = \left(\sum_{i=1}^{n} \cos\theta_i, \sum_{i=1}^{n} \sin\theta_i\right)$$

In the former case, the maximum likelihood estimate of α is given by the direction of $\boldsymbol{R}$. Tables in Pearson and Hartley (1972) facilitate the calculations. The derivative $I_0'(\hat{\kappa})$ may also be written $I_1(\hat{\kappa})$, another Bessel function. The estimate of κ is necessarily positive, and when κ and n are small it may be seriously biased. Modified estimates have been suggested; see Fisher (1993), p88, and references cited there.

Modified forms of the distribution are possible. In particular, replacing $\cos(\theta - \alpha)$ by $\cos 2(\theta - \alpha)$ gives a symmetrical bimodal distribution. This may be used for modelling bimodal data sets, such as, for example, annual data with modes in spring and autumn. A more important application is to angles defined by undirected lines, when θ is in $(0, \pi)$ and the normalising constant is doubled.

Distributions on the sphere

14.29 Spherical polar coordinates are usually defined by

$$\left.\begin{aligned} x &= r\sin\theta\cos\psi \\ y &= r\sin\theta\sin\psi \\ z &= r\cos\theta, \end{aligned}\right\} \qquad (14.36)$$

but other systems are possible; see Mardia (1972).

It is also often convenient to express directional data in terms of a random unit vector $\boldsymbol{l}$, where $\boldsymbol{l}'\boldsymbol{l} = 1$. In three dimensions, the components of the unit vector are the *direction cosines*, the cosines of the angles between the vector and the coordinate axes,

$$(l_1, l_2, l_3) = (\sin\theta\cos\psi, \sin\theta\sin\psi, \cos\theta)$$

so that a density on the sphere may take the form

$$f(\boldsymbol{l})dS = f(\sin\theta\cos\psi, \sin\theta\sin\psi, \cos\theta)dS$$

where dS is the surface element of the unit sphere, equal to $\sin\theta d\theta d\psi$.

The uniform distribution on the sphere is given by

$$f(\theta, \psi) = \frac{1}{4\pi}\sin\theta, \quad (0 \leqslant \theta < \pi), (0 \leqslant \psi < 2\pi). \tag{14.37}$$

This gives a uniform distribution over the surface of the sphere; the angle ψ, but not the angle θ, is uniformly distributed.

The spherical distribution analogous to the von Mises distribution was introduced by R.A.Fisher in the study of geomagnetism (Fisher, 1953). It is given by

$$f(\theta, \psi) = \frac{\kappa}{4\pi \sinh \kappa} e^{\kappa \cos\phi} \sin\theta, \quad (0 \leqslant \theta < \pi, 0 \leqslant \phi, \psi < 2\pi, 0 \leqslant \kappa < \infty). \tag{14.38}$$

Here, ϕ is the angle with some fixed, preferred direction, say (θ_0, ψ_0). Now, considering the spherical triangle defined by the z-axis, the vector (θ, ψ), and the preferred direction (θ_0, ψ_0), a standard formula of spherical trigonometry gives

$$\cos\phi = \cos\theta\cos\theta_0 + \sin\theta\sin\theta_0\cos(\psi - \psi_0).$$

As with the von Mises distribution, the resultant vector is sufficient for θ_0, ψ_0, κ.

The modified distribution

$$f(\theta, \psi) = C(\kappa)e^{\kappa \cos 2\phi} \tag{14.39}$$

$$(0 \leqslant \theta < \pi), (0 \leqslant \phi, \psi < 2\pi), (-\infty < \kappa < \infty)$$

$$C(\kappa) = \left\{2\pi e^{\kappa} \int_0^1 e^{-2\kappa z^2} dz\right\}^{-1}$$

is generally known as the Dimroth–Watson distribution (Dimroth, 1962; Watson, 1965). The distribution with $\kappa > 0$ is bimodal, with directions clustering symmetrically about an axis; like the modified von Mises distribution, it is often used to represent clustered undirected lines. When $\kappa < 0$, the directions avoid the axis corresponding to ϕ, and cluster round a great circle on the sphere, giving a *girdle distribution.* Baldwin *et al* (1971) use this distribution to represent the behaviour of roots in soil. Species with roots growing nearly vertically, such as onions, have positive values of κ, while species such as couch grass that form a nearly horizontal mat of roots have negative κ.

The Bingham distribution (Bingham 1964, 1974) is a slightly more general distribution with axial symmetry on the sphere. It involves clustering about (or avoiding) *two* orthogonal axes. In the bimodal form, the distribution is clustered about two poles, but the contours of equal density are not small circles but ovals. In the girdle form of the distribution, the density is concentrated around a great circle, but has a symmetrical bimodal distribution on the circle and the neighbouring small circles.

The distribution may be written in the form

$$f(\boldsymbol{l}) = c(\boldsymbol{\kappa})\exp\{\mathrm{tr}\,(\boldsymbol{\kappa}\boldsymbol{A}'\boldsymbol{l}\boldsymbol{l}'\boldsymbol{A})\} \tag{14.40}$$

or, taking new axes (a_1, a_2, a_3)

$$f(\theta, \psi) = 4\pi c(\boldsymbol{\kappa})^{-1}[\exp\{(\kappa_1 \cos^2 \psi + \kappa_2 \sin^2 \psi) \sin^2 \theta + \kappa_3 \cos^2 \theta\}] \sin \theta$$

where $\boldsymbol{\kappa}$ is a 3×3 diagonal matrix, $\boldsymbol{l}$ is a random unit vector, $\boldsymbol{A}$ is an orthogonal 3×3 matrix, defining three orthogonal axes, and $c(\boldsymbol{\kappa})$ is a normalising constant dependent only on $\boldsymbol{\kappa}$. In fact, the constraint that $\boldsymbol{l}$ is a unit vector implies that the sum of the values of $\boldsymbol{\kappa}$ is arbitrary, and usually κ_3 is set to zero.

The constant $c(\boldsymbol{\kappa})$ can be expressed as a confluent hypergeometric function of matrix argument

$$c(\boldsymbol{\kappa}) = {}_1F_1\left(\frac{1}{2}; \frac{3}{2}; \boldsymbol{\kappa}\right)$$

(Herz, 1955), or in series form

$$c(\boldsymbol{\kappa}) = (2\pi)^{-1} \sum_{i,j,k=0}^{\infty} \frac{\Gamma(i+\frac{1}{2})\Gamma(j+\frac{1}{2})\Gamma(k+\frac{1}{2})}{\Gamma(i+j+k+\frac{3}{2})} \frac{\kappa_1^i \kappa_2^j \kappa_3^k}{i!j!k!}$$

When $\kappa_1 = \kappa_2 = \kappa_3 = 0$, the distribution is uniform. The case $\kappa_2 = \kappa_3 = 0$ is the Dimroth–Watson distribution, bimodal or girdle according to the sign of κ_1. If κ_1 is much larger than κ_2 in absolute value, the distribution is bimodal ($\kappa_1 > 0$) with oval clusters about the ends of the axis, or girdle ($\kappa_1 < 0$) with clustering about a diameter of the great circle.

See Mardia (1972) for a full account of the distribution.

Inference for angular distributions

14.30 Inferential procedures for angular distributions are fully discussed in Mardia (1972), Pearson and Hartley (1972), Fisher *et al* (1987) and Fisher (1993). Only a brief account is given here.

The von Mises and Fisher distributions have sufficient statistics, and for large samples maximum likelihood theory shows that the estimates are approximately normally distributed with known covariance matrix. The estimates of the concentration parameter are biased, and for small samples the asymptotic theory may be misleading. Stephens (1962, 1967, 1969a,b, 1972) discusses the problems, and gives tables; see also Pearson and Hartley (1972). The tables make it possible to calculate the following.

(1) Confidence intervals for the concentration parameter κ of the von Mises distribution.
(2) Confidence intervals for the direction of the modal vector in the von Mises distribution.
(3) Two-sample tests comparing the concentration parameters and modal vectors from samples drawn from two von Mises distributions.
(4) Confidence intervals for the concentration parameter κ of the Fisher distribution.
(5) Confidence intervals for the direction of the modal vector in the Fisher distribution.

(6) Two-sample tests comparing the concentration parameters and modal vectors from samples drawn from two Fisher distributions.

These confidence intervals include, of course, tests of uniformity against the alternative of a unimodal distribution (Rayleigh's tests).

The use of Watson's U^2 statistic for tests of uniformity on the circle has been extended by Lockhart and Stephens (1985) to the problem of testing the goodness-of-fit of the von Mises distribution. As parameters are fitted, special tables are needed for the critical points of the distribution.

Watson (1956) and Watson and Williams (1956) introduced an analysis of variance technique for comparing groups of measurements consisting of unit vectors. This is based on the assumption that the groups follow von Mises–Fisher type distributions, in general in p dimensions, with a common concentration parameter κ but possibly different modal directions. Now suppose s samples are based on numbers $N_1, \ldots, N_s$, with $N = \sum_{i=1}^{s} N_s$. Suppose further that the resultant vectors for the groups have lengths $R_1, \ldots, R_s$, and the resultant vector for the n observations has length R. Then the 'analysis of variance' takes the form

$$2\kappa(N - R) = 2\kappa(N_1 - R_1) + 2\kappa(N_2 - R_2) + \cdots + 2\kappa\left(\sum_{i=1}^{s} R_i - R\right)$$

Now, on the assumption of a common distribution of von Mises type, each of these terms has asymptotically a χ^2 distribution, with degrees of freedom given by one fewer than the number of terms involved, multiplied by $p - 1$. Thus, the asymptotic distributions are

$$\chi^2_{(N-1)r} = \chi^2_{(N_1-1)r} + \chi^2_{(N_2-1)r} \cdots + \chi^2_{(s-1)r}$$

where $r = p - 1$.

This leads immediately to the test

$$\frac{(N - s)(\sum_{i=1}^{s} R_i - R)}{(s - 1)(N - \sum_{i=1}^{s} R_i)} \sim F_{(s-1)r,(N-s)r}$$

A detailed account of the procedure, with justification, is given by Stephens (1992). Stephens (1982) discusses possible applications to compositional data.

Example 14.2 (Stephens, 1992)
The data in Table 14.1 represent coal-cleat attitudes—vectors in three dimensions—and are adapted by Stephens from data from Jeran and Mashey (1970) to illustrate the extension of Watson's analysis of variance to a two-way (nested) classification. The resultant vector of the 124 observations has length $R_{..} = 84.15$. It is then easy to follow Watson's analysis to obtain, first, a test for the difference between rows, and then tests for variation among columns within each row. The results are shown in Table 14.2.

The bimodal and girdle distributions are less easy to handle. In the planar case, Stephens (1972) suggests representing a bimodal distribution by a

Table 14.1 Resultant lengths and sample sizes

		Columns				
		1	2	3	4	Total
Row 1	N	19	19	16	15	69
	R	18.93	18.94	15.93	14.95	$R_{1.} = 68.37$
Row 2	N	21	20	7	7	55
	R	20.89	19.92	6.97	6.97	$R_{2.} = 54.22$

Table 14.2 Two-way analysis of variance table: sums of squares, degrees of freedom and F statistic values

Between rows	38.45	2	$F_{2,232} = 8907$
Columns in row 1	0.39	6	$F_{6,232} = 30.1$
Columns in row 2	0.52	6	$F_{6,232} = 40.06$
Within groups	0.50	232	

mixture of two von Mises distributions, with equal probabilities and concentrations, and opposite modal vectors. He shows how the calculations 1–3 above could be adapted for the distribution mixture.

Estimation for the Dimroth–Watson distribution depends on the *orientation matrix*

$$T = \begin{pmatrix} \sum x_i^2 & \sum x_i y_i & \sum x_i z_i \\ \sum x_i y_i & \sum y_i^2 & \sum x_i y_i \\ \sum x_i z_i & \sum y_i z_i & \sum z_i^2 \end{pmatrix}$$

The three eigenvectors of this matrix define three directions. If the data follow a Dimroth–Watson distribution of the bimodal type, it is likely that there is one large eigenvalue and two smaller, approximately equal, eigenvalues. The eigenvector corresponding to the largest eigenvalue is then the axis of the bimodal distribution. In the case of a girdle distribution, there is typically one small eigenvalue and two, approximately equal, larger eigenvalues. The eigenvector corresponding to the smallest eigenvalue is then the diameter normal to the preferred great circle. The maximum likelihood estimate of the axis is the eigenvector corresponding to the largest eigenvalue in the bimodal case, and to the smallest eigenvalue in the case of the girdle distribution.

Exercises

14.1 Suppose that π_1 and π_2 represent two p-variate normal populations with means $\boldsymbol{\mu}_1$, $\boldsymbol{\mu}_2$ and dispersion matrices $\boldsymbol{\Sigma}_1$, $\boldsymbol{\Sigma}_2$ respectively. Show that the distance Δ_6 between these populations is given by

$$\Delta_6 = \frac{1}{2}\text{tr}[(\boldsymbol{\Sigma}_1 - \boldsymbol{\Sigma}_2)(\boldsymbol{\Sigma}_1^{-1} - \boldsymbol{\Sigma}_2^{-1})] + \frac{1}{2}\text{tr}[(\boldsymbol{\Sigma}_1^{-1} + \boldsymbol{\Sigma}_2^{-1})(\boldsymbol{\mu}_1 - \boldsymbol{\mu}_2)(\boldsymbol{\mu}_1 - \boldsymbol{\mu}_2)']$$

(Kullback, 1959, p190), and that the affinity ρ between them is given by

$$\rho = |\frac{1}{2}(\boldsymbol{\Sigma}_1^{-1} + \boldsymbol{\Sigma}_2^{-1})|^{-1/2}|\boldsymbol{\Sigma}_1\boldsymbol{\Sigma}_2|^{-1/4}\exp\{-\frac{1}{4}[\boldsymbol{\mu}_1'\boldsymbol{\Sigma}_1^{-1}\boldsymbol{\mu}_1 + \boldsymbol{\mu}_2'\boldsymbol{\Sigma}_2^{-1}\boldsymbol{\mu}_2 - (\boldsymbol{\mu}_1'\boldsymbol{\Sigma}_1^{-1} + \boldsymbol{\mu}_2'\boldsymbol{\Sigma}_2^{-1})(\boldsymbol{\Sigma}_1^{-1} + \boldsymbol{\Sigma}_2^{-1})^{-1}(\boldsymbol{\Sigma}_1^{-1}\boldsymbol{\mu}_1 + \boldsymbol{\Sigma}_2^{-1}\boldsymbol{\mu}_2)]\}$$

(Matusita, 1967). Hence, deduce that, in the special case $\boldsymbol{\Sigma}_1 = \boldsymbol{\Sigma}_2 = \boldsymbol{\Sigma}$, these expressions reduce to $\Delta_6 = (\boldsymbol{\mu}_1 - \boldsymbol{\mu}_2)'\boldsymbol{\Sigma}^{-1}(\boldsymbol{\mu}_1 - \boldsymbol{\mu}_2)$ and $\rho = \exp(-\frac{1}{8}\Delta_6)$.

If now $\boldsymbol{\Sigma}_1$ and $\boldsymbol{\Sigma}_2$ follow the common principal component model, i.e. $\boldsymbol{\Sigma}_i = \boldsymbol{L}\boldsymbol{D}_i\boldsymbol{L}'$ for $i = 1, 2$ where $\boldsymbol{L}$ is an orthogonal matrix of common eigenvectors and $\boldsymbol{D}_1, \boldsymbol{D}_2$ are diagonal matrices of eigenvalues in π_1 and π_2, show that

$$\rho = 2^{p/2}\frac{|\boldsymbol{D}_1\boldsymbol{D}_2|^{1/4}}{|\boldsymbol{D}_1 + \boldsymbol{D}_2|^{1/2}}\exp\{-\frac{1}{4}(\boldsymbol{\mu}_1 - \boldsymbol{\mu}_2)'\boldsymbol{L}(\boldsymbol{D}_1 + \boldsymbol{D}_2)^{-1}\boldsymbol{L}'(\boldsymbol{\mu}_1 - \boldsymbol{\mu}_2)\}$$

(Krzanowski, 1990).

14.2 Suppose that p variables $\boldsymbol{x}' = (x_1, \ldots, x_p)$ have been observed on each of n individuals, which are divided *a priori* into g groups. Let $\boldsymbol{B}$ and $\boldsymbol{W}$ denote the usual $(p \times p)$ between-group and within-group covariance matrices, $\boldsymbol{G}$ the $(g \times p)$ matrix containing the group means for each of the variables, and $\boldsymbol{R}$ a matrix such that $\boldsymbol{R}'\boldsymbol{R} = \boldsymbol{W}^{-1}$. The k columns of the matrix $\boldsymbol{K}$ represent growth vectors if the elements of each column give the coefficients of a linear combination of the p variables x_i that measure some facet of growth of the individuals. Show that *growth-free canonical variables*, i.e. variables $y_i = \boldsymbol{a}_i'\boldsymbol{x}$ with coefficients $\boldsymbol{a}_i$ satisfying $(\boldsymbol{B} - \lambda\boldsymbol{W})\boldsymbol{a}_i = \boldsymbol{0}$ and $\boldsymbol{a}_i'\boldsymbol{K} = \boldsymbol{0}$, can be obtained either as $(\boldsymbol{QR})'\boldsymbol{m}_i$, where $\boldsymbol{Q} = \boldsymbol{I} - (\boldsymbol{RK})[(\boldsymbol{RK})'(\boldsymbol{RK})]^{-1}(\boldsymbol{RK})'$ and $\boldsymbol{m}_i$ is a latent vector of $(\boldsymbol{QR})\boldsymbol{B}(\boldsymbol{QR})'$ (Burnaby, 1966), or as latent vectors of $\boldsymbol{Q}(\boldsymbol{QWQ})^-\boldsymbol{QG}'\boldsymbol{G}$ where $\boldsymbol{Q}$ is as before and $(\boldsymbol{QWQ})^-$ is a generalised inverse of $\boldsymbol{QWQ}$ (Gower, 1976).

14.3 Three points fall at random on the circumference of a circle. What is the probability that it is possible to connect them with a semicircle?

14.4 (Buffon's problem). A needle of length $l < 1$ falls on a floor ruled with parallel lines 1 unit apart. Show that the probability that it intersects a line is $2l/\pi$.

Solve the corresponding three-dimensional problem, in which the needle is randomly orientated in space, and the lines are replaced by a system of parallel planes.

[For history, developments and applications see Kendall and Moran (1963); Miles and Serra (1978).]

14.5 If θ is uniformly distributed on $(0, 2\pi)$ and $T = \sum_{i=1}^{n}\cos\theta_i$, where $\theta_1, \cdots, \theta_n$ is a sample of n independent observations. Show that $\mathrm{E}(T) = 0$, $\mathrm{var}(T) \approx n/2$.

(Cox and Hinkley, 1974).

14.6 The following data, from Edwards (1961), show the monthly numbers of cases of a congenital birth defect (Patent Ductus Arteriosus) over several years in the Birmingham area. Investigate the existence of a seasonal effect.

Jan	Feb	Mar	Apr	May	Jun	Jul	Aug	Sep	Oct	Nov	Dec
13	7	21	13	32	21	22	24	16	26	19	22

15

Review: Strategic Aspects and Future Prospects

15.1 This brings us to the end of these two volumes outlining the current state of multivariate methodology. Right from the outset, we were only too conscious of the size of our task. Given the growth of the subject in the last quarter of this century, providing a full treatment of each possible topic would have resulted in a vast and totally indigestible work. Inevitably, therefore, many of the possible topics have received only cursory mention, while some of the more esoteric or theoretical aspects have been entirely bypassed. By and large, we have concerned ourselves specifically with those features of the subject that have some practical importance.

Perhaps of more importance, however, is the way in which a subject is presented, and here we encounter a problem. Demands of conciseness have almost inevitably dictated a *technique-orientated* presentation, in which we have described in turn a succession of multivariate techniques (albeit grouped in what we hope to be a sensible fashion). The statistician who wishes to use any of these techniques must thus know *a priori* which ones are appropriate. In other words, he or she must already have decided on how the analysis of a data set is to be conducted. However, such a decision may not always be an easy one to make. Large multivariate data sets are now routinely amassed in surveys or through automatic data recording devices, and the typical statistical adviser is often unsure as to how best to proceed with an analysis. Also, statisticians in industry may be faced with frequent demands for the analysis of multivariate data sets, without the luxury of being able to set aside a lot of time for dwelling on each set. Such consultants thus need some general guidelines about a *strategy* to adopt when faced with an unfamiliar multivariate situation, which will narrow down the options and help them to home in quickly on a sensible analysis. The main purpose of this final chapter is thus to consider such strategic aspects of multivariate analysis. Having done so, we will then conclude by speculating briefly on how we see the subject developing in the future.

15.2 The question of a general strategy for multivariate analysis seems not to have been addressed explicitly in the literature, other than in passing fashion, until fairly recently. The impetus for setting down some systematic ideas has come from interest in the design of an online multivariate analysis computing system. One such system is the *OMEGA* pipeline, which is described by Weihs and Schmidli (1990) with an accompanying discussion by a number of prominent researchers. These contributions provide a useful framework for development of strategic considerations, so we now summarise some key points.

15.3 Assuming adequate computing facilities exist, a typical multivariate data set will need some mix of graphical display, preliminary filtering or preprocessing, exploratory dimension-reducing inspection, formal analysis, investigation of stability, search for unusual features, modification of data and return to the start for re-analysis, or conclusions and summary. The initial data inspection will clearly be governed by the sophistication of the graphical software available. As a baseline, marginal distributions of single variables and bivariate scatterplots of pairs of variables should be easy to obtain on all systems. At the simplest level these scatterplots are called up singly, but a more sophisticated system will allow display of the *scatterplot matrix*. This is a 'matrix' of graphs, having the scatterplot of the ith variable against the jth as its (i,j)th entry for $i \neq j$, with the diagonal free for additional information about the marginal distributions of the separate variables (e.g. histograms of the single variables, or their names and scale limits, or the different variable values). Further enhancements of such displays are facilities such as 'brushing', mentioned in Chapter 3 (Part 1). At the most sophisticated level are the dynamic graphical facilities which include 3-D displays, rotation, spinning, and so on (Becker *et al*, 1987). All of these graphical displays are geared towards identification of the general features of the data, and help to decide not only which multivariate technique might be potentially usable but also whether any systematic trends are evident that would necessitate a preliminary transformation of the data (see also Part 1, Chapter 3). If necessary, different transformations of each variable should be tried and the results re-subjected to the various graphical techniques until the overall features appear satisfactory.

15.4 Turning next to dimension-reducing inspection of the data, various projection techniques are available (Part 1, Chapters 4 and 5) for a more detailed investigation of patterns among the units of the sample or relationships among the variables. Here, it is important to be aware of any *a priori* external structure that may be present, in order to choose the most appropriate projection technique to use. Thus, unstructured quantitative data may require either principal component analysis or one of the variants of projection pursuit; quantitative data grouped by units will generally benefit from canonical variate analysis, while canonical correlation analysis may yield useful projections if the grouping is on the variables; categorical data will need some form of correspondence analysis, and any form of data can be subjected to scaling methods if the raw data are first converted to distances or dissimilarities.

15.5 One question that arises with all these techniques is that of stability of solution, and this can be tackled very effectively by a combination of data resampling and application of Procrustes analysis (Part 1, Chapter 5). The idea here is to see how much variation there is among the data projections obtained from repeated subsamples of the data. However, since any projection is only defined to within an arbitrary rotation of axes, it is important to ensure that all configurations for consideration have, as far as possible, matching coordinate axes. This can be achieved by rotating each subset projection to match as closely as possible the corresponding points from the complete-data projection.

The details are as follows. For a given integer p, where $0 < p < 100$, randomly choose $(100 - p)$% of the objects in the observed sample, apply the appropriate projection method to this subsample, project the omitted objects into the k-dimensional space defined by this method, and match (by Procrustes analysis) this configuration of points to the corresponding configuration extracted from application of the projection method to the whole sample. Repeat the complete procedure a large number, N say, of times, so that each point is represented in approximately $pN/100$ of the replicates. The envelope of the $pN/100$ optimally matched positions then gives an idea of the 'variability' of each point in the configuration over repeated sampling, and hence the collection of such envelopes reflects the stability of the particular projection method used: all envelopes very small suggests a very stable solution, while some or all envelopes large indicates much variability over repeated sampling.

Data resampling or cross-validation is also a very useful procedure for checking on applicability of a technique to a very large data set. For example, one might randomly divide the data set into two, conduct a projection technique such as principal component analysis on each subset, and then compare the two sets of principal component coefficients (Part 1, Chapter 4). Again, the process can be repeated a large number of times. If the data set is homogeneous then there should be little difference between the two sets of coefficients for each subdivision, but if there are some (hidden) heterogeneities then there will be some big differences for at least some of the replicates. Note, however, that any comparison of coefficients must allow for such problems as arbitrary orientation of a component (i.e. invariance under multiplication of all its coefficients by -1), and the possibility of inversions in the order of importance of components (particularly for those components whose corresponding eigenvalues are nearly equal). See Milan and Whittaker (1995) for further discussion of these specific aspects; other uses for resampling may suggest themselves in particular situations.

15.6 Once the data have been fully investigated in these ways then there may be a call to conduct a more formal inferential technique (Part 1, Chapters 6 and 7) or to fit a specific model to the data and test some consequent hypotheses about the model. This model may concern discriminating between groups of individuals (Chapter 9), explaining observed associations between variables (Chapters 11 and 12), or investigating relations among subgroups of individuals under repeated measurements (Chapter 13). An important feature of model fitting is that of model validation, so one may want to apply general procedures of outlier detection (Part 1, Chapter 3) to the residuals from a fitted model as well as conducting any of the special goodness-of-fit tests for the given models. Presence of outliers, or poorly fitting models, may then suggest modifications that need to be made to the model or data, and the whole process of analysis may need to be restarted.

15.7 An extended case study illustrating some of the above ideas on a large industrial data set concerned with the quality of dyestuffs is discussed in detail by Weihs and Schmidli (1990). The considerations arising from, and approaches demonstrated on, this set are relevant to many multivariate data sets arising in other fields.

15.8 A prime consideration in all these various data-exploratory activities is the dimensionality of data that can be handled by present-day computers. For the more analytically based techniques, such as principal components or canonical variates, there should be no problem, as extremely high dimensionalities can be handled comfortably even with several levels of data re-sampling. However, some of the more data-based techniques are much more restricted in scope. For example, a graphical grand tour to identify an optimal projection (Part 1, Chapter 4) may take about an hour in four dimensions, and the best part of a day in five dimensions on a modern workstation. High dimensionalities are thus still clearly infeasible. Again, estimating multivariate densities non-parametrically, e.g. by the kernel method (**9.27**), is perfectly feasible for up to four dimensions (Scott, 1992) but computationally prohibitive much beyond this value. A suitable compromise here is preliminary projection of the data into a low dimensionality, followed by density estimation in the projected subspace. It is evident that much remains to be done in bringing all these techniques fully within the scope of everyday use.

15.9 Looking more generally at the development of multivariate techniques, it is evident that computation has played a major role in the directions of progress. Mathematical tractability is often cited as a major reason for the pre-eminence of the multivariate normal distribution in multivariate analysis, but the relative simplicity of computations that result from techniques based around this distribution was also a strong attraction in the days before extensive computer power was available. This simplicity arises in several distinct ways. First, the direct correspondence between linear combinations of the variables in a multivariate situation and the corresponding univariate one led to the strong early emphasis on such parametric generalisations of familiar univariate inferential techniques as Hotelling's T^2, multivariate analysis of variance, and all those other methods discussed in Chapters 6 and 7 (Part 1). Secondly, the fact that the sample mean and covariance matrix form jointly sufficient statistics for the unknown parameters of the multivariate normal distribution meant that a large body of linear descriptive multivariate techniques could be built around fairly simple computational procedures such as inversion of, and extraction of latent roots and vectors from, a symmetric positive definite matrix. These techniques include all the familiar classical ones such as principal component analysis, canonical analysis, discriminant analysis and classical scaling methods. Early computer power made all these techniques easy to apply, but did not necessarily facilitate development of new methods. What might be termed the 'classical' (linear-algebra based) techniques thus held sway till the late 1960s.

15.10 At this stage, the increase in computer power led to greater emphasis on iterative numerical rather than algebraic methods for obtaining useful multivariate techniques. Non-metric multidimensional scaling, as proposed by Shepard (1962) and refined by Kruskal (1964), was arguably the first stage of a process that saw the development of methods based on alternating least squares (Part 1, **5.12**) for displaying multivariate data, and on iterative relocation (**10.14**) for partitioning and clustering such data. A short step from these techniques led to implementation of iterative methods for dealing with outliers and hence to the 'robustification' of many classical descriptive methods, and thence to the nonlinear generalisations of these techniques as described in Part 1, Chapter 8. All these developments took place in the 1970s and early 1980s. The sudden increase in computer power achieved at this time then led to the next major impetus, into computer-intensive methods such as resampling for data-based inference and model validation, and to the heavily computer-based non-parametric methods such as projection pursuit and its legion of offshoots, adaptive regression splines, non-parametric density estimation, and Monte Carlo Markov Chain methods.

15.11 So what for the future? There is clearly still much scope for computer-orientated methodological research in such areas as 'black-box' rule-based classification methods (e.g. neural networks and genetic algorithms, see Chapters 4 and 9), nonlinear generalisations of familiar linear techniques (see Part 1, Chapter 8) and associated function optimisation methods such as majorization (see Heiser, 1995), analysis of high-dimensional and functional data (see Chapter 14), image processing and spatially constrained data (see Chapter 9), and non-parametric data processing techniques. One topic that has generated considerable recent research interest is wavelet analysis, which brings together aspects of nonlinear, non-parametric, Bayesian functional analysis and which offers much scope for both theoretical and practical development (see, for example, Daubechies, 1988; Meyer, 1990; Donoho and Johnstone, 1991). Many of the areas just mentioned still lack a firm theoretical basis, so providing this is one obvious direction for theoretical research. Another is to widen the scope of existing methodology to non-normal data, continuing the development that has seen the gradual extension of normal distribution theory to samples from more general elliptical distributions.

Whatever the future brings, the above brief review means that one thing is certain: multivariate analysis continues to be a vigorous area for research, as well as an indispensible part of statistics from the point of view of the practitioner.

References

Adhikari, B. P. and Joshi, D. D. (1956). Distance, discrimination et résumé exhaustif. *Publications de l'Institut de Statistique de l'Univesité de Paris*, **5**, 57–74.

Agrawala, A. K. (Ed.) (1977). *Machine Recognition of Patterns*. IEEĖ Press, New York, USA.

Ahmad, I. A. (1982). Matusita's distance. In *Encyclopedia of Statistical Sciences, Volume 5* (eds S. Kotz and N. L. Johnson), pp. 334–336. Wiley, New York, USA.

Ahmed, S. W. and Lachenbruch, P. A. (1975). Discriminant analysis when one or both of the initial samples is contaminated: large sample results. *EDV in Medizin und Biologie*, **6**, 35–42.

Ahmed, S. W. and Lachenbruch, P. A. (1977). Discriminant analysis when scale contamination is present in the initial sample. In *Classification and Clustering* (ed J. van Ryzin), pp. 331–353. Academic Press, New York, USA.

Aigner, D. J., Hsiao, C., Kapteyn, A. and Wansbeek, T. (1984). Latent variable models in econometrics. In *Handbook of Econometrics, Volume 2* (eds Z. Griliches and M. D. Intriligator), pp. 1321–1393. North-Holland, Amsterdam, The Netherlands.

Aitchison, J. (1975). Goodness of prediction fit. *Biometrika*, **62**, 547–554.

Aitchison, J. and Aitken, C. G. G. (1976). Multivariate binary discrimination by the kernel method. *Biometrika*, **63**, 413–420.

Aitchison, J. and Dunsmore, I. R. (1975). *Statistical Prediction Analysis*. Cambridge University Press, Cambridge, England.

Aitchison, J., Habbema, J. D. F. and Kay, J. W. (1977). A critical comparison of two methods of statistical discrimination. *Applied Statistics*, **26**, 15–25.

Aitken, C. G. G. (1978). Methods of discrimination in multivariate binary data. *Compstat 1978, Proceedings of Computational Statistics*, pp. 155–161. Physica-Verlag, Vienna, Austria.

Aitkin, M., Anderson, D. A. and Hinde, J. P. (1981). Statistical modelling of data on teaching styles (with discussion). *Journal of the Royal Statistical Society, Series A*, **144**, 419–461.

Aitkin, M., Anderson, D. A., Francis, B. and Hinde, J. P. (1989). *Statistical Modelling with GLIM*. Oxford University Press, Oxford, England.

Akaike, H. (1973). Information theory and an extension of the maximum likelihood principle. In *Second International Symposium on Information Theory* (eds B. N. Petrov and F. Czáki), pp. 267-281. Akademiai Kiadó, Budapest, Hungary.

Akaike, H. (1974). A new look at statistical model identification. *IEEE Transactions on Automatic Control*, **AC-19**, 716–722.

Akaike, H. (1983). Information measures and model selection. *Bulletin of the International Statistical Institute*, **50**(1), 277–290.

Akaike, H. (1987). Factor analysis and AIC. *Psychometrika*, **52**, 317–332.

Aladjem, M. (1991a). Parametric and nonparametric linear mappings of multidimensional data. *Pattern Recognition*, **24**, 543–553.

Aladjem, M. (1991b). PNM: a program for parametric and nonparametric mapping of multidimensional data. *Computers in Biology and Medicine*, **21**, 321–343.

Ali, S. M. and Silvey, S. D. (1966). A general class of coefficients of divergence of one distribution from another. *Journal of the Royal Statistical Society, Series B*, **28**, 131–142.

Ambartzumian, R. V. (1982). Random shapes by factorisation. In *Statistics in Theory and Practice* (ed B. Ranneby), pp. 35–41. Swedish University of Agricultural Sciences, Umea, Sweden.

Amemiya, Y. (1993). Instrumental variable estimation for nonlinear factor analysis. In *Multivariate Analysis: Future Directions 2* (eds C. M. Cuadras and C. R. Rao), pp. 113–129. Elsevier Science Publishers, Amsterdam, The Netherlands.

Amit, D. J. (1989). *Modeling Brain Function*. Cambridge University Press, Cambridge, England.

Anderson, B. M., Anderson, T. W. and Olkin, I. (1986). Maximum likelihood estimators and likelihood ratio criteria in multivariate components of variance. *Annals of Statistics*, **14**, 405–417.

Anderson, J. A. (1969). Discrimination between *k* populations with constraints on the probabilities of misclassification. *Journal of the Royal Statistical Society, Series B*, **31**, 123–139.

Anderson, J. A. (1972). Separate sample logistic discrimination. *Biometrika*, **59**, 19–35.

Anderson, J. A. (1974). Diagnosis by logistic discriminant functions: further practical problems and results. *Applied Statistics*, **23**, 397-404.

Anderson, J. A. (1975). Quadratic logistic discrimination. *Biometrika*, **62**, 149–154.

Anderson, J. A. (1979). Multivariate logistic compounds. *Biometrika*, **66**, 17–26.

Anderson, J. A. (1982). Logistic discrimination. In *Handbook of Statistics, Volume 2* (eds P. R. Krishnaiah and L. N. Kanal), pp. 169–191. North-Holland, Amsterdam, The Netherlands.

Anderson, J. A. and Richardson, S. C. (1979). Logistic discrimination and bias correction in maximum likelihood estimation. *Technometrics*, **21**, 71–78.

Anderson, T. W. (1951). Classification by multivariate analysis. *Psychometrika*, **16**, 31–50.
Anderson, T. W. (1958). *An Introduction to Multivariate Statistical Analysis*, 1st edition. Wiley, New York, USA.
Anderson, T. W. (1984). *An Introduction to Multivariate Statistical Analysis*, 2nd edition. Wiley, New York, USA.
Anderson, T. W. and Bahadur, R. R. (1962). Classification into two multivariate normal populations with different covariance matrices. *Annals of Mathematical Statistics*, **33**, 420–431.
Anderson, T. W. and Rubin, H. (1956). Statistical inference in factor analysis. *Proceedings of the Third Berkeley Symposium in Mathematical Statistics and Probability*, **5**, 111–150.
Arabie, P. and Carroll, J. D. (1980). MAPCLUS: a mathematical programming approach to fitting the ADCLUS model. *Psychometrika*, **45**, 211–235.
Arabie, P., Carroll, J. D. and DeSarbo, W. S. (1987). *Three-way Scaling and Clustering*. Sage Publications, Newbury Park, California, USA.
Ashton, E. H., Healy, M. J. R. and Lipton, S. (1957). The descriptive use of discriminant functions in physical anthropology. *Proceedings of the Royal Society, Series B*, **146**, 552–572.
Atkinson, C. and Mitchell, A. F. S. (1981). Rao's distance measure. *Sankhyā A*, **43**, 345–365.
Aubin, J.-P. (1979). *Applied Functional Analysis*. Wiley Interscience, New York, USA.
Badsberg, J. H. (1991). *A Guide to CoCo*. Research Report R 91–43, Department of Mathematics and Computer Science, University of Aalborg, Denmark.
Badsberg, J. H. (1992). Model search in contingency tables by CoCo. In *Computational Statistics*, COMPSTAT 1992, Neûchatel (eds Y. Dodge and J. Whittaker), vol 1, pp. 251–256. Physica Verlag, Heidelberg, Germany.
Badsberg, J. H. (1993). *A Graphical Tool for Association Models*. Report No IR 93–2006, Institute of Electronic Systems, Aalborg, Denmark.
Bahadur, R. R. (1961). A representation of the joint distribution of responses to *n* dichotomous items. In *Studies in Item Analysis and Prediction* (ed H. Solomon), pp. 158–168. Stanford University Press, Stanford, California, USA.
Baksalary, J. K., Corsten, L. C. A. and Kala, R. (1978). Reconciliation of two different views on estimation of growth curve parameters. *Biometrika*, **65**, 662–665.
Balakrishnan, N. and Kocherlakota, K. (1985). Robustness to nonnormality of the linear discriminant function: mixtures of normal distributions. *Communications in Statistics, Theory and Methods*, **14**, 465–478.
Balakrishnan, N. and Tiku, M. L. (1988). Robust classification procedures. In *Classification and Related Methods of Data Analysis* (ed H. H. Bock), pp. 269–276. North-Holland, Amsterdam, The Netherlands.
Balakrishnan, V. and Sanghvi, L. D. (1968). Distance between populations on the basis of attribute data. *Biometrics*, **24**, 859–865.

Baldwin, J. P., Tinker, P. B. and Marriott, F. H. C. (1971). The measurement of length and distribution of onion roots in the field and in the laboratory. *Journal of Applied Ecology*, **8**, 543–554.

Banfield, C. F. and Bassill, S. (1977). A transfer algorithm for non-hierarchical classification. Algorithm 133. *Applied Statistics*, **26**, 206–210.

Banfield, J. D. and Raftery, A. E. (1993). Model-based Gaussian and non-Gaussian clustering. *Biometrics*, **49**, 803–821.

Bar-Hen, A. and Daudin, J. J. (1995). Generalization of the Mahalanobis distance in the mixed case. *Journal of Multivariate Analysis*, **53**, 332–342.

Bartholomew, D. J. (1980). Factor analysis for categorical data (with discussion). *Journal of the Royal Statistical Society, Series B*, **42**, 293–321.

Bartholomew, D. J. (1987). *Latent Variable Models and Factor Analysis*. Griffin, Oxford, England.

Bartlett, M. S. (1937). The statistical conception of mental factors. *British Journal of Psychology*, **28**, 97–104.

Bartlett, M. S. (1938). Methods of estimating mental factors. *Nature*, **141**, 609–610.

Bartlett, M. S. (1950). Tests of significance in factor analysis. *British Journal of Psychology, Statistical Section*, **3**, 77–85.

Bartlett, M. S. (1951). An inverse matrix adjustment arising in discriminant analysis. *Annals of Mathematical Statistics*, **22**, 107–111.

Bartlett, M. S. (1954). A note on the multiplying factors for various χ^2 approximations. *Journal of the Royal Statistical Society, Series B*, **16**, 296–298.

Bartlett, M. S. and Please, N. W. (1963). Discrimination in the case of zero mean differences. *Biometrika*, **50**, 17–21.

Basu, A. P. and Gupta, A.K. (1974). Classification rules for exponential populations. In *Reliability and Biometry: Statistical Analysis of Lifelength* (eds F. Proschan and J. Serfling), pp. 637–650. SIAM, Philadelphia, USA.

Batschelet, E. (1981). *Circular Statistics in Biology*. Academic Press, London, England.

Bayne, C. K., Beauchamp, J. J., Kane, V. E. and McCabe, G. P. (1983). Assessment of Fisher and logistic linear and quadratic discrimination models. *Computational Statistics and Data Analysis*, **1**, 257–273.

Beale, E. M. L. and Little, R. J. A. (1975). Missing values in multivariate analysis. *Journal of the Royal Statistical Society, Series B*, **37**, 129–145.

Becker, R. A., Chambers, J. M. and Wilks, A. R. (1988). *The New S Language*. Wadsworth, Belmont, California, USA.

Becker, R. A.,Cleveland, W. S. and Wilks, A. R. (1987). Dynamic graphics for data analysis. *Statistical Science*, **2**, 355-383.

Bello, A. L. (1993a). Choosing among imputation techniques for incomplete multivariate data: a simulation study. *Communications in Statistics, Theory and Methods*, **22**, 853–877.

Bello, A. L. (1993b). A comparative study of imputation techniques in linear, quadratic and kernel discriminant analyses. *Journal of Statistical Computation and Simulation*, **48**, 167–180.

Ben-Bassat, M. (1982). Use of distance measures, information matrices and error bounds in feature evaluation. In *Handbook of Statistics, Volume 2* (eds P. R. Krishnaiah and L. N. Kanal), pp. 773–791. North-Holland, Amsterdam, The Netherlands.

Bentler, P. M. (1980). Multivariate analysis with latent variables: causal modeling. *Annual Review of Psychology*, **31**, 419–456.

Bentler, P. M. (1986). Structural modeling and Psychometrika: a historical perspective on growth and achievements. *Psychometrika*, **51**, 35–51.

Bentler, P. M. (1989). *EQS. Structural Equations Manual.* BMDP Statistical Software, Los Angeles, California, USA.

Bentler, P. M. (1990). Comparative fit indices in structural models. *Psychological Bulletin*, **107**, 238–246.

Bentler, P. M. and Bonett, D. G. (1980). Significance tests and goodness of fit in the analysis of covariance structures. *Psychological Bulletin*, **88**, 588–606.

Bentler, P. M. and Lee, S. Y. (1983). Covariance structures under polynomial constraints: applications to correlation and alpha-type structural models. *Journal of Educational Statistics*, **8**, 207–222.

Bentler, P. M. and Weeks, D. G. (1980). Multivariate analysis with latent variables. In *Handbook of Statistics, Volume 2* (eds P. R. Krishnaiah and L. N. Kanal), pp. 747–771. North-Holland, Amsterdam, The Netherlands.

Bentler, P. M. and Tanaka, J. S. (1983). Problems with EM algorithms for ML factor analysis. *Psychometrika*, **48**, 371–375.

Bertolino, F. (1988). On the classification of observations structured into groups. *Applied Stochastic Models and Data Analysis*, **4**, 239–251.

Besag, J. E. (1986). On the statistical analysis of dirty pictures (with discussion). *Journal of the Royal Statistical Society, Series B*, **48**, 259–302.

Besse, P. (1988). Spline functions and optimal metric in linear principal component analysis. In *Component and Correspondence Analysis: Dimensional Reduction by Functional Approximation* (eds J. L. A. van Rijckevorsel and J. de Leeuw), pp. 81–101. Wiley, New York, USA.

Besse, P. and Ramsay, J. O. (1986). Principal components analysis of sampled functions. *Psychometrika*, **51**, 285–311.

Bezdek, J. C. (1974). Cluster validity with fuzzy sets. *Journal of Cybernetics*, **3**, 58–72.

Bhattacharya, P. K. and Das Gupta, S. (1964). Classification into exponential populations. *Sankhyā A*, **26**, 17–24.

Bhattacharyya, A. (1943). On a measure of divergence between two statistical populations defined by their probability distributions. *Bulletin of the Calcutta Mathematical Society*, **35**, 99–109.

Bhattacharyya, A. (1946). On a measure of divergence between two multinomial populations. *Sankhyā A*, **7**, 401–406.

Bielby, W. T. and Hauser, R. M. (1977). Structural equation models. *Annual Review of Sociology*, **3**, 137–161.

Bingham, C. (1964). *Distributions on the sphere and on the projective plane.* Unpublished Ph.D. Thesis, Yale University, USA.

Bingham, C. (1974). An antipodally symmetric distribution on the sphere. *Annals of Statistics*, **2**, 1201–1225.

Bishop, C. (1992). Exact calculation of the Hessian matrix for the multilayer perceptron. *Neural Computation*, **4**, 494–501.

Bishop, Y. M. M. (1971). Effects of collapsing multi-dimensional contingency tables. *Biometrics*, **27**, 119–128.

Bishop, Y. M. M., Fienberg, S. E. and Holland, P. W. (1975). *Discrete Multivariate Analysis: Theory and Practice*. MIT Press, Cambridge, Massachusetts, USA.

Bock, H. H. (1984). Statistical testing and evaluation methods in cluster analysis. In *Proceedings of the Golden Jubilee Conference on Statistics: Applications and New Directions* (eds J. K. Ghosh and J. Roy), pp. 116–146. Indian Statistical Institute, Calcutta, India.

Bock, H. H. (1985). On some significance tests in cluster analysis. *Journal of Classification*, **2**, 77–108.

Bock, R. D. and Aitkin, M. (1981). Marginal maximum likelihood estimation of item parameters: application of an EM algorithm. *Psychometrika*, **46**, 443–459.

Bock, R. D. and Lieberman, M. (1970). Fitting a response model for n dichotomously scored items. *Psychometrika*, **35**, 179–197.

Bollen, K. A. (1979). Political democracy and the timing of development. *American Sociological Review*, **44**, 572–587.

Bollen, K. A. (1980). Issues in the comparative measurement of political democracy. *American Sociological Review*, **45**, 370–390.

Bollen, K. A. (1989). *Structural Equations with Latent Variables*. Wiley, New York, USA.

Bookstein, F. L. (1978). *The Measurement of Biological Shape and Shape Change*. Lecture Notes in Mathematics, Volume 24; Springer, New York, USA.

Bookstein, F. L. (1986). Size and shape spaces for landmark data in two dimensions. *Statistical Science*, **1**, 181–221.

Bookstein, F. L. (1991). *Morphometric Tools for Landmark Data*. Cambridge University Press, New York, USA.

Bookstein, F. L. and Sampson, P. D. (1990). Statistical models for geometric components of shape change. *Communications in Statistics, Theory and Methods*, **19**, 1939–1972.

Bookstein, F. L., Chernoff, B., Elder, R., Humphries, J., Smith, G. and Strauss, R. (1985). *Morphometrics in Evolutionary Biology. The Geometry of Size and Shape Change, with Examples from Fishes*. Academy of Natural Sciences of Philadelphia, Philadelphia, USA.

Bower, F. O. (1930). *Size and Form in Plants*. Macmillan, London, England.

Bowker, A. H. (1961). A representation of Hotelling's T^2 and Anderson's classification statistic W in terms of simple statistics. In *Studies in Item Analysis and Prediction* (ed H. Solomon), pp. 285–292. Stanford University Press, Stanford, California, USA.

Box, G. E. P. (1949). A general distribution theory for a class of likelihood criteria. *Biometrika*, **40**, 318–335.

Box, G. E. P. and Jenkins, G. M. (1976). *Time Series Analysis, Forecasting and Control* (Revised Edition). Holden-Day, San Francisco, USA.

Box, G. E. P. and Tiao, G. C. (1973). *Bayesian Inference in Statistical Analysis*. Addison-Wesley, Reading, Massachusetts, USA.

Bray, J. R. and Curtis, J. T. (1957). An ordination of the upland forest communities of S. Wisconsin. *Ecological Monographs*, **27**, 325–349.

Breiman, L., Friedman, J. H., Olshen, R. A. and Stone, C. J. (1984). *Classification and Regression Trees*. Wadsworth, Belmont, California, USA.

Brennan, R. L. and Light, R. J. (1974). Measuring agreement when two observers classify people into categories not defined in advance. *British Journal of Mathematical and Statistical Psychology*, **27**, 154–163.

Broemeling, L. D. and Son, M. S. (1987). The classification problem with autoregressive processes. *Communications in Statistics, Theory and Methods*, **16**, 927–936.

Broffitt, J. D. (1982). Nonparametric classification. In *Handbook of Statistics, Volume 2* (eds P. R. Krishnaiah and L. N. Kanal), pp. 139–168. North-Holland, Amsterdam, The Netherlands.

Broffitt, J. D., Clarke, W. R. and Lachenbruch, P. A. (1980). The effect of Huberizing and trimming on the quadratic discriminant function. *Communications in Statistics, Theory and Methods*, **9**, 13–25.

Broffitt, J. D., Clarke, W. R. and Lachenbruch, P. A. (1981). Measurement errors—a location contamination problem in discriminant analysis. *Communications in Statistics, Simulation and Computation*, **10**, 129–141.

Brooks, C. A., Clark, R. R., Hodgu, A. and Jones, A. M. (1988). The robustness of the logistic risk function. *Communications in Statistics, Simulation and Computation*, **17**, 1–24.

Brown, P. J. (1982). Multivariate calibration (with discussion). *Journal of the Royal Statistical Society, Series B*, **44**, 287–321.

Brown, P. J. (1993). *Measurement, Regression and Calibration*. Oxford University Press, Oxford, England.

Browne, M. W. (1982). Covariance structures. In *Topics in Applied Multivariate Analysis* (ed D. M. Hawkins), pp. 72–141. Cambridge University Press, Cambridge, England.

Browne, M. W. (1984). Asymptotically distribution-free methods for the analysis of covariance structures. *British Journal of Mathematical and Statistical Psychology*, **37**, 62–83.

Browne, M. W. (1993). Structured latent curve models. In *Multivariate Analysis: Future Directions 2* (eds C. M. Cuadras and C. R. Rao), pp. 171–197. Elsevier Science Publishers, Amsterdam, The Netherlands.

Browne, M. W. and Cudeck, R. (1989). Single sample cross-validation indices for covariance structures. *Multivariate Behavioral Research*, **24**, 445–455.

Browne, M. W. and Cudeck, R. (1992). Alternative ways of assessing model fit. *Sociological Methods and Research*, **21**, 230–258.

Bryan, J. G. (1951). The generalized discriminant function: mathematical foundation and computational routine. *Harvard Educational Review*, **21**, 90–95.

Bryant, P. G. (1988). On characterizing optimization-based clustering methods. *Journal of Classification*, **5**, 81–84.

Bryant, P. G. (1991). Large-scale results for optimization-based clustering methods. *Journal of Classification*, **8**, 31–44.

Buck, S. F. (1960). A method of estimation of missing values in multivariate data suitable for use with an electronic computer. *Journal of the Royal Statistical Society, Series B*, **22**, 302–306.

Bull, S. B. and Donner, A. (1987). The efficiency of multinomial logistic regression compared with multiple group discriminant analysis. *Journal of the American Statistical Association*, **82**, 1118–1122.

Burbea, J. and Rao, C. R. (1982). Entropy differential metric, distance and divergence in probability spaces: a unified approach. *Journal of Multivariate Analysis*, **12**, 575–596.

Burbea, J. and Rao, C. R. (1984). Differential metrics in probability spaces. *Probability and Mathematical Statistics*, **3**, 241–258.

Burnaby, T. P. (1966). Growth-invariant discriminant functions and generalized distances. *Biometrics*, **22**, 96–110.

Busemeyer, J. R. and Jones, L. E. (1983). Analysis of multiplicative combination rules when the causal variables are measured with error. *Psychological Bulletin*, **93**, 549–562.

Cacoullos, T. (1965a). Comparing Mahalanobis distances, I: comparing distances between k normal populations and another unknown. *Sankhyā A*, **27**, 1–22.

Cacoullos, T. (1965b). Comparing Mahalanobis distances, II: Bayes procedures when the mean vectors are unknown. *Sankhyā A*, **27**, 23–32.

Calvo, M. and Oller, J. M. (1990). A distance between multivariate normal distributions based in an embedding into a Siegel group. *Journal of Multivariate Analysis*, **35**, 223–242.

Campbell, N. A. (1980). Shrunken estimators in discriminant and canonical variate analysis. *Applied Statistics*, **29**, 5–14.

Campbell, N. A. (1984). Canonical variate analysis with unequal covariance matrices. Generalizations of the usual solution. *Mathematical Geology*, **16**, 109–124.

Campbell, N. A. (1985). Updating formulae for allocation of individuals. *Applied Statistics*, **34**, 235–236.

Campbell, N. A. and Atchley, W. R. (1981). The geometry of canonical variate analysis. *Systematic Zoology*, **30**, 268–280.

Carlson, M. and Mulaik, S. A. (1993). Trait ratings from descriptions of behavior as mediated by components of meaning. *Multivariate Behavioral Research*, **28**, 111–159.

Carne, T. K. (1990). The geometry of shape spaces. *Proceedings of the London Mathematical Society*, **61**, 407–432.

Carroll, J. D. and Arabie, P. (1983). INDCLUS: an individual differences generalization of the ADCLUS model and the MAPCLUS algorithm. *Psychometrika*, **48**, 157–169.

Carroll, J. D. and Chang, J. J. (1970). Analysis of individual differences in multidimensional scaling via an N-way generalisation of the 'Eckart-Young' decomposition. *Psychometrika*, **35**, 283–319.

Carroll, J. D., Clark, L. A. and DeSarbo, W. S. (1984). The representation of three-way proximity data by single and multiple tree structure models. *Journal of Classification*, **1**, 25–74.

Castagliola, P. and Dubuisson, B. (1989). Two classes linear discrimination. A min-max approach. *Pattern Recognition Letters*, **10**, 281-287.

Celeux, G. and Mkhadri, A. (1992). Discrete regularized discriminant analysis. *Statistics and Computing*, **2**, 143–151.

Chaddha, R. L. and Marcus, L. F. (1968). An empirical comparison of distance statistics for populations with unequal covariance matrices. *Biometrics*, **24**, 683–694.

Chang, W. C. (1983). On using principal components before separating a mixture of multivariate normal distributions. *Applied Statistics*, **32**, 276–286.

Chatterjee, S. and Chatterjee, S. (1983). Estimation of misclassification probabilities. *Communications in Statistics, Simulation and Computation*, **12**, 645–656.

Cheng, B. and Titterington, D. M. (1994). Neural networks: a review from a statistical perspective (with discussion). *Statistical Science*, **9**, 2–54.

Chernick, M. R., Murthy, V. K. and Nealy, C. D. (1985). Application of bootstrap and other resampling techniques: evaluation of classifier performance. *Pattern Recognition Letters*, **3**, 167–178.

Chernick, M. R., Murthy, V. K. and Nealy, C. D. (1986a). Correction note to Application of bootstrap and other resampling techniques: evaluation of classifier performance. *Pattern Recognition Letters*, **4**, 133–142.

Chernick, M. R., Murthy, V. K. and Nealy, C. D. (1986b). Estimation of error rate for linear discriminant functions by resampling: non-Gaussian populations. *Computers and Mathematics with Applications*, **13**, 29–37.

Chernoff, H. (1952). A measure of asymptotic efficiency for tests of a hypothesis based on the sum of observations. *Annals of Mathematical Statistics*, **23**, 493–507.

Chhikara, R. S. and Folks, J. L. (1989). *The Inverse Gaussian Distribution: Theory, Methodology, and Applications.* Marcel Dekker, New York, USA.

Chhikara, R. S. and Odell, P. L. (1973). Discriminant analysis using certain normed exponential densities with emphasis on remote sensing application. *Pattern Recognition*, **5**, 259–272.

Child, D. (1970). *The Essentials of Factor Analysis.* Holt, Reinhart and Winston, London, England.

Chinchilli, V. M. and Carter, W. H. (1984). A likelihood ratio test for a patterned covariance matrix in a multivariate growth-curve model. *Biometrics*, **40**, 151–156.

Chinganda, E. F. and Subrahmaniam, K. (1979). Robustness of the linear discriminant function to nonnormality: Johnson's system. *Journal of Statistical Planning and Inference*, **3**, 69–77.

Choi, S. C. (1972). Classification of multiply observed data. *Biometrical Journal*, **14**, 8–11.

Chou, P. A. (1991). Optimal partitioning for classification and regression trees. *IEEE Transactions on Pattern Analysis and Machine Intelligence*, **PAMI-13**, 340–354.

Christofferson, A. (1975). Factor analysis of dichotomized variables. *Psychometrika*, **40**, 5–32.

Clark, M. R. B. (1970). A rapidly convergent method for maximum likelihood factor analysis. *British Journal of Mathematical and Statistical Psychology*, **23**, 43–52.

Clarke, W. R., Lachenbruch, P. A. and Broffitt, J. D. (1979). How nonnormality affects the quadratic discriminant function. *Communications in Statistics, Theory and Methods*, **8**, 1285–1301.

Clogg, C. C. (1981). Latent structure models of mobility. *American Journal of Sociology*, **86**, 836–868.

Clogg, C. C. (1992). The impact of sociological methodology on statistical methodology. *Statistical Science*, **7**, 183–196.

Clunies-Ross, C. W. and Riffenburgh, R. H. (1960). Geometry and linear discrimination. *Biometrika*, **47**, 185–189.

Coomans, D. and Broeckaert, I. (1986). *Potential Pattern Recognition in Chemical and Medical Decision Making*. Research Studies Press, Letchworth, England.

Cooper, P. W. (1963). Statistical classification with quadratic forms. *Biometrika*, **50**, 439–448.

Cormack, R. M. (1971). A review of classification (with discussion). *Journal of the Royal Statistical Society, Series A*, **134**, 321–367.

Cornfield, J. (1962). Joint dependence of risk of coronary heart disease on serum cholesterol and systolic blood pressure: a discriminant function approach. *Federation Proceedings. Federation of American Societies for Experimental Biology*, **11**, 58–61.

Costanza, M. C. and Afifi, A. A. (1979). Comparison of stopping rules in forward stepwise discriminant analysis. *Journal of the American Statistical Association*, **74**, 777–785.

Cox, D. R. (1966). Some procedures associated with the logistic qualitative response curve. In *Research Papers on Statistics: Festschrift for J. Neyman* (ed F. David), pp. 55–71. Wiley, New York, USA.

Cox, D. R. and Hinkley, D. V. (1974). *Theoretical Statistics*. Chapman and Hall, London, England.

Crawley, D. R. (1979). Logistic discrimination as an alternative to Fisher's linear discriminant function. *New Zealand Statistician*, **14**, 21–25.

Cressie, N. A. (1991). *Statistics for Spatial Data*. Wiley, New York, USA.

Crichton, N. J. and Hinde, J. P. (1989). Correspondence analysis as a screening method for indicants for clinical diagnosis. *Statistics in Medicine*, **8**, 1351–1362.

Critchley, F., Ford, I. and Rijal, O. (1987). Uncertainty in discrimination. *Proceedings of Conference DIANA II*, pp. 83–106. Mathematical Institute of the Czechoslovak Academy of Sciences, Prague, Czech Republic.

Croon, M. (1990). Latent class analysis with ordered classes. *British Journal of Mathematical and Statistical Psychology*, **43**, 171–192.

Crowder, M. J. and Hand, D. J. (1990). *Analysis of Repeated Measures*. Chapman and Hall, London, England.

Cuadras, C. M. (1989). Distance analysis in discrimination and classification using both continuous and categorical variables. In *Statistical Data Analysis and Inference* (ed Y. Dodge), pp. 459–473. North-Holland, Amsterdam, The Netherlands.

Cuadras, C. M. (1991). A distance based approach to discriminant analysis and its properties. *Mathematics Preprint Series, No. 90*, Second version, University of Barcelona, Spain.

Cuadras, C. M. (1992). Some examples of distance based discrimination. *Biometrical Letters*, **29**, 3–20.
Cudeck, R. and Browne, M.W. (1983). Cross-validation of covariance structures. *Multivariate Behavioral Research*, **18**, 147–167.
Cudeck, R. and Henly, S. J. (1991). Model selection in covariance structure analysis and the "problem" of sample size: a clarification. *Psychological Bulletin*, **109**, 512–519.
Cureton, E. E. and D'Agostino, R. B. (1983). *Factor Analysis: An Applied Approach*. Lawrence Erblaum, New Jersey, USA.
Cutler, A. and Breiman, L. (1994). Archetypal analysis. *Technometrics*, **36**, 338–347.
Dargahi-Noubari, G. R. (1981). An application of discrimination when covariance matrices are proportional. *Australian Journal of Statistics*, **23**, 38–44.
Dasarathy, B. V. (1991). *Nearest Neighbor (NN) Norms: NN Pattern Classification Techniques*. IEEE Computer Society Press, Los Alamos, California, USA.
Daubechies, I. (1988). Orthonormal bases of compactly supported wavelets. *Communications in Pure and Applied Mathematics*, **41**, 909–996.
Daudin, J. J. (1986). Selection of variables in mixed-variable discriminant analysis. *Biometrics*, **42**, 473–481.
Davison, A. C. and Hall, P. (1992). On the bias and variability of bootstrap and cross-validation estimates of error rate in discrimination problems. *Biometrika*, **79**, 279–284.
Dawid, A. P. (1979). Conditional independence in statistical theory (with discussion). *Journal of the Royal Statistical Society, Series B*, **41**, 1–31.
Day, N. E. (1969a). Linear and quadratic discrimination in pattern recognition. *IEEE Transactions on Information Theory*, **IT-15**, 419–420.
Day, N. E. (1969b). Estimating the components of a mixture of normal distributions. *Biometrika*, **56**, 463–474.
Day, N. E. and Kerridge, D. F. (1967). A general maximum likelihood discriminant. *Biometrics*, **23**, 313–324.
Day, W. H. E. (1986). Foreword: comparison and consensus of classifications. *Journal of Classification*, **3**, 183–185.
Dear, R. E. (1959). *A Principal-component Missing Data Method for Multiple Regression Models*. Report SP-86, System Development Corporation, Santa Monica, California, USA.
De Boor, C. (1978). *A Practical Guide to Splines*. Springer-Verlag, New York, USA.
DeGroot, M. H. (1970). *Optimal Statistical Decisions*. McGraw-Hill, New York, USA.
De Leeuw, J. and van der Heijden, P. G. M. (1991). Reduced rank models for contingency tables. *Biometrika*, **78**, 229–232.
De Leeuw, J., van der Heijden, P. G. M. and Verboon, P. (1990). A latent time budget model. *Statistica Neerlandica*, **44**, 1-22.
Dempster, A. P. (1972). Covariance selection. *Biometrics*, **28**, 157–175.
Dempster, A. P., Laird, N. M. and Rubin, D. B. (1977). Maximum likelihood from incomplete data via the EM algorithm (with discussion). *Journal of the Royal Statistical Society, Series B*, **39**, 1–38.

Denham, M. C. and Brown, P. J. (1993). Calibration with many variables. *Applied Statistics*, **42**, 515–528.

DeSarbo, W. S., Ramaswamy, V. and Lenk, P. (1993). A latent class procedure for the structural analysis of two-way compositional data. *Journal of Classification*, **10**, 159–193.

De Soete, G. and Winsberg, S. (1993). A latent class vector model for preference ratings. *Journal of Classification*, **10**, 195–218.

De Soete, G., DeSarbo, W. S. and Carroll, J. D. (1985). Optimal variable weighting for hierarchical clustering: an alternating least-squares algorithm. *Journal of Classification*, **2**, 173–192.

Devroye, L. P. (1987). *A Course in Density Estimation*. Birkhauser, Boston, USA.

Devroye, L. P. (1988). Automatic pattern recognition: a study of the probability of error. *IEEE Transactions on Pattern Analysis and Machine Intelligence*, **PAMI-10**, 530–543.

Dietterich, T. G. (1990). Machine learning. *Annual Review of Computer Science*, **4**, 255–306.

Diggle, P. J. (1988). An approach to the analysis of repeated measurements. *Biometrics*, **44**, 959–971.

Diggle, P. J. and Kenward, M. G. (1994). Informative drop-out in longitudinal data analysis (with discussion). *Applied Statistics*, **43**, 49–93.

Dillon, W. R. and Goldstein, M. (1978). On the performance of some multinomial classification rules. *Journal of the American Statistical Association*, **73**, 305–313.

Dimroth, E. (1962). Untersuchungen zum Mechanismus von Blastesis und Syntexis in Phylliten und Hornfelsen des südwestlichen Fichtelgebirges. I. Die statische Auswertung einfacher Gürteldiagramme. *Tschermaks Mineralogische und Petrographische Mitteilungen*, **8**, 248–274.

Di Pillo, P. J. (1976). The application of bias to discriminant analysis. *Communications in Statistics, Theory and Methods*, **A5**, 843–854.

Di Pillo, P. J. (1977). Further applications of bias to discriminant analysis. *Communications in Statistics, Theory and Methods*, **A6**, 933–943.

Donoho, D. L. and Johnstone, I. M. (1991). Minimax estimation via wavelet shrinkage. Paper presented to Annual Meeting of the Institute of Mathematical Statistics, Atlanta, Georgia, August 1991.

Dunn, J. C. (1974). A fuzzy relative of the ISODATA process and its use in detecting compact well-separated clusters. *Journal of Cybernetics*, **3**, 32–57.

Dunn, J. C. (1976). Indices of partition fuzziness and the detection of clusters in large data sets. In *Fuzzy Automata and Decision Processes* (ed M. M. Gupta), pp. 271–283. Elsevier, New York, USA.

Eckart, C. and Young, G. (1936). The approximation of one matrix by another of lower rank. *Psychometrika*, **1**, 211–218.

Edwards, A. W. F. (1971). Distances between populations on the basis of gene frequencies. *Biometrics*, **27**, 873–881.

Edwards, A. W. F. and Cavalli-Sforza, L. L. (1965). A method for cluster analysis. *Biometrics*, **21**, 362–375.

Edwards, D. (1990). Hierarchical interaction models (with discussion). *Journal of the Royal Statistical Society, Series B*, **52**, 3–20.

Edwards, D. (1993). *Graphical Modelling with MIM (version 2.1)*. Hypergraph Software, Roskilde, Denmark.
Edwards, J. H. (1961). The recognition and estimation of cyclic trends. *Annals of Human Genetics*, **25**, 83–86.
Efron, B. (1975). The efficiency of logistic regression compared to normal discriminant analysis. *Journal of the American Statistical Association*, **70**, 892–898.
Efron, B. (1982). *The Jackknife, the Bootstrap, and Other Resampling Plans*. SIAM, Philadelphia, USA.
Efron, B. (1983). Estimating the error of a prediction rule: improvement on cross-validation. *Journal of the American Statistical Association*, **78**, 316–331.
Efron, B. and Gong, G. (1983). A leisurely look at the bootstrap, the jackknife and cross-validation. *The American Statistician*, **37**, 36–48.
Efron, B. and Hinkley, D. V. (1978). Assessing the accuracy of the maximum likelihood estimators: observed versus expected Fisher information. *Biometrika*, **65**, 457–487.
Efron, B. and Tibshirani, R. J. (1986). Bootstrap methods for standard errors, confidence intervals, and other measures of statistical accuracy (with discussion). *Statistical Science*, **1**, 54–77.
Egan, J. P. (1975). *Signal Detection Theory and ROC Analysis*. Academic Press, New York, USA.
Emerson, P. A., Russell, N. J., Wyatt, J., Crichton, N. J., Pantin, C. F. A., Morgan, A. D. and Fleming, P. R. (1989). An audit of doctors' management of patients with chest pain in the accident and emergency department. *The Quarterly Journal of Medicine*, **70**, 213–220.
Engelman, L. and Hartigan, J. A. (1969). Percentage points for a test for clusters. *Journal of the American Statistical Association*, **64**, 1647–1648.
Enis, P. and Geisser, S. (1970). Sample discriminants which minimize posterior squared error loss. *South African Statistical Journal*, **4**, 85–93.
Enis, P. and Geisser, S. (1974). Optimal predictive linear discriminants. *Annals of Statistics*, **2**, 403–410.
Etezadi-Amoli, J. and McDonald, R. P. (1983). A second generation nonlinear factor analysis. *Psychometrika*, **48**, 315–342.
Everitt, B. S. (1974). *Cluster Analysis*. Heinemann, London, England.
Everitt, B. S. (1977). *The Analysis of Contingency Tables*. Chapman and Hall, London, England.
Everitt, B. S. (1981). A Monte Carlo investigation of the likelihood ratio test for the number of components in a mixture of normal distributions. *Multivariate Behavioral Research*, **16**, 171–180.
Everitt, B. S. (1984a). *An Introduction to Latent Variable Models*. Chapman and Hall, London, England.
Everitt, B. S. (1984b). Maximum likelihood estimation of the parameters in a mixture of two univariate normal distributions; a comparison of different algorithms. *The Statistician*, **33**, 205–215.
Everitt, B. S. (1984c). A note on parameter estimation for Lazarsfeld's latent class model using the EM algorithm. *Multivariate Behavioral Research*, **19**, 79–89.

Everitt, B. S. (1987) *Introduction to Optimization Methods and their Application in Statistics*. Chapman and Hall, London, England.
Everitt, B. S. (1988). A finite mixture model for the clustering of mixed-mode data. *Statistics and Probability Letters*, **6**, 305–309.
Everitt, B. S. (1993). *Cluster Analysis*, 3rd edn. Edward Arnold, London.
Everitt, B. S. and Dunn, G. (1991). *Applied Multivariate Data Analysis*. Edward Arnold, London, England.
Everitt, B. S. and Hand, D. J. (1981). *Finite Mixture Distributions*. Chapman and Hall, London, England.
Everitt, B. S. and Merette, C. (1990). The clustering of mixed-mode data: a comparison of possible approaches. *Journal of Applied Statistics*, **17**, 283–297.
Fachel, J. M. G. (1986). *The C-type distribution as an underlying model for categorical data and its use in factor analysis*. Unpublished Ph.D. Thesis, London University, England.
Felling, A., Peters, J. and Schreuder, O. (1987). *Religion in Dutch Society. Documentation of a National Survey on Religious and Secular Attitudes in 1985*. Steinmetz Archives, Amsterdam, The Netherlands.
Feltz, C. J. and Dykstra, R. L. (1985). Maximum likelihood estimation of the survival functions of N stochastically ordered random variables. *Journal of the American Statistical Association*, **80**, 1012–1019.
Fienberg, S. E. (1980). *The Analysis of Cross-Classified Categorical Data*, 2nd edition. MIT Press, Cambridge, Massachusetts, USA.
Fingleton, B. (1984). *Models of Category Counts*. Cambridge University Press, Cambridge, England.
Fisher, N. I. (1993). *Statistical Analysis of Circular Data*. Cambridge University Press, Cambridge, England.
Fisher, N. I., Lewis, T. and Embleton, B. J. J. (1987). *Statistical Analysis of Spherical Data*. Cambridge University Press, Cambridge, England.
Fisher, R. A. (1936). The use of multiple measurements in taxonomic problems. *Annals of Eugenics*, **7**, 179–184.
Fisher, R. A. (1938). The statistical utilization of multiple measurements. *Annals of Eugenics*, **8**, 376–386.
Fisher, R. A. (1953). Dispersion on a sphere. *Proceedings of the Royal Society of London, Series A*, **217**, 295–305.
Fitzmaurice, G. M. and Hand, D. J. (1987). A comparison of two average conditional error rate estimators. *Pattern Recognition Letters*, **6**, 221–224.
Fitzmaurice, G. M., Krzanowski, W. J. and Hand, D. J. (1991). A Monte Carlo study of the 632 bootstrap estimator of error rate. *Journal of Classification*, **8**, 239–250.
Fix, E. and Hodges, J. L. (1951). *Discriminatory Analysis—Nonparametric Discrimination: Consistency Properties*. Report No. 4, US Air Force School of Aviation Medicine, Randolph Field, Texas, USA. (Reprinted as pp. 261–279 of Agrawala, 1977).
Fix, E. and Hodges, J. L. (1952). *Discriminatory Analysis. Nonparametric Discrimination: Small Sample Performance*. Report No. 11, US Air Force School of Aviation Medicine, Randolph Field, Texas, USA. (Reprinted as pp. 280–322 of Agrawala, 1977).

Florek, K., Łukaszewicz, J., Perkal, J., Steinhaus, H. and Zubrzycki, S. (1951). Sur la liaison et la division des points d'un ensemble fini. *Colloquium Mathematicum*, **2**, 282–285.
Flury, B. D. (1988). *Common Principal Components and Related Models*. Wiley, New York, USA.
Flury, B. D. (1990). Principal points. *Biometrika*, **77**, 33–41.
Flury, B. D. (1993). Estimation of principal points. *Applied Statistics*, **42**, 139–151.
Flury, B. D. and Schmid, M. J. (1992). Quadratic discriminant functions with constraints on the covariance matrices: some asymptotic results. *Journal of Multivariate Analysis*, **40**, 244–261.
Flury, B. D. and Tarpey, T. (1993). Representing a large collection of curves: a case for principal points. *The American Statistician*, **47**, 304–306.
Folks, J. L. and Chhikara, R. S. (1978). The inverse Gaussian distribution and its statistical applications—a review. *Journal of the Royal Statistical Society, Series B*, **40**, 263–289.
Fowlkes, E. B., Gnanadesikan, R. and Kettenring, J. R. (1988). Variable selection in clustering. *Journal of Classification*, **5**, 205–288.
Frank, I. E. (1987). Intermediate least squares regression method. *Journal of Chemometrics*, **1**, 233–242.
Frank, I. E. and Friedman, J. H. (1989). Classification: oldtimers and newcomers. *Journal of Chemometrics*, **3**, 463–475.
Freed, N. and Glover, F. (1981). A linear programming approach to the discriminant problem. *Decision Sciences*, **12**, 68–74.
Friedman, H. P. and Rubin, J. (1967). On some invariant criteria for grouping data. *Journal of the American Statistical Association*, **62**, 1159–1178.
Friedman, J. H. (1989). Regularized discriminant analysis. *Journal of the American Statistical Association*, **84**, 165–175.
Friedman, J. H. (1991). Multivariate adaptive regression splines (with discussion). *Annals of Statistics*, **19**, 1–141.
Friedman, J. H. and Rafsky, L. C. (1981). Graphics for the multivariate two-sample problem. *Journal of the American Statistical Association*, **76**, 277–287.
Friedman, J. H. and Stuetzle, W. (1981). Projection pursuit regression. *Journal of the American Statistical Association*, **78**, 817–823.
Fryer, M. J. (1977). A review of some methods of density estimation. *Journal of the Institute of Mathematics and its Applications*, **20**, 335–354.
Fujikoshi, Y. (1985). Selection of variables in two-group discriminant analysis by error rate and Akaike's information criteria. *Journal of Multivariate Analysis*, **17**, 27–37.
Fujikoshi, Y. and Rao, C. R. (1991). Selection of covariables in the growth curve model. *Biometrika*, **78**, 779–785.
Fukunaga, K. and Kessel, D. L. (1971). Estimation of classification error. *IEEE Transactions on Computers*, **C-20**, 1521–1527.
Fukunaga, K. and Hayes, R. R. (1989). Estimation of classifier performance. *IEEE Transactions on Pattern Analysis and Machine Intelligence*, **PAMI-11**, 1087–1101.

Fukunaga, K. and Hummels, D. M. (1989). Leave-one-out procedures for nonparametric error estimates. *IEEE Transactions on Pattern Analysis and Machine Intelligence*, **PAMI-11**, 421–423.
Fuller, W. A. (1987). *Measurement Error Models*. Wiley, New York, USA.
Furnival, G. M. (1971). All possible regressions with less computation. *Technometrics*, **13**, 403–408.
Gabriel, K. R. (1961). The model of ante-dependence for data of biological growth. *Bulletin of the International Statistical Institute*, 33rd Session (Paris), 253–264.
Gabriel, K. R. (1962). Antedependence analysis of a set of ordered variables. *Annals of Mathematical Statistics*, **33**, 201–212.
Ganesalingam, S. and McLachlan, G. J. (1978). The efficiency of a linear discriminant function based on unclassified initial samples. *Biometrika*, **65**, 658–662.
Ganesalingam, S. and McLachlan, G. J. (1979). Small sample results for a linear discriminant function estimated from a mixture of normal populations. *Journal of Statistical Computation and Simulation*, **9**, 151–158.
Ganeshanandam, S. and Krzanowski, W. J. (1989). On selecting variables and assessing their performance in linear discriminant analysis. *Australian Journal of Statistics*, **32**, 443–447.
Ganeshanandam, S. and Krzanowski, W. J. (1990). Error-rate estimation in two-group discriminant analysis using the linear discriminant function. *Journal of Statistical Computation and Simulation*, **36**, 157–175.
Garside, M. J. (1971). Some computational procedures for the best subset problem. *Applied Statistics*, **20**, 8–15.
Garthwaite, P. H. (1994). An interpretation of partial least squares. *Journal of the American Statistical Association*, **89**, 122–127.
Geisser, S. (1967). Estimation associated with linear discriminants. *Annals of Mathematical Statistics*, **38**, 807–817.
Geisser, S. (1977). Discrimination, allocatory, and separatory, aspects. In *Classification and Clustering* (ed J. van Ryzin), pp. 303–330. Academic Press, New York, USA.
Geisser, S. (1982). Bayesian discrimination. In *Handbook of Statistics, Volume 2* (eds P. R. Krishnaiah and L. N. Kanal), pp. 101–120. North-Holland, Amsterdam, The Netherlands.
Gelfand, S. B. and Delp, E. J. (1991). On tree structured classifiers. In *Artificial Neural Networks and Statistical Pattern Recognition. Old and New Connections* (eds I. K. Sethi and A. K. Jain), pp. 51–70. North-Holland, Amsterdam, The Netherlands.
Geman, S. and Geman, D. (1984). Stochastic relaxation, Gibbs distribution and the Bayesian restoration of images. *IEEE Transactions on Pattern Analysis and Machine Intelligence*, **PAMI-6**, 721–741.
Gilbert, E. S. (1969). The effect of unequal variance covariance matrices on Fisher's linear discriminant function. *Biometrics*, **25**, 505–515.
Gill, R. D. (1977). Consistency of maximum likelihood estimators of the factor analysis model, when the observations are not multivariate normal. In *Recent Developments in Statistics* (eds J. R. Barra, F. Brodeau, G. Romier and B. van Cutsem), pp. 437–440. North-Holland, Amsterdam, The Netherlands.

Glasbey, C. A. (1987). Complete linkage as a multiple stopping rule for single linkage clustering. *Journal of Classification*, **4**, 103–109.
Glick, N. (1973). Separation and probability of correct classification among two or more distributions. *Annals of the Institute of Statistical Mathematics*, **25**, 373–382.
Glick, N. (1976). Sample-based classification procedures related to empirical distributions. *IEEE Transactions on Information Theory*, **IT-22**, 454–461.
Glick, N. (1978). Additive estimators for probabilities of correct classification. *Pattern Recognition*, **10**, 211–222.
GLIM (1986). *Release 3.77 of GLIM System*. Numerical Algorithm Group Ltd, Oxford, England.
Golub, G. H. and van Loan, C. F. (1983). *Matrix Computations*. John Hopkins University Press, Baltimore, USA.
Gong, G. (1982). *Cross-Validation, the Jackknife, and the Bootstrap*. Unpublished Ph.D. Thesis, Stanford University, USA.
Gong, G. (1986). Cross-validation, the jackknife, and the bootstrap: excess error estimation in forward logistic regression. *Journal of the American Statistical Association*, **81**, 108–113.
Goodall, C. R. (1991). Procrustes methods in the statistical analysis of shape (with discussion). *Journal of the Royal Statistical Society, Series B*, **53**, 285–339.
Goodall, C. R. and Mardia, K. V. (1993). Multivariate aspects of shape theory. *Annals of Statistics*, **21**, 848–866.
Goodman, I. R. and Kotz, S. (1973). Multivariate θ-generalized normal distributions. *Journal of Multivariate Analysis*, **3**, 204–219.
Goodman, L. A. (1978). *Analyzing Qualitative/Categorical Data* (ed J. Magidson). Abt Books, Cambridge, Massachusetts, USA.
Gordon, A. D. (1981). *Classification*. Chapman and Hall, London, England.
Gordon, A. D. (1994). Identifying genuine clusters in a classification. *Computational Statistics and Data Analysis*, **18**, 561–581.
Gower, J. C. (1966). A Q-technique for the calculation of canonical variates. *Biometrika*, **53**, 588–589.
Gower, J. C. (1967). A comparison of some methods of cluster analysis. *Biometrics*, **23**, 623–628.
Gower, J. C. (1971). A general coefficient of similarity and some of its properties. *Biometrics*, **27**, 857–872.
Gower, J. C. (1976). Growth-free canonical variates and generalized inverses. *Bulletin of the Geological Institution of the University of Uppsala, N.S.*, **7**, 1-10.
Gower, J. C. (1985). Measures of similarity, dissimilarity, and distance. In *Encyclopedia of Statistical Sciences, Volume 5* (eds S. Kotz and N. L. Johnson), pp. 397–405. Wiley, New York, USA.
Gower, J. C. (1989). Generalised canonical analysis. In *Multiway Data Analysis* (eds R. Coppi and S. Bolasco), pp. 221–232. North-Holland, Amsterdam, The Netherlands.
Gower, J. C. and Banfield, C. F. (1975). Goodness-of-fit criteria for hierarchical classification, and their empirical distributions. In *Proceedings of the 8th International Biometric Conference* (eds L. C. A. Corsten and T.

Postelnicu), pp. 347–361. Editura Academici Republicii Socialiste Romania.

Gower, J. C. and Legendre, P. (1986). Metric and Euclidean properties of dissimilarity coefficients. *Journal of Classification*, **3**, 5–48.

Gower, J. C. and Ross, G. J. S. (1969). Minimum spanning trees and single linkage cluster analysis. *Applied Statistics*, **18**, 54–64.

Green, P. E., Carmone, F. J. and Kim, J. (1990). A preliminary study of optimal variable weighting in K-means clustering. *Journal of Classification*, **7**, 271–285.

Green, P. J. and Silverman, B. W. (1993). *Nonparametric Regression and Generalized Linear Models: A Roughness Penalty Approach*. Chapman and Hall, London, England.

Greenhouse, S. W. and Geisser, S. (1959). On methods in the analysis of profile data. *Psychometrika*, **24**, 95–112.

Greer, R. L. (1984). *Trees and Hills: Methodology for Maximizing Functions of Systems of Linear Relations*. North-Holland, Amsterdam, The Netherlands.

Grizzle, J. E. and Allen, D. M. (1969). Analysis of growth and dose response curves. *Biometrics*, **25**, 357–381.

Guénoche, A., Hansen, P. and Jaumard, B. (1991). Efficient algorithms for divisive hierarchical clustering with the diameter criterion. *Journal of Classification*, **8**, 5–30.

Gupta, A. K. (1980). On a multivariate statistical classification model. In *Multivariate Statistical Analysis* (ed R. P. Gupta), pp. 83–93. North-Holland, Amsterdam, The Netherlands.

Gupta, A. K. (1986). On a classification rule for multiple measurements. *Computers and Mathematics with Applications*, **12A(2)**, 301–308.

Gupta, A. K. and Logan, T. P. (1990). On a multiple observations model in discriminant analysis. *Journal of Statistical Computation and Simulation*, **34**, 119–132.

Gweke, J. F. and Singleton, K. J. (1980). Interpreting the likelihood ratio statistic in factor models when sample size is small. *Journal of the American Statistical Association*, **75**, 133–137.

Habbema, J. D. F. and Hermans, J. (1977). Selection of variables in discriminant analysis by F-statistic and error rate. *Technometrics*, **19**, 487-493.

Habbema, J. D. F., Hermans, J. and van der Burgt, A. T. (1974). Cases of doubt in allocation problems. *Biometrika*, **61**, 313–324.

Habbema, J. D. F., Hilden, J. and Bjerregaard, B. (1981). The measurement of performance in probabilistic diagnosis V. General recommendations. *Methods of Information in Medicine*, **20**, 97–100.

Haberman, S. J. (1974). *The Analysis of Frequency Data*. University of Chicago Press, Chicago, USA.

Hakstian, A. R., Rogers, W. T. and Cattell, R. B. (1982). The behavior of number-of-factors rules with simulated data. *Multivariate Behavioral Research*, **17**, 193–219.

Hall, P. (1981). Optimal near neighbour estimator for use in discriminant analysis. *Biometrika*, **68**, 572–575.

Hall, P. (1983) Orthogonal series methods for both qualitative and quantitative data. *Annals of Statistics*, **11**, 1004–1007.

Hall, P. (1986). On the number of bootstrap simulations required to construct a confidence interval. *Annals of Statistics*, **14**, 1453–1462.
Hall, P., Sheather, S. J., Jones, M. C. and Marron, J. S. (1991). On optimal data-based bandwidth selection in kernel density estimation. *Biometrika*, **78**, 263–269.
Halperin, M., Blackwelder, W. C. and Verter, J. I. (1971). Estimation of the multivariate logistic risk function: a comparison of the discriminant and maximum likelihood approaches. *Journal of Chronic Diseases*, **24**, 125–158.
Halperin, M., Abbott, R. D., Blackwelder, W. C., Jacobowitz, R., Lan, K. K. G., Verter, J. I. and Wedel, H. (1985). On the use of the logistic model in prospective studies. *Statistics in Medicine*, **4**, 227–235.
Han, C. P. (1968). A note on discrimination in the case of unequal covariance matrices. *Biometrika*, **55**, 586–587.
Han, C. P. (1979). Alternative methods of estimating the likelihood ratio inclassification of multivariate normal observations. *The American Statistician*, **33**, 204–206.
Hand, D. J. (1981). *Discrimination and Classification*. Wiley, Chichester, England.
Hand, D. J. (1982). *Kernel Discriminant Analysis*. Research Studies Press, Letchworth, England.
Hand, D. J. (1983). A comparison of two methods of discriminant analysis applied to binary data. *Biometrics*, **39**, 683–694.
Hand, D. J. (1986). Recent advances in error rate estimation. *Pattern Recognition Letters*, **4**, 335–346.
Hand, D. J. (1987). A shrunken leaving-one-out estimator of error rate. *Computers and Mathematics with Applications*, **14**, 161–167.
Hand, D. J. (1994). Assessing classification rules. *The Journal of Applied Statistics*, **21**, 3–16.
Härdle, W. (1991). *Smoothing Techniques*. Springer-Verlag, Berlin, Germany.
Harman, H. H. (1976). *Modern Factor Analysis*, 3rd edition. University of Chicago Press, Chicago, USA.
Harshman, R. A. and Lundy, M. E. (1984). The PARAFAC model for three-way factor analysis and multidimensional scaling. In *Research Methods in Multimode Data Analysis* (eds H. G. Law, C. W. Snyder Jr., J. A. Hattie and R. P. McDonald). Praeger, New York, USA.
Hartigan, J. A. (1975a). *Clustering Algorithms*. Wiley, New York, USA.
Hartigan, J. A. (1975b). Printer graphics for clustering. *Journal of Statistical Computation and Simulation*, **4**, 187-213.
Hartigan, J. A. (1977). Distribution problems in clustering. In *Classification and Clustering* (ed J. van Ryzin), pp. 45–71. Academic Press, New York, USA.
Hartigan, J. A. (1985). Statistical theory in clustering. *Journal of Classification*, **2**, 63–76.
Hartigan, J. A. and Hartigan, P. M. (1985). The dip test of multimodality. *Annals of Statistics*, **13**, 70–84.
Hartley, H. O. (1967). Expectations, variances and covariances of Anova mean squares by 'synthesis'. *Biometrics*, **23**, 105–114.

Hastie, T. J. and Tibshirani, R. J. (1990). *Generalized Additive Models*. Chapman and Hall, London, England.
Hawkins, D. M. (1976). The subset problem in multivariate analysis of variance. *Journal of the Royal Statistical Society, Series B*, **38**, 132–139.
Hawkins, D. M. and Raath, E. L. (1982). An extension of Geisser's discrimination model to proportional covariance matrices. *Canadian Journal of Statistics*, **10**, 261–270.
Heise, D. R. (1986). Estimating nonlinear models correcting for measurement error. *Sociological Methods and Research*, **14**, 447–472.
Heiser, W. J. (1995). Convergent computation by iterative majorization: theory and applications in multidimensional data analysis. In *Recent Advances in Descriptive Multivariate Analysis* (ed W. J. Krzanowski), pp. 157–189. Clarendon Press, Oxford, England.
Heitjan, D. F. (1991a). Generalized Norton-Simon models of tumour growth. *Statistics in Medicine*, **10**, 1075–1088.
Heitjan, D. F. (1991b). Nonlinear modeling of serial immunologic data; a case study. *Journal of the American Statistical Association*, **86**, 891–898.
Helland, I. S. (1990). Partial least squares regression and statistical methods. *Scandinavian Journal of Statistics*, **17**, 97–114.
Hellman, M. E. (1970). The nearest neighbor classification rule with a reject option. *IEEE Transactions on Systems, Science and Cybernetics*, **SSC-6**, 179–185.
Hermans, J. and Habbema, J. D. F. (1975). Comparison of five methods to estimate posterior probabilities. *EDV in Medizin und Biologie*, **6**, 14–19.
Hermans, J., Habbema, J. D. F. and Schaefer, J. R. (1982). The ALLOC80 package for discriminant analysis. *Statistical Software Newsletter*, **8**, 15–20.
Hernandez Avila, A. (1979). *Problems in Cluster Analysis*. D. Phil. Thesis (unpubl.), Oxford University, England.
Hertz, J., Krogh, A. and Palmer, R. G. (1991). *Introduction to the Theory of Neural Computation*. Addison-Wesley, Reading, Massachusetts, USA.
Herz, C. S. (1955). Bessel functions of matrix argument. *Annals of Mathematics, Series 2*, **61**, 474–523.
Heywood, H. B. (1931). On finite sequences of real numbers. *Proceedings of the Royal Society, Series A*, **134**, 486–510.
Highleyman, W. H. (1962). The design and analysis of pattern recognition experiments. *Bell Systems Technical Journal*, **41**, 723–744.
Hills, M. (1966). Allocation rules and their error rates (with discussion). *Journal of the Royal Statistical Society, Series B*, **28**, 1–31.
Hills, M. (1967). Discrimination and allocation with discrete data. *Applied Statistics*, **16**, 237–250.
Hinde, J. P. (1985). Descriptive classification of Cetacea: a latent class solution. In *Data Analysis in Real Life Environment: Ins and Outs of Solving Problems* (eds J.-F. Marcotorchino, J.-M. Proth and J. Jansen), pp. 37–43. North-Holland, Amsterdam, The Netherlands.
Hopkins, J. W. (1966). Some considerations in multivariate allometry. *Biometrics*, **22**, 747–760.

Horn, J. and Knapp, J. R. (1973). On the subjective character of the empirical base of Guilford's structure-of-intellect model. *Psychological Bulletin*, **80**, 33–43.

Höskuldsson, A. (1988). PLS regression methods. *Journal of Chemometrics*, **2**, 211–228.

Hosmer, D. W., Jr. (1973). A comparison of iterative maximum likelihood estimates of the parameters of a mixture of two normal distributions under three different types of sample. *Biometrics*, **29**, 761–770.

Hosmer, T. A., Hosmer, D. W. and Fisher, L. (1983a). A comparison of the maximum likelihood and discriminant function estimators of the coefficients of the logistic regression model for mixed continuous and discrete variables. *Communications in Statistics, Simulation and Computation*, **12**, 23–43.

Hosmer, T. A., Hosmer, D. W. and Fisher, L. (1983b). A comparison of three methods of estimating the logistic regression coefficients. *Communications in Statistics, Simulation and Computation*, **12**, 577–593.

Hotelling, H. (1951). A generalized T test and measure of multivariate dispersion. *Proceedings of the 2nd Berkeley Symposium*, pp. 23–41. University of California Press, Berkeley, California, USA.

Howe, W. G. (1955). *Some Contributions to Factor Analysis*. Oak Ridge National Laboratory, Oak Ridge, Tennessee, USA.

Huber, P. J. (1981). *Robust Statistics*. Wiley, New York, USA.

Hubert, L. J. and Arabie, P. (1985). Comparing partitions. *Journal of Classification*, **2**, 193–218.

Hufnagel, G. (1988). On estimating missing values in linear discriminant analysis: Part I. *Biometrical Journal*, **30**, 69–75.

Huitson, A. (1989). Problems with Procrustes analysis. *Journal of Applied Statistics*, **16**, 39–45.

Huynh, H. and Feldt, L. S. (1976). Estimation of the Box correction for degrees of freedom from sample data in the randomized block and split-plot designs. *Journal of Educational Statistics*, **1**, 69–82.

Jaccard, P. (1908). Nouvelles recherches sur la distribution florale. *Bulletin de la Société Vaudoise des Sciences Naturelles*, **44**, 223–270.

Jackson, M. (1989). *Michael Jackson's Malt Whisky Companion: a Connoisseur's Guide to the Malt Whiskies of Scotland*. Dorling Kindersley, London, England.

Jacobs, R. A. (1988). Increased rates of convergence through learning rate adaptation. *Neural Networks*, **1**, 295–307.

Jain, A. K. and Chandrasekaran, B. (1982). Dimensionality and sample size considerations in pattern recognition practice. In *Handbook of Statistics, Volume 2* (eds P. R. Krishnaiah and L. N. Kanal), pp. 835–855. North-Holland, Amsterdam, The Netherlands.

Jain, A. K., Dubes, R. C. and Chen, C.-C. (1987). Bootstrap techniques for error estimation. *IEEE Transactions on Pattern Analysis and Machine Intelligence*, **PAMI-9**, 628–633.

Jardine, N. and Sibson, R. (1968). The construction of hierarchic and nonhierarchic classifications. *Computer Journal*, **11**, 177–194.

Jardine, N. and Sibson, R. (1971). *Mathematical Taxonomy*. Wiley, London, England.

Jeffreys, H. (1948). *Theory of Probability*, 2nd edition. Clarendon Press, Oxford, England.
Jennrich, R. I. and Robinson, S. M. (1969). A Newton–Raphson algorithm for maximum likelihood factor analysis. *Psychometrika*, **34**, 111–123.
Jennrich, R. I. and Thayer, D. T. (1973). A note on Lawley's formulas for standard errors in maximum likelihood factor analysis. *Psychometrika*, **38**, 571–580.
Jeran, P. W. and Mashey, J. R. (1970). A computer program for the stereographic analysis of coal fractures and cleats. US Bureau of Mines Information Circular No. 8454.
Joachimsthaler, E. A. and Stam, A. (1988). Four approaches to the classification problem in discriminant analysis: an experimental study. *Decision Sciences*, **19**, 322–333.
John, S. (1959). The distribution of Wald's classification statistic when the dispersion matrix is known. *Sankhyā*, **21**, 371–375.
John, S. (1960). On some classification statistics. *Sankhyā*, **22**, 309–316.
John, S. (1961). Errors in discrimination. *Annals of Mathematical Statistics*, **32**, 1125–1144.
Jolicoeur, P. (1963). The multivariate generalization of the allometry equation. *Biometrics*, **19**, 497–499.
Jolicoeur, P. (1984). Principal components, factor analysis and multivariate allometry: a small-sample direction test. *Biometrics*, **40**, 685–690.
Jolliffe, I. T. (1986). *Principal Component Analysis*. Springer-Verlag, New York, USA.
Jones, M. C. and Rice, J. A. (1992). Displaying the important features of a large collection of similar curves. *The American Statistician*, **46**, 140–145.
Jöreskog, K. G. (1963). *Statistical Estimation in Factor Analysis*. Almqvist and Wiksell, Stockholm, Sweden.
Jöreskog, K. G. (1967). Some contributions to maximum likelihood factor analysis. *Psychometrika*, **32**, 443–482.
Jöreskog, K. G. (1969). A general approach to confirmatory maximum likelihood factor analysis. *Psychometrika*, **34**, 183–220.
Jöreskog, K. G. (1970). A general method for analysis of covariance structures. *Biometrika*, **57**, 239–251.
Jöreskog, K. G. (1973a). A general method for estimating a linear structural equation system. In *Structural Equation Models in the Social Sciences* (eds A. S. Goldberger and O. D. Duncan), pp. 85–112. Academic Press, New York, USA.
Jöreskog, K. G. (1973b). Analysis of covariance structures. In *Multivariate Analysis III* (ed P. R. Krishnaiah), pp. 263–285. Academic Press, New York, USA.
Jöreskog, K. G. and Sorbom, D. (1986). *LISREL VI: Analysis of Linear Structural Relationships by Maximum Likelihood and Least Squares Methods*. Scientific Software, Inc, Mooresville, Indiana, USA.
Jöreskog, K. G. and Sorbom, D. (1988). *LISREL 7: A Guide to the Program and Applications*. SPSS, Inc, Chicago, Illinois, USA.
Kaiser, H. F. (1958). The varimax criterion for analytical rotation in factor analysis. *Psychometrika*, **23**, 187–200.

Kang, K. M. and Seneta, E. (1980). Path analysis: an exposition. In *Developments in Statistics, Volume 3* (ed P. R. Krishnaiah), pp. 217–246. Academic Press, New York, USA.
Kaufman, L. and Rousseeuw, P. J. (1990). *Finding Groups in Data: An Introduction to Cluster Analysis*. Wiley, New York, USA.
Kendall, D. G. (1977). The diffusion of shape. *Advances in Applied Probability*, **9**, 428–430.
Kendall, D. G. (1984). Shape manifolds, procrustean metrics, and complex projective spaces. *Bulletin of the London Mathematical Society*, **16**, 81–121.
Kendall, D. G. (1989). A survey of the statistical theory of shape (with discussion). *Statistical Science*, **4**, 87–120.
Kendall, M. G. (1966). Discrimination and clustering. In *Multivariate Analysis* (ed P. R. Krishnaiah), pp. 165-185, Academic Press, New York, USA.
Kendall, M. G. (1980). *Multivariate Analysis, 2nd Edition*. Griffin, London, England.
Kendall, M. G. and Moran, P. A. P. (1963). *Geometrical Probability*. Griffin, London, England.
Kendall, M. G., Stuart, A. S. and Ord, J. K. (1983). *The Advanced Theory of Statistics, Volume 3*, 4th edition. Griffin, London, England.
Kenny, D. A. and Judd, C. M. (1984). Estimating the non-linear and interactive effects of latent variables. *Psychological Bulletin*, **96**, 201–210.
Kent, J. T. (1994). The complex Bingham distribution and shape analysis. *Journal of the Royal Statistical Society, Series B*, **56**, 285–299.
Kenward, M. G. (1985). The use of fitted higher-order polynomial coefficients as covariates in the analysis of growth curves. *Biometrics*, **41**, 19–28.
Kenward, M. G. (1986). The distribution of a generalized least squares estimator with covariance adjustment. *Journal of Multivariate Analysis*, **3**, 244–250.
Kenward, M. G. (1987). A method for comparing profiles of repeated measurements. *Applied Statistics*, **36**, 296–308.
Kiiveri, H. T. (1992). Canonical variate analysis of high-dimensional spectral data. *Technometrics*, **34**, 321–331.
Kiiveri, H. and Speed, T. P. (1982). Structural analysis of multivariate data: a review. In *Sociological Methodology* (ed S. Leinhardt), pp. 209–289. Jossey-Buss, San Francisco, USA.
Kimura, F., Takashina, K., Tsuruoka, S. and Miyake, Y. (1987). Modified quadratic discriminant functions and their application to Chinese character recognition. *IEEE Transactions on Pattern Analysis and Machine Intelligence*, **PAMI-9**, 149–153.
Knoke, J. D. (1982). Discriminant analysis with discrete and continuous variables. *Biometrics*, **38**, 191–200.
Knoke, J. D. (1986). The robust estimation of classification error rates. *Computers and Mathematics with Applications*, **12A(2)**, 253–260.
Kohonen, T. (1990). The self-organizing map. *Proceedings of the IEEE*, **78**, 1464–1480.
Krane, W. R. and McDonald, R. P. (1978). Scale invariance and the factor analysis of correlation matrices. *British Journal of Mathematical and Statistical Psychology*, **31**, 218–228.

Krishnan, T. and Nandy, S. C. (1987). Discriminant analysis with stochastic supervisor. *Pattern Recognition*, **20**, 379–384.
Kroonenberg, P. M. and ten Berge, J. M. F. (1989). Three-mode principal component analysis and perfect congruence analysis for sets of covariance matrices. *British Journal of Mathematical and Statistical Psychology*, **42**, 62–80.
Krusinska, E. (1988). Variable selection in location model for mixed variable discrimination: a procedure based on total probability of misclassification. *EDV in Medizin und Biologie*, **19**, 14–18.
Krusinska, E. (1989a). New procedure for selection of variables in location model for mixed variable discrimination. *Biometrical Journal*, **31**, 511–523.
Krusinska, E. (1989b). Two step semi-optimal branch and bound algorithm for feature selection in mixed variable discrimination. *Pattern Recognition*, **22**, 455–459.
Krusinska, E. (1990). Suitable location model selection in the terminology of graphical models. *Biometrical Journal*, **32**, 817–826.
Kruskal, J. B. (1964). Multidimensional scaling by optimising goodness-of-fit to a nonmetric hypothesis. *Psychometrika*, **29**, 1–27.
Kruskal, J. B. and Shepard, R. N. (1974). A nonmetric variety of linear factor analysis. *Psychometrika*, **39**, 123–157.
Krzanowski, W. J. (1971). A comparison of some distance measures applicable to multinomial data, using a rotational fit technique. *Biometrics*, **27**, 1062–1068.
Krzanowski, W. J. (1975). Discrimination and classification using both binary and continuous variables. *Journal of the American Statistical Association*, **70**, 782–790.
Krzanowski, W. J. (1977). The performance of Fisher's linear discriminant function under non-optimal conditions. *Technometrics*, **19**, 191–200.
Krzanowski, W. J. (1980). Mixtures of continuous and categorical variables in discriminant analysis. *Biometrics*, **36**, 493–499.
Krzanowski, W. J. (1982). Mixtures of continuous and categorical variables in discriminant analysis: a hypothesis-testing approach. *Biometrics*, **38**, 991–1002.
Krzanowski, W. J. (1983a). Stepwise location model choice in mixed-variable discrimination. *Applied Statistics*, **32**, 260–266.
Krzanowski, W. J. (1983b). Distance between populations using mixed continuous and categorical variables. *Biometrika*, **70**, 235–243.
Krzanowski, W. J. (1986). Multiple discriminant analysis in the presence of mixed continuous and categorical data. *Computers and Mathematics with Applications*, **12A(2)**, 179–185.
Krzanowski, W. J. (1987). A comparison between two distance-based discriminant principles. *Journal of Classification*, **4**, 73–84.
Krzanowski, W. J. (1988a). *Principles of Multivariate Analysis; a User's Perspective*. Clarendon Press, Oxford, England.
Krzanowski, W. J. (1988b). Missing value imputation in multivariate data using the singular value decomposition of a matrix. *Biometrical Letters*, **25**, 31–39.

Krzanowski, W. J. (1989). On confidence regions in canonical variate analysis. *Biometrika*, **76**, 107–116.
Krzanowski, W. J. (1990). Between-group analysis with heterogeneous covariance matrices: the common principal component model. *Journal of Classification*, **7**, 81–98.
Krzanowski, W. J. (1992). Ranking principal components to reflect group structure. *Journal of Chemometrics*, **6**, 97–102.
Krzanowski, W. J. (1993a). The location model for mixtures of categorical and continuous variables. *Journal of Classification*, **10**, 25–49.
Krzanowski, W. J. (1993b). Antedependence modelling in discriminant analysis of high-dimensional spectroscopic data. In *Multivariate Analysis: Future Directions 2* (eds C. M. Cuadras and C. R. Rao), pp. 87–95. Elsevier Science Publishers, Amsterdam, The Netherlands.
Krzanowski, W. J. (1994a). Ordination in the presence of group structure, for general multivariate data. *Journal of Classification*, **11**, 195–207.
Krzanowski, W. J. (1994b). Quadratic location discriminant functions for mixed categorical and continuous data. *Statistics and Probability Letters*, **19**, 91–95.
Krzanowski, W. J. (1995). Selection of variables, and assessment of their performance, in mixed-variable discriminant analysis. *Computational Statistics and Data Analysis*, **19**, 419–431.
Krzanowski, W. J. and Radley, D. (1989). Nonparametric confidence and tolerance regions in canonical variate analysis. *Biometrics*, **45**, 1163–1173.
Krzanowski, W. J., Jonathan, P., McCarthy, W. V. and Thomas, M. R. (1995). Discriminant analysis with singular covariance matrices: methods and applications to spectroscopic data. *Applied Statistics*, **44**, 101–115.
Kullback, S. (1959). *Information Theory and Statistics*. Wiley, New York, USA.
Kullback, S. and Leibler, R. (1951). On information and sufficiency. *Annals of Mathematical Statistics*, **22**, 79–86.
Kurczynski, T. W. (1970). Generalised distance and discrete variables. *Biometrics*, **26**, 525–534.
Lachenbruch, P. A. (1967). An almost unbiased method of obtaining confidence intervals for the probability of misclassification in discriminant analysis. *Biometrics*, **23**, 639–645.
Lachenbruch, P. A. (1975). *Discriminant Analysis*. Hafner, New York, USA.
Lachenbruch, P. A. (1979). Note on initial misclassification effects on the quadratic discriminant function. *Technometrics*, **21**, 129–132.
Lachenbruch, P. A. and Mickey, M. R. (1968). Estimation of error rates in discriminant analysis. *Technometrics*, **10**, 1–11.
Lachenbruch, P. A., Sneeringer, C. and Revo, L. T. (1973). Robustness of the linear and quadratic discriminant function to certain types of non-normality. *Communications in Statistics*, **1**, 39–57.
Lancaster, H. O. (1969). *The Chi-Squared Distribution*. Wiley, New York, USA.
Lance, G. N. and Williams, W. T. (1967a). Mixed-data agglomerative programs. I. Agglomerative systems. *Australian Computing Journal*, **1**, 15–20.
Lance, G. N. and Williams, W. T. (1967b). A general theory of classificatory sorting strategies: 1. Hierarchical systems. *Computer Journal*, **9**, 373–380.
Lapointe, F.-J. and Legendre, P. (1990). A statistical framework to test the consensus of two nested classifications. *Systematic Zoology*, **39**, 1–13.

Lapointe, F.-J. and Legendre, P. (1991). The generation of random ultrametric matrices representing dendrograms. *Journal of Classification*, **8**, 177–200.

Lapointe, F.-J. and Legendre, P. (1992). Statistical significance of the matrix correlation coefficient for comparing independent phylogenetic trees. *Systematic Biology*, **41**, 378–384.

Lapointe, F.-J. and Legendre, P. (1994). A classification of pure malt Scotch whiskies. *Applied Statistics*, **43**, 237–257.

Lauritzen, S. L. (1993). *Graphical Association Models*. (Draft version, University of Aalborg, Denmark).

Lauritzen, S. L. and Wermuth, N. (1984). *Mixed Interaction Models*, Research Report, Institute of Electronic Systems, University of Aalborg, Denmark.

Lauritzen, S. L. and Wermuth, N. (1989). Graphical models for associations between variables, some of which are qualitative and some quantitative. *Annals of Statistics*, **17**, 31–57.

Lawley, D. N. (1940). The estimation of factor loadings by the method of maximum likelihood. *Proceedings of the Royal Society of Edinburgh*, **60**, 64–82.

Lawley, D. N. (1941). Further investigations in factor estimation. *Proceedings of the Royal Society of Edinburgh*, **61**, 176–185.

Lawley, D. N. and Maxwell, A. E. (1971). *Factor Analysis as a Statistical Method*, 2nd edition. Butterworth, London, England.

Lazarsfeld, P. F. (1950). The logical and mathematical foundation of latent structure analysis. In *Measurement and Prediction* (eds S. A. Stouffer, L. Guttman, E. A. Suchman, P. F. Lazarsfeld, S. A. Star and J. A. Clausen), pp. 362–412. Princeton University Press, USA.

Lazarsfeld, P. F. (1961). The algebra of dichotomous systems. In *Studies in Item Analysis and Prediction* (ed H. Solomon), pp. 111–157. Stanford University Press, Stanford, California, USA.

Lazarsfeld, P. F. and Henry, N. W. (1968). *Latent Structure Analysis*. Houghton-Miflin, Boston, USA.

Le, H. and Kendall, D. G. (1993). The Riemannian structure of Euclidean shape spaces: a novel environment for statistics. *Annals of Statistics*, **21**, 1225–1271.

Le Cam, L. (1970). On the assumptions used to prove asymptotic normality of maximum likelihood estimates. *Annals of Mathematical Statistics*, **41**, 802–828.

Lee, J. C. (1982). Classification of growth curves. In *Handbook of Statistics, Volume 2* (eds P. R. Krishnaiah and L. N. Kanal), pp. 121–137. North-Holland, Amsterdam, The Netherlands.

Lee, S. Y. (1980). Estimation of covariance structure models with parameters subject to functional restraints. *Psychometrika*, **45**, 309–324.

Lee, S. Y. and Jennrich, R. I. (1979). A study of algorithms for covariance structure analysis with specific comparisons using factor analysis. *Psychometrika*, **44**, 99–113.

Lee, T. C., Judge, G. G. and Zellner, A. (1977). *Estimating the Parameters of the Markov Probability Model from Aggregate Time Data*. North-Holland, Amsterdam, The Netherlands.

Leech, F. B. and Healy, M. J. R. (1959). The analysis of experiments on growth rate. *Biometrics*, **15**, 98–106.

Leurgans, S. and Ross, R. T. (1992). Multilinear models: applications in spectroscopy. *Statistical Science*, **7**, 289–319.
Levin, J. (1965). Three-mode factor analysis. *Psychological Bulletin*, **64**, 442–452.
Lindsey, J. K. (1989). *The Analysis of Categorical Data Using GLIM*. Springer-Verlag, Berlin, Germany.
Lindsey, J. K. (1992). *The Analysis of Stochastic Processes Using GLIM*. Springer-Verlag, Berlin, Germany.
Lindsey, J. K. (1993). *Models for Repeated Measurements*. Oxford University Press, Oxford, England.
Ling, R. F. (1973). A probability theory for cluster analysis. *Journal of the American Statistical Association*, **68**, 159–164.
Lissack, T. and Fu, K. S. (1976). Error estimation in pattern recognition via L^{α}-distance between posterior density functions. *IEEE Transactions on Information Theory*, **IT-22**, 34–45.
Little, R. J. A. (1988). Robust estimation of the mean and covariance matrix from data with missing values. *Applied Statistics*, **37**, 23–38.
Little, R. J. A. and Rubin, D. B. (1987). *Statistical Analysis with Missing Data*. Wiley, New York, USA.
Little, R. J. A. and Schluchter, M. D. (1985). Maximum likelihood estimation for mixed continuous and categorical data with missing values. *Biometrika*, **72**, 497–512.
Lockhart, R. A. and Stephens, M. A. (1985). Tests of fit for the von Mises distribution. *Biometrika*, **72**, 647–652.
Logan, T. P. and Gupta, A. K. (1993). Bayesian discrimination using multiple observations. *Communications in Statistics, Theory and Methods*, **22**, 1735–1754.
Loh, W.-Y. and Vanichsetakul, N. (1988). Tree-structured classification via generalized discriminant analysis (with discussion). *Journal of the American Statistical Association*, **83**, 715–728.
Lord, F. M. and Novick, M. R. (1968). *Statistical Theories of Mental Test Scores*. Addison-Wesley, Reading, Massachusetts, USA.
Louis, T. A. (1982). Finding the observed information matrix when using the EM algorithm. *Journal of the Royal Statistical Society, Series B*, **44**, 226–233.
Lubischew, A. A. (1962). On the use of discriminant functions in taxonomy. *Biometrics*, **18**, 455–477.
Luk, A. and MacLeod, J. E. S. (1986). An alternative nearest neighbor classification scheme. *Pattern Recognition Letters*, **4**, 375–381.
Macdonald, P. D. M. (1987). Analysis of length-frequency distributions. In *Age and Growth of Fish* (eds R. C. Summerfelt and G. E. Hall), pp. 371–384. Iowa State University Press, Ames, USA.
MacNaughton-Smith, P., Williams, W. T., Dale, M. B. and Mockett, L. G. (1964). Dissimilarity analysis: a new technique of hierarchical subdivision. *Nature*, **202**, 1034–1035.
MacQueen, J. B. (1967). Some methods for classification and analysis of multivariate observations. In *Proceedings of the 5th Berkeley Symposium in Mathematical Statistics and Probability*, pp. 281–297. University of California Press, Berkeley, USA.

Macready, G. B. and Dayton, C. M. (1977). The use of probabilistic models in the assessment of mastery. *Journal of Educational Statistics*, **2**, 99–120.
Madansky, A. (1959). *Partitioning Methods in Latent Class Analysis*. Report P-1644, RAND Corporation, USA.
Magnus, J. R. and Neudecker, H. (1988). *Matrix Differential Calculus*. Wiley, New York, USA.
Mahalanobis, P. C. (1930). On tests and measures of group divergence. I. Theoretical formulae. *Journal of the Asiatic Society of Bengal*, **26**, 541–588.
Mahalanobis, P. C. (1936). On the generalized distance in statistics. *Proceedings of the National Institute of Science India*, **2**, 49–55.
Manly, B. F. J. (1986). *Multivariate Statistical Methods*. Chapman and Hall, London, England.
Manly, B. F. J. and Rayner, J. C. W. (1987). The comparison of sample covariance matrices using likelihood ratio tests. *Biometrika*, **74**, 841–847.
Mantel, N. (1967). The detection of disease clustering and a generalized regression approach. *Cancer Research*, **27**, 209–220.
Marco, V. R., Young, D. M. and Turner, D. W. (1988). Predictive discrimination for autoregressive processes. *Pattern Recognition Letters*, **7**, 145–149.
Mardia, K. V. (1970). Measures of multivariate skewness and kurtosis with applications. *Biometrika*, **57**, 519–520.
Mardia, K. V. (1972). *Statistics of Directional Data*. Academic Press, London, England.
Mardia, K. V. (1974). Applications of some measures of multivariate skewness and kurtosis for testing normality and robustness studies. *Sankhyā B*, **36**, 115–128.
Mardia, K. V. (1984). Spatial discrimination and classification maps. *Communications in Statistics, Theory and Methods*, **13**, 2181–2197. (Correction **16**, 3749.)
Mardia, K. V. and Dryden, I. L. (1989). Shape distributions for landmark data. *Advances in Applied Probability*, **21**, 742–755.
Mardia, K. V., Kent, J. T. and Bibby, J. M. (1979). *Multivariate Analysis*. Academic Press, London, England.
Marks, S. and Dunn, O. J. (1974). Discriminant functions when covariance matrices are unequal. *Journal of the American Statistical Association*, **69**, 555–559.
Marriott, F. H. C. (1971). Practical problems in a method of cluster analysis. *Biometrics*, **27**, 501–514.
Marriott, F. H. C. (1982). Optimization methods of cluster analysis. *Biometrika*, **69**, 417–422.
Martens, H. and Naes, T. (1989). *Multivariate Calibration*. Wiley, Chichester, England.
Martin, D. C. and Bradley, R. A. (1972). Probability models, estimation and classification for multivariate dichotomous populations. *Biometrics*, **28**, 203–221.
Martinson, E. O. and Hamdan, M. A. (1975). Calculation of the polychoric estimate of correlation in contingency tables. Algorithm AS87, *Applied Statistics*, **24**, 272–278.

Matusita, K. (1956). Decision rule, based on distance, for the classification problem. *Annals of the Institute of Statistical Mathematics*, **8**, 67–77.
Matusita, K. (1967). Classification based on distance in multivariate Gaussian cases. *Proceedings of the Fifth Berkeley Symposium in Mathematical Statistics and Probability*, **1**, 299–304.
Matusita, K. (1971). Some properties of affinity and applications. *Annals of the Institute of Statistical Mathematics*, **23**, 137–155.
Matusita, K. (1973). Discrimination and the affinity of distributions. In *Discriminant Analysis and Applications* (ed T. Cacoullos), pp. 213–223. Academic Press, New York, USA.
Mauchley, J. W. (1940). Significance test for sphericity of a normal n-variate distribution. *Annals of Mathematical Statistics*, **11**, 204–209.
McArdle, J. J. and McDonald, R. P. (1984). Some algebraic properties of the reticular action model for moment structures. *British Journal of Mathematical and Statistical Psychology*, **37**, 234–251.
McCabe, G. P. (1975). Computations for variable selection in discriminant analysis. *Technometrics*, **17**, 103–109.
McCullagh, P. and Nelder, J. A. (1989). *Generalized Linear Models*, 2nd edition. Chapman and Hall, London, England.
McDonald, R. P. (1962). A general approach to nonlinear factor analysis. *Psychometrika*, **27**, 397–415.
McDonald, R. P. (1965). Difficulty factors and nonlinear factor analysis. *British Journal of Mathematical and Statistical Psychology*, **18**, 11–23.
McDonald, R. P. (1967a). Nonlinear factor analysis. *Psychometrika Monograph*, No. 15.
McDonald, R. P. (1967b). Numerical methods for polynomial models in nonlinear factor analysis. *Psychometrika*, **32**, 77–112.
McDonald, R. P. (1967c). Factor interaction in nonlinear factor analysis. *British Journal of Mathematical and Statistical Psychology*, **20**, 205–215.
McDonald, R. P. (1978). A simple comprehensive model for the analysis of covariance structures. *British Journal of Mathematical and Statistical Psychology*, **31**, 59–72.
McDonald, R. P. (1979). Simultaneous estimation of factor loadings and scores. *British Journal of Mathematical and Statistical Psychology*, **32**, 212–228.
McDonald, R. P. (1980). A simple comprehensive model for the analysis of covariance structures: some remarks on applications. *British Journal of Mathematical and Statistical Psychology*, **33**, 161–168.
McDonald, R. P. (1986). Comments on D. J. Bartholomew, Foundations of factor analysis: some practical implications. *British Journal of Mathematical and Statistical Psychology*, **38**, 134–137.
McDonald, R. P. and Marsh, H. W. (1990). Choosing a multivariate model: noncentrality and goodness of fit. *Psychological Bulletin*, **107**, 247–255.
McGilchrist, C. A. (1989). Bias of ML and REML estimators in regression models with ARMA errors. *Journal of Statistical Computation and Simulation*, **32**, 127–136.
McGilchrist, C. A. and Cullis, B. R. (1991). REML estimation for repeated measures analysis. *Journal of Statistical Computation and Simulation*, **38**, 151–163.

McHugh, R. B. (1956). Efficient estimation and local identification in latent class analysis. *Psychometrika*, **21**, 331–347.
McHugh, R. B. (1958). Note on 'Efficient estimation and local identification in latent class analysis'. *Psychometrika*, **23**, 273–274.
McKay, R. J. (1976). Simultaneous procedures in discriminant analysis involving two groups. *Technometrics*, **18**, 47–53.
McKay, R. J. (1977). Simultaneous procedures for variable selection in multiple discriminant analysis. *Biometrika*, **64**, 283–290.
McKay, R. J. and Campbell, N. A. (1982a). Variable selection techniques in discriminant analysis I. Description. *British Journal of Mathematics and Statistics in Psychology*, **35**, 1–29.
McKay, R. J. and Campbell, N. A. (1982b). Variable selection techniques in discriminant analysis II. Allocation. *British Journal of Mathematics and Statistics in Psychology*, **35**, 30–41.
McLachlan, G. J. (1974). An asymptotic unbiased technique for estimating the error rates in discriminant analysis. *Biometrics*, **30**, 239–249.
McLachlan, G. J. (1975). Iterative reclassification procedure for constructing an asymptotically optimal rule of allocation in discriminant analysis. *Journal of the American Statistical Association*, **70**, 365–369.
McLachlan, G. J. (1976). A criterion for selecting variables for the linear discriminant function. *Biometrics*, **32**, 529–534.
McLachlan, G. J. (1977). Constrained sample discrimination with the studentized classification statistic *W*. *Communications in Statistics, Theory and Methods*, **6**, 575–583.
McLachlan, G. J. (1980). On the relationship between the *F*-test and the overall error rate for variable selection in two-group discriminant analysis. *Biometrics*, **36**, 501–510.
McLachlan, G. J. (1986). Assessing the performance of an allocation rule. *Computers and Mathematics with Applications*, **12A(2)**, 261–272.
McLachlan, G. J. (1987). On bootstrapping the likelihood ratio statistic for the number of components in a normal mixture. *Applied Statistics*, **36**, 318–324.
McLachlan, G. J. (1992). *Discriminant Analysis and Statistical Pattern Recognition*. Wiley, New York, USA.
McLachlan, G. J. and Basford, K. E. (1988). *Mixture Models: Inference and Applications to Clustering*. Marcel Dekker, New York, USA.
McLachlan, G. J. and Byth, K. (1979). Expected error rates for logistic regression versus normal discriminant analysis. *Biometrical Journal*, **21**, 47–56.
Mendell, N. R., Finch, S. J. and Thode, H. C. (1993). Where is the likelihood ratio test powerful for detecting two component normal mixtures? *Biometrics*, **49**, 907–915.
Mendell, N. R., Thode, H. C. and Finch, S. J. (1991). The likelihood ratio test for the two-component normal mixture problem: power and sample size analysis. *Biometrics*, **47**, 1143–1148.
Meng, X.-L. and Rubin, D. B. (1993). Maximum likelihood estimation via the ECM algorithm: a general framework. *Biometrika*, **80**, 267–278.
Meredith, W. (1977). On weighted Procrustes and hyperplane fitting in factor analysis rotation. *Psychometrika*, **42**, 491–522.

Meyer, Y. (1990). *Ondelettes*. Hermann, Paris, France.
Michie, D., Spiegelhalter, D. J. and Taylor, C. C. (Eds) (1994). *Machine Learning, Neural and Statistical Classification*. Ellis Horwood, New York, USA.
Milan, L. and Whittaker, J. (1995). Application of the parametric bootstrap to models that incorporate a singular value decomposition. *Applied Statistics*, **44**, 31–49.
Miles, R. E. and Serra, J. (1978). En matière d'introduction ... In *Geometrical Probability and Biological Structures: Buffon's 200th Anniversary* (eds R. E. Miles and J. Serra), pp. 3–28. Springer-Verlag, Berlin, Germany.
Miller, A. J. (1990). *Subset Selection in Regression*. Chapman and Hall, London, England.
Miller, G. A. and Nicely, P. E. (1955). An analysis of perceptual confusions among some English consonants. *Journal of the Acoustical Society of America*, **27**, 338–352.
Milligan, G. W. (1989). A validation study of a variable weighting algorithm for cluster analysis. *Journal of Classification*, **6**, 53–71.
Milligan, G. W. and Cooper, M. C. (1988). A study of standardization of variables in cluster analysis. *Journal of Classification*, **5**, 181–204.
Minsky, M. L. and Papert, S. (1969). *Perceptrons*. MIT Press, Cambridge, USA.
Mitchell, A. F. S. (1988). Statistical manifolds of univariate elliptic distributions. *International Statistical Review*, **56**, 1–16.
Mitchell, A. F. S. and Krzanowski, W. J. (1985). The Mahalanobis distance and elliptic distributions. *Biometrika*, **72**, 464–467.
Moran, M. A. (1975). On the expectation of errors of allocation associated with a linear discriminant function. *Biometrika*, **62**, 141–148.
Moran, M. A. and Murphy, B. J. (1979). A closer look at two alternative methods of statistical discrimination. *Applied Statistics*, **28**, 223–232.
Morgan, J. N. and Messenger, R. C. (1973). *THAID: a Sequential Search Program for the Analysis of Nominal Scale Dependent Variables*. Institute for Social Research, University of Michigan, Ann Arbor, USA.
Morgan, J. N. and Sonquist, J. A. (1963). Problems in the analysis of survey data. *Journal of the American Statistical Association*, **58**, 415–434.
Morrison, D. F. (1976). *Multivariate Statistical Methods*, 2nd edition. McGraw-Hill, New York, USA.
Mosimann, J. E. (1970). Size allometry; size and shape variables with characterizations of the lognormal and gamma distributions. *Journal of the American Statistical Association*, **65**, 930–945.
Mosimann, J. E. and James, F. C. (1979). New statistical methods for allometry with application to Florida Redwing Blackbirds. *Evolution*, **33**, 444–459.
Mulaik, S. A., James, L. R., van Alstine, J., Bennett, N., Lind, S. and Stilwell, C. D. (1989). Evaluation of goodness-of-fit indices for structural equation models. *Psychological Bulletin*, **105**, 430–445.
Murphy, B. J. and Moran, M. A. (1986). Parametric and kernel density methods in discriminant analysis: another comparison. *Computers and Mathematics with Applications*, **12A(2)**, 197–207.
Murray, G. D. (1977a). A note on the estimation of probability density functions. *Biometrika*, **64**, 150–152.

Murray, G. D. (1977b). A cautionary note on selection of variables in discriminant analysis. *Applied Statistics*, **26**, 246–250.

Murray, G. D. and Titterington, D. M. (1978). Estimation problems with data from a mixture of normal populations. *Applied Statistics*, **27**, 325–334.

Muthen, B. O. (1978). Contributions to factor analysis of dichotomous variables. *Psychometrika*, **43**, 551–560.

Muthen, B. O. (1993). Latent variable modelling of growth with missing data and multilevel data. In *Multivariate Analysis: Future Directions 2* (eds C. M. Cuadras and C. R. Rao), pp. 199–210. Elsevier Science Publishers, Amsterdam, The Netherlands.

Myles, J. P. and Hand, D. J. (1990). The multi-class metric problem in nearest neighbour discrimination rules. *Pattern Recognition*, **23**, 1291–1297.

Nelder, J. A. (1961). The fitting of a generalization of the logistic curve. *Biometrika*, **17**, 89–100.

Nelder, J. A. (1962). An alternative form of a generalized logistic equation. *Biometrics*, **18**, 614–616.

Nelder, J. A. (1977). A reformulation of linear models (with discussion). *Journal of the Royal Statistical Society, Series A*, **140**, 48–77.

Neuhaus, J. and Wrigley, C. (1954). The quartimax method: an analytical approach to orthogonal simple structure. *British Journal of Statistical Psychology*, **7**, 81–91.

O'Hara, T. F., Hosmer, D. W., Lemeshow, S. and Hartz, S. C. (1982). A comparison of discriminant function and maximum likelihood estimates of logistic coefficients for categorical-scaled data. *Journal of Statistical Computation and Simulation*, **14**, 169–178.

Okamoto, M. (1963). An asymptotic expansion for the distribution of the linear discriminant function. *Annals of Mathematical Statistics*, **34**, 1286-1301 (with correction in **39**, 1358–1359).

Oliver, J. J. and Hand, D. J. (1995). Averaging over decision trees. *Journal of Classification*, **12**, in press.

Olkin, I. and Tate, R. F. (1961). Multivariate correlation models with mixed discrete and continuous variables. *Annals of Mathematical Statistics*, **32**, 448-465 (with correction in **39**, 1358–1359).

Olsson, U. (1979). Maximum likelihood estimation of the polychoric coefficient. *Psychometrika*, **44**, 443–459.

O'Neill, T. J. (1978). Normal discrimination with unclassified observations. *Journal of the American Statistical Association*, **73**, 821–826.

O'Neill, T. J. (1986). The variance of the error rates of classification rules. *Computers and Mathematics with Applications*, **12A(2)**, 273–287.

Ott, J. and Kronmal, R. A. (1976). Some classification procedures for multivariate binary data using orthogonal functions. *Journal of the American Statistical Association*, **71**, 391–399.

Owen, A. (1984). A neighbourhood-based classifier for LANDSAT data. *Canadian Journal of Statistics*, **12**, 191–200.

Patterson, H. D. and Thompson, R. (1971). Recovery of interblock information when the block sizes are unequal. *Biometrika*, **58**, 545–554.

Pearson, E. S. and Hartley, H. O. (1972). *Biometrika Tables for Statisticians, Volume 2*. Cambridge University Press, Cambridge, England.

Pearson, K. (1894). Contribution to the mathematical theory of evolution. *Philosophical Transactions of the Royal Society of London, Series A*, **185**, 71–110.

Pearson, K. (1904). On the theory of contingency and its relation to association and normal correlation. *Drapers' Company Research Memoirs, Biometric Series I*, Dulau and Co, London, England.

Pearson, K. (1926). On the coefficient of racial likeness. *Biometrika*, **18**, 105–117.

Peck, R. and van Ness, J. W. (1982). The use of shrinkage estimators in linear discriminant analysis. *IEEE Transactions on Pattern Analysis and Machine Intelligence*, **PAMI-4**, 530–537.

Potthoff, R. F. and Roy, S. N. (1964). A generalized multivariate analysis model useful especially for growth curve problems. *Biometrika*, **51**, 313–326.

Press, S. J. and Wilson, S. (1978). Choosing between logistic regression and discriminant analysis. *Journal of the American Statistical Association*, **73**, 699–705.

Quenouille, M. (1956). Notes on bias in estimation. *Biometrika*, **43**, 353–360.

Quinlan, J. R. (1986). Induction of decision trees. *Machine Learning*, **1**, 81–106.

Quinlan, J. R. (1990). Decision trees and decision making. *IEEE Transactions on Systems, Man, and Cybernetics*, **20**, 339–346.

Quinlan, J. R. (1993). *C4.5: Programs for Machine Learning*. Morgan Kaufman, San Mateo, California, USA.

Ramsay, J. O. (1982). When the data are functions. *Psychometrika*, **47**, 379–396.

Ramsay, J. O. (1995). Some tools for the multivariate analysis of functional data. In *Recent Advances in Descriptive Multivariate Analysis* (ed W. J. Krzanowski), pp. 269–282. Clarendon Press, Oxford, England.

Ramsay, J. O. and Dalzell, C. J. (1991). Some tools for functional data analysis (with discussion). *Journal of the Royal Statistical Society, Series B*, **53**, 539–572.

Ramsay, J. O., Flanagan, R. and Wang, X. (1995). The functional data analysis of pinch force. *Applied Statistics*, **44**, 17–30.

Rand, W. M. (1971). Objective criteria for the evaluation of clustering methods. *Journal of the American Statistical Association*, **66**, 846–850.

Randles, R. H., Broffitt, J. D., Ramberg, J. S. and Hogg, R. V. (1978). Generalized linear and quadratic discriminant functions using robust estimates. *Journal of the American Statistical Association*, **73**, 564–568.

Rao, C. R. (1945). Information and the accuracy attainable in the estimation of statistical parameters. *Bulletin of the Calcutta Mathematical Society*, **37**, 81–91.

Rao, C. R. (1948). The utilization of multiple measurements in problems of biological classification. *Journal of the Royal Statistical Society, Series B*, **10**, 159–203.

Rao, C. R. (1949). On the distance between two populations. *Sankhyā*, **9**, 246–248.

Rao, C. R. (1965). The theory of least squares when the parameters are stochastic and its application to the analysis of growth curves. *Biometrika*, **52**, 447–458.

Rao, C. R. (1966a). Discriminant function between composite hypotheses and related problems. *Biometrika*, **53**, 339–345.

Rao, C. R. (1966b). Covariance adjustment and related topics in multivariate analysis. In *Multivariate Analysis I*, (ed P. R. Krishnaiah), pp. 87–103. Academic Press, New York, USA.

Rao, C. R. (1973). *Linear Statistical Inference and its Applications*, 2nd edition. Wiley, New York, USA.

Rao, C. R. (1982). Diversity and dissimilarity coefficients: a unified approach. *Theoretical Population Biology*, **21**, 24–43.

Rao, J. N. K. (1968). Expectations, variances and covariances of Anova mean squares by 'synthesis'. *Biometrics*, **24**, 963–978.

Rasch, G. (1960). *Probabilistic Models for Some Intelligence and Attainment Tests*. Paedogogiske Institut, Copenhagen, Denmark.

Raudys, S. J. and Pikelis, V. (1980). On dimensionality, sample size, classification error, and complexity of classification algorithm in pattern recognition. *IEEE Transactions on Pattern Analysis and Machine Intelligence*, **PAMI-2**, 242–252.

Rawlings, R. R. and Faden, V. B. (1986). A study on discriminant analysis techniques applied to multivariate lognormal data. *Journal of Statistical Computation and Simulation*, **26**, 79–100.

Ray, S. (1989a). On a theoretical property of the Bhattacharyya coefficient as a feature evaluation criterion. *Pattern Recognition Letters*, **9**, 315–319.

Ray, S. (1989b). On looseness of error bounds provided by the generalized separability measures of Lissack and Fu. *Pattern Recognition Letters*, **9**, 321–325.

Rayens, W. and Greene, T. (1991). Covariance pooling and stabilization for classification. *Computational Statistics and Data Analysis*, **11**, 17–42.

Rayleigh, Lord (1919). On the problem of random vibrations, and of random flights, in one, two or three dimensions. *Philosophical Magazine*, **37**, 321–347.

Rayner, J. H. (1966). Classification of soils by numerical methods. *Journal of Soil Science*, **17**, 79–92.

Redner, R. A. and Walker, H. F. (1984). Mixture densities, maximum likelihood and the EM algorithm. *SIAM Review*, **26**, 195–239.

Reinsel, G. (1982). Multivariate repeated-measurement or growth curve models with multivariate random-effects covariance structure. *Journal of the American Statistical Association*, **77**, 190–195.

Reinsel, G. (1984). Estimation and prediction in a multivariate random-effects generalized linear model. *Journal of the American Statistical Association*, **79**, 406–414.

Reinsel, G. (1985). Mean squared error properties of empirical Bayes estimators in a multivariate random effects general linear model. *Journal of the American Statistical Association*, **80**, 642–650.

Rencher, A. C. (1993). The contribution of individual variables to Hotelling's T^2, Wilks' Λ, and R^2. *Biometrics*, **49**, 479–489.

Renner, R. M. (1993). The resolution of a compositional data set into mixtures of fixed source compositions. *Applied Statistics*, **42**, 615–631.

Reyment, R. A., Blackith, R. E. and Campbell, N. A. (1984). *Multivariate Morphometrics*, 2nd edition. Academic Press, New York, USA.

Rice, J. A. and Silverman, B. W. (1991). Estimating the mean and covariance structure nonparametrically when the data are curves. *Journal of the Royal Statistical Society, Series B*, **53**, 233–243.

Richards, F. J. (1959). A flexible growth function for empirical use. *Journal of Experimental Botany*, **10**, 290–300.

Ridout, M. S. (1988). Algorithm AS 233. An improved branch and bound algorithm for feature subset selection. *Applied Statistics*, **37**, 139–147.

Riffenburgh, R. H. and Clunies-Ross, C. W. (1960). Linear discriminant analysis. *Pacific Science*, **14**, 251–256.

Ringrose, T. J. and Krzanowski, W. J. (1991). Simulation study of confidence regions for canonical variate analysis. *Statistics and Computing*, **1**, 41–46.

Ripley, B. D. (1986). Statistics, images and pattern recognition (with discussion). *Canadian Journal of Statistics*, **14**, 83–111.

Ripley, B. D. (1988). *Statistical Inference for Spatial Processes*. Cambridge University Press, Cambridge, England.

Ripley, B. D. (1993). Statistical aspects of neural networks. In *Chaos and Networks—Statistical and Probabilistic Aspects* (eds O. E. Barndorff-Nielsen, D. R. Cox, J. L. Jensen and W. S. Kendall), pp. 40–123. Chapman and Hall, London, England.

Ripley, B. D. (1994). Neural networks and related methods for classification (with discussion). *Journal of the Royal Statistical Society, Series B*, **56**, 409–456.

Roberts, E. A. and Raison, J. M. (1983). An analysis of moisture content of soil cores in a designed experiment. *Biometrics*, **39**, 1097–1105.

Robinson, P. M. (1974). Identification, estimation, and large sample theory for regressions containing unobservable variables. *International Economic Review*, **15**, 680–692.

Roger, J. H. (1971). Algorithm AS 20. Updating a minimum spanning tree. *Applied Statistics*, **20**, 204–206.

Roger, J. H. and Carpenter, R. G. (1971). The cumulative construction of minimum spanning trees. *Applied Statistics*, **20**, 192–194.

Rohlf, F. J. (1974). Graphs implied by the Jardine–Sibson overlapping clustering methods, B_k. *Journal of the American Statistical Association*, **69**, 705–710.

Rohlf, F. J. and Bookstein, F. L. (1987). A comment on shearing as a method for 'size correction'. *Systematic Zoology*, **36**, 356–367.

Romesburg, H. C. (1984). *Cluster Analysis for Researchers*. Lifetime Learning Publications, Belmont, California, USA.

Rosenblatt, F. (1962). *Principles of Neurodynamics*. Spartan Books, New York, USA.

Ross, G. J. S. (1969). Algorithm AS 13. Minimum spanning tree. *Applied Statistics*, **18**, 103–104.

Rost, J. (1985). A latent class model for rating data. *Psychometrika*, **50**, 37–49.

Rost, J. (1988). Rating scale analysis with latent class models. *Psychometrika*, **53**, 327–348.

Roy, S. N., Gnanadesikan, R. and Srivastava, J. N. (1971). *Analysis and Design of Certain Quantitative Multiresponse Experiments, with an Appendix on*

Computer Algorithms and Programs by E. B. Fowlkes II and E. T. Lee. Pergamon Press, Oxford, England.
Royall, M. R. (1986). Model robust confidence intervals using maximum likelihood estimators. *International Statistical Review*, **54**, 221–226.
Rubin, D. B. and Thayer, D. T. (1982). EM algorithms for ML factor analysis. *Psychometrika*, **47**, 69–76.
Rubin, D. B. and Thayer, D. T. (1983). More on EM for ML factor analysis. *Psychometrika*, **48**, 253–257.
Rumelhart, D. E., Hinton, G. E. and Williams, R. J. (1986). Learning internal representations by error propagation. In *Neural Information Processing Systems* (ed D. Z. Anderson), pp. 602–611. American Institute of Physics, New York, USA.
Safavian, S. R. and Landgrebe, D. (1991). A survey of decision tree classifier methodology. *IEEE Transactions on Systems, Man, and Cybernetics*, **21**, 660–674.
Sanathanan, L. and Blumenthal, S. (1978). The logistic model and estimation of latent structure. *Journal of the American Statistical Association*, **73**, 794–799.
Sandland, R. L. and McGilchrist, C. A. (1979). Stochastic growth curve analysis. *Biometrics*, **35**, 255–271.
Sarle, W. S. (1991). Book review of Kaufman and Rousseeuw (1990). *Journal of the American Statistical Association*, **86**, 830–832.
SAS Institute, Inc. (1990). *SAS User's Guide*. Cary, North Carolina, USA.
Sayre, J. W. (1980). The distribution of the actual error rates in linear discriminant analysis. *Journal of the American Statistical Association*, **75**, 201–205.
Schaafsma, W. (1982). Selecting variables in discriminant analysis for improving upon classical procedures. In *Handbook of Statistics, Volume 2* (eds P. R. Krishnaiah and L. N. Kanal), pp. 857–881. North-Holland, Amsterdam, The Netherlands.
Schaafsma, W. and van Vark, G. N. (1977). Classification and discrimination problems with applications. Part I. *Statistica Neerlandica*, **31**, 25–45.
Schmitz, P. I. M., Habbema, J. D. F. and Hermans, J. (1983). The performance of logistic discrimination on myocardial infarction data in comparison with some other discriminant analysis methods. *Statistics in Medicine*, **2**, 199-205.
Schmitz, P. I. M., Habbema, J. D. F. and Hermans, J. (1985). A simulation study of the performance of five discriminant analysis methods for mixtures of continuous and binary variables. *Journal of Statistical Computation and Simulation*, **23**, 69–95.
Schönemann, P. H. A. (1981). Power as a function of communality in factor analysis. *Bulletin of the Psychonomic Society*, **17**, 57–60.
Schott, J. R. (1990). Canonical mean projections and confidence regions in canonical variate analysis. *Biometrika*, **77**, 587–596.
Scott, A. J. and Symons, M. J. (1971). Clustering methods based on maximum likelihood. *Biometrics*, **27**, 387–398.
Scott, D. W. (1992). *Multivariate Density Estimation*. Wiley, New York, USA.
Seber, G. A. F. (1984). *Multivariate Observations*. Wiley, New York, USA.

Shapiro, A. (1987). Robustness properties of the MDF analysis of moment structures. *South African Statistical Journal*, **21**, 39–62.

Shepard, R. N. (1962). The analysis of proximities: multidimensional scaling with an unknown distance function. *Psychometrika*, **27**, 125–140 and 219-246.

Shepard, R. N. (1972). Psychological representation of speech sounds. In *Human Communication: A Unified View* (eds E. E. David and P. B. Denes). McGraw-Hill, New York, USA.

Shepard, R. N. and Arabie, P. (1979). Additive clustering: representation of similarities as combinations of discrete overlapping properties. *Psychological Review*, **86**, 87–123.

Shumway, R. H. (1982). Discriminant analysis for time series. In *Handbook of Statistics, Volume 2* (eds P. R. Krishnaiah and L. N. Kanal), pp. 1–46. North-Holland, Amsterdam, The Netherlands.

Sibson, R. (1969). Information radius. *Zeitschrift für Wahrscheinlichkeitstheorie und Verwandte Gebeite*, **14**, 149–160.

Siddiqi, J. S. (1991). *Mixture and latent class models for discrete multivariate data*. Unpublished Ph.D. Thesis, Exeter University, England.

Silverman, B. W. (1981). Using kernel densities to investigate multimodality. *Journal of the Royal Statistical Society, Series B*, **43**, 97–99.

Silverman, B. W. (1986). *Density Estimation for Statistics and Data Analysis*. Chapman and Hall, London, England.

Silverman, B. W. and Jones, M. C. (1989). E. Fix and J. L. Hodges (1951): An important contribution to nonparametric discriminant analysis and density estimation. *International Statistical Review*, **57**, 233–247.

Siotani, M. (1982). Large sample approximations and asymptotic expansions of classification statistics. In *Handbook of Statistics, Volume 2* (eds P. R. Krishnaiah and L. N. Kanal), pp. 61–100. North-Holland, Amsterdam, The Netherlands.

Siotani, M. and Wang, R.-H. (1975). *Further Expansion Formulae for Error Rates and Comparison of the W- and Z-procedures in Discriminant Analysis*. Technical Report No. 33, Department of Statistics, Kansas State University, Kansas, USA.

Siotani, M. and Wang, R.-H. (1977). Asymptotic expansions for error rates and comparison of the W-procedure and Z-procedure in discriminant analysis. In *Multivariate Analysis IV* (ed P. R. Krishnaiah), pp. 523–545. North-Holland, Amsterdam, The Netherlands.

Sitgreaves, R. (1961). Some results on the W-classification statistic. In *Studies in Item Analysis and Prediction* (ed H. Solomon), pp. 158–168. Stanford University Press, Stanford, California, USA.

Skovgaard, I. M. (1985). A second order investigation of asymptotic ancillarity. *Annals of Statistics*, **13**, 534–557.

Small, C. G. (1981). *Distribution of shape and maximal invariant statistics*. Unpublished Ph.D. Thesis, Cambridge University, England.

Small, C. G. (1988). Techniques of shape analysis on sets of points. *International Staistical Review*, **56**, 243–257.

Smith, C. A. B. (1947). Some examples of discrimination. *Annals of Eugenics*, **13**, 272–282.

Smith, D. J. (1993). *The application of artificial neural networks to the classification of underwater cetacean sounds.* Unpublished Ph.D. Thesis, Exeter University, England.

Smith, D. J., Bailey, T. C. and Munford, A. G. (1993). Robust classification of high-dimensional data using artificial neural networks. *Statistics and Computing*, **3**, 71–81.

Snapinn, S. M. and Knoke, J. D. (1985). An evaluation of smoothed classification error-rate estimators. *Technometrics*, **27**, 199–206.

Snapinn, S. M. and Knoke, J. D. (1988). Bootstrapped and smoothed classification error rate estimators. *Communications in Statistics, Simulation and Computation*, **17**, 1135–1153.

Snapinn, S. M. and Knoke, J. D. (1989). Estimation of error rates in discriminant analysis with selection of variables. *Biometrics*, **45**, 289–299.

Sobel, M. E. and Bohrnstedt, G. W. (1985). Use of null models in evaluating the fit of covariance structure models. In *Sociological Methodology* (ed N. B. Tuma). Jossy-Bass, San Francisco, California, USA.

Sokal, R. R. and Michener, C. D. (1958). A statistical method for evaluating systematic relationships. *University of Kansas Science Bulletin*, **38**, 1409–1438.

Sokal, R. R. and Sneath, P. H. A. (1963). *Principles of Numerical Taxonomy.* Freeman, San Francisco, USA.

Solla, S. A., Levin, E. and Fleisher, M. (1988). Accelerated learning in layered neural networks. *Complex Systems*, **2**, 625–639.

Solomon, H. (ed) (1961). *Studies in Item Analysis and Prediction.* Stanford University Press, Stanford, California, USA.

Somers, K. M. (1986). Multivariate allometry and removal of size with principal components analysis. *Systematic Zoology*, **35**, 359–368.

Somers, K. M. (1989). Allometry, isometry and shape in principal components analysis. *Systematic Zoology*, **38**, 169–173.

Sonquist, J. A. (1970). *Multivariate Model Building.* Institute for Survey Research, Michigan, USA.

Sprent, P. (1972). The mathematics of size and shape. *Biometrics*, **28**, 23–37.

Srivastava, J. N. and Zaatar, M. K. (1972). On the maximum likelihood classification rule for incomplete multivariate samples and its admissibility. *Journal of Multivariate Analysis*, **2**, 115–126.

Srivastava, M. S. (1967). Comparing distances between multivariate populations—the problem of minimum distance. *Annals of Mathematical Statistics*, **38**, 550–556.

Stapleton, D. C. (1977). Analyzing political participation data with a MIMIC model. In *Sociological Methodology 1978* (ed K. Schuessler), pp. 52–74. Jossey-Bass, San Francisco, California, USA.

Stephens, M. A. (1962). Exact and approximate tests for directions. II. *Biometrika*, **49**, 547–552.

Stephens, M. A. (1967). Tests for the dispersion and for the modal vector of a distribution on the sphere. *Biometrika*, **54**, 211–223.

Stephens, M. A. (1969a). Multi-sample tests for the Fisher distribution for directions. *Biometrika*, **56**, 169–181.

Stephens, M. A. (1969b). *Techniques for Directional Data.* Stanford University Department of Statistics Technical Report No. 150. Stanford University Press, Stanford, California, USA.

Stephens, M. A. (1972). Multisample tests for the von Mises distribution. *Journal of the American Statistical Association*, **67**, 456–461.

Stephens, M. A. (1982). Use of the von Mises distribution to analyse continuous proportions. *Biometrika*, **69**, 197–203.

Stephens, M. A. (1992). On Watson's ANOVA for directions. In *The Art of Statistical Science* (ed K. V. Mardia), pp. 75–85. Wiley, Chichester, England.

Stone, C. J. (1977). Consistent nonparametric regression (with discussion). *Annals of Statistics*, **5**, 595–645.

Stone, M. (1974). Cross-validatory choice and assessment of statistical predictions (with discussion). *Journal of the Royal Statistical Society, Series B*, **36**, 111–148.

Stone, M. and Brooks, R. J. (1990). Continuum regression: cross-validated sequentially constructed prediction embracing ordinary least squares, partial least squares and principal components regression (with discussion). *Journal of the Royal Statistical Society, Series B*, **52**, 237–269.

Streit, F. (1979). Multivariate linear discrimination when the covariance matrix is unknown. *South African Statistical Journal*, **14**, 76.

Stuart, A. S. and Ord, J. K. (1991). *Kendall's Advanced Theory of Statistics, Volume II*, 5th edition. Edward Arnold, London, England.

Subrahmaniam, K. and Chinganda, E. F. (1978). Robustness of the linear discriminant function to nonnormality: Edgeworth series distribution. *Journal of Statistical Planning and Inference*, **2**, 79–91.

Sundberg, R. and Brown, P.J. (1989). Multivariate calibration with more variables than observations. *Technometrics*, **31**, 365–371.

Sutradhar, B. C. (1990). Discrimination of observations into one of two *t* populations. *Biometrics*, **46**, 827–835.

Switzer, P. (1980). Extensions of linear discriminant analysis for statistical classification of remotely sensed satellite imagery. *Mathematical Geology*, **12**, 367–376.

Switzer, P. (1983). Some spatial statistics for the interpretation of satellite data (with discussion). *Bulletin of the International Statistical Institute*, **50**, 962–972.

Symons, M. J. (1981). Clustering criteria and multivariate normal mixtures. *Biometrics*, **37**, 35–43.

Takane, Y., Bozdogan, H. and Shibayama, T. (1987). Ideal point discriminant analysis. *Psychometrika*, **52**, 371–392.

Tanaka, J. S. and Huba, G. J. (1985). A fit index for covariance structure models under arbitrary GLS. *British Journal of Mathematical and Statistical Psychology*, **38**, 197–201.

Thode, H. C., Finch, S. J. and Mendell, N. R. (1989). Simulated percentage points for the null distribution of the likelihood ratio test for a mixture of two normals. *Biometrics*, **44**, 1195–1201.

Thompson, D. W. (1917). *On Growth and Form*, (2 volumes). Cambridge University Press, Cambridge, England.

Thomson, G. H. (1951). *The Factorial Analysis of Human Ability*. London University Press, London, England.
Thurstone, L. L. (1947). *Multiple Factor Analysis*. University of Chicago Press, Chicago, USA.
Tiku, M. L. and Balakrishnan, N. (1984). Robust multivariate classification procedures based on the MML estimators. *Communications in Statistics, Theory and Methods*, **13**, 967–986.
Tiku, M. L. and Balakrishnan, N. (1989). Robust classification procedures based on the MML estimators. *Communications in Statistics, Theory and Methods*, **18**, 1047–1066.
Titterington, D. M. (1989). An alternative stochastic supervisor in discriminant analysis. *Pattern Recognition*, **22**, 91–95.
Titterington, D. M., Murray, G. D., Murray, L. S., Spiegelhalter, D. J., Skene, A. M., Habbema, J. D. F. and Gelpke, G. J. (1981). Comparison of discrimination techniques applied to a complex set of head injured patients (with discussion). *Journal of the Royal Statistical Society, Series A*, **144**, 145–175.
Titterington, D. M., Smith, A. F. M. and Makov, U. E. (1985). *Statistical Analysis of Finite Mixture Distributions*. Wiley, New York, USA.
Todeschini, R. (1989). *k*-nearest neighbour method: the influence of data transformations and metrics. *Chemometrics and Intelligent Laboratory Systems*, **6**, 213–220.
Todorov, V., Neykov, N. and Neytchev, P. (1994). Robust two-group discrimination by bounded influence regression. A Monte Carlo simulation. *Computational Statistics and Data Analysis*, **17**, 289–302.
Toussaint, G. T. (1974). Bibliography on estimation of misclassification. *IEEE Transactions on Information Theory*, **IT-20**, 472–479.
Truett, J., Cornfield, J. and Kannel, W. B. (1967). A multivariate analysis of the risk of coronary heart disease in Framingham. *Journal of Chronic Diseases*, **20**, 511–524.
Tucker, L.R. (1966). Some mathematical notes on three-mode factor analysis. *Psychometrika*, **31**, 279–311.
Tucker, L. R. (1972). Relations between multidimensional scaling and three-mode factor analysis. *Psychometrika*, **37**, 3–27.
Tucker, L. R. and Lewis, C. (1973). The reliability coefficient for maximum likelihood factor analysis. *Psychometrika*, **38**, 197–201.
Tucker, L. R. and Messick, S. (1963). An individual differences model for multidimensional scaling. *Psychometrika*, **28**, 333–367.
Tyteca, D. and Dufrêne, M. (1993). On the use of distances in the taxonomic study of critical plant groups — case studies of Western European Orchidaceae. *Annals of Botany*, **71**, 257–277.
Vach, W. and Degens, P. O. (1991). A new approach to isotonic agglomerative hierarchical clustering. *Journal of Classification*, **8**, 217–237.
van Buuren, S. and Heiser, W. J. (1989). Clustering *N* objects into *K* groups under optimal scaling of variables. *Psychometrika*, **54**, 699–706.
van der Heijden, P. G. M., Mooijaart, A. and de Leeuw, J. (1992). Constrained latent budget analysis. In *Sociological Methodology 22* (ed C. C. Clogg), pp. 279–320. Blackwell, Cambridge, England.

van Driel, O. P. (1978). On various cases of improper solutions in maximum likelihood factor analysis. *Psychometrika*, **43**, 225–243.

van Ness, J. W. (1979). On the effects of dimension in discriminant analysis for unequal covariance populations. *Technometrics*, **21**, 119–127.

van Ness, J. W. and Simpson, C. (1976). On the effects of dimension in discriminant analysis. *Technometrics*, **18**, 175–187.

van Ooyen, A. and Nienhuis, B. (1992). Improving the convergence of the back-propagation algorithm. *Neural Networks*, **5**, 465–471.

Vasdekis, V. G. S. (1993). *An investigation of certain methods in the analysis of growth curves.* Unpublished D. Phil. Thesis, University of Oxford, England.

Verbyla, A. P. (1986). Conditioning in the growth curve model. *Biometrika*, **75**, 129–138.

Verbyla, A. P. and Cullis, B. R. (1990). Modelling in repeated measures experiments. *Applied Statistics*, **39**, 129–138.

Vlachonikolis, I. G. (1985). On the asymptotic distribution of the lócation linear discriminant function. *Journal of the Royal Statistical Society, Series B*, **47**, 498–509.

Vlachonikolis, I. G. (1990). Predictive discrimination and classification with mixed binary and continuous variables. *Biometrika*, **77**, 657–662.

Vlachonikolis, I. G. and Marriott, F. H. C. (1982). Discrimination with mixed binary and continuous data. *Applied Statistics*, **31**, 23–31.

Vlachonikolis, I. G. and Vasdekis, V. G. S. (1994). On a class of change-point models in covariance structures for growth curves and repeated measurements. *Communications in Statistics, Part A—Theory and Methods*, **23**, 1087–1102.

von Mises, R. (1918). Über die "Ganzzahligkeit" der Atomgewichte und Verwandte Fragen. *Physikal Zeitung*, **19**, 490–500.

von Mises, R. (1945). On the classification of observation data into distinct groups. *Annals of Mathematical Statistics*, **16**, 68–73.

Wahl, P. W. and Kronmal, R. A. (1977). Discriminant functions when covariances are unequal and sample sizes are moderate. *Biometrics*, **33**, 479–484.

Wakaki, H. (1990). Comparison of linear and quadratic discriminant functions. *Biometrika*, **77**, 227–229.

Wald, A. (1944). On a statistical problem arising in the classification of an individual into one of two groups. *Annals of Mathematical Statistics*, **15**, 145–162.

Walker, S. H. and Duncan, D. B. (1967). Estimation of the probability of an event as a function of several independent variables. *Biometrika*, **54**, 167–169.

Ward, J. H. (1963). Hierarchical grouping to optimize an objective function. *Journal of the American Statistical Association*, **58**, 236–244.

Watson, G. S. (1956). Analysis of dispersion on a sphere. *Monthly Notices of the Royal Astronomical Society; Geophysics Supplement*, **7**, 153–159.

Watson, G. S. (1965). Equatorial distributions on a sphere. *Biometrika*, **52**, 193–201.

Watson, G. S. and Williams, E. J. (1956). On the construction of significance tests on the circle and the sphere. *Biometrika*, **43**, 344–352.

Webster, R. and McBratney, A. B. (1981). Soil segment overlap in character space and its implication for soil classification. *Journal of Soil Science*, **32**, 133–147.
Wegman, E. J. (1972a). Non-parametric probability density estimation. I. *Technometrics*, **14**, 533–546.
Wegman, E. J. (1972b). Non-parametric probability density estimation. II. *Journal of Statistical Computation and Simulation*, **1**, 225–246.
Weihs, C. and Schmidli, H. (1990). OMEGA: online multivariate exploratory graphical analysis: routine searching for structure (with discussion). *Statistical Science*, **5**, 175–226.
Welch, B. L. (1939). Note on discriminant functions. *Biometrika*, **31**, 218–220.
Wermuth, N. (1980). Linear recursive equations, covariance selection and path analysis. *Journal of the American Statistical Association*, **75**, 963–972.
Wermuth, N. and Lauritzen, S. L. (1990). On substantive research hypotheses, independence graphs and graphical chain models (with discussion). *Journal of the Royal Statistical Society, Series B*, **52**, 21–50.
Wertz, W. and Schneider, B. (1979). Statistical density estimation: a bibliography. *International Statistical Review*, **47**, 155–175.
Whittaker, J. (1990). *Graphical Models in Applied Multivariate Statistics*. John Wiley and Son, Chichester, England.
Whittle, P. (1952). On principal components and least square methods of factor analysis. *Skandinavisk Aktuarietidskrift*, **35**, 223–239.
Wilks, S. S. (1932). Certain generalizations in the analysis of variance. *Biometrika*, **24**, 471–494.
Windham, M. P. (1986). A unification of optimization-based clustering algorithms. In *Classification as a Tool of Research* (eds W. Gaul and M. Schader), pp. 447–451. North Holland, Amsterdam, Netherlands.
Wishart, D. (1969). Mode analysis. In *Numerical Taxonomy* (ed A. J. Cole). Academic Press, New York, USA.
Wishart, D. (1987). *CLUSTAN User Manual*, University of St. Andrews, Scotland.
Wishart, J. (1938). Growth rate determinations in nutrition studies with the bacon pig, and their analysis. *Biometrika*, **30**, 16–28.
Wojciechowski, T. J. (1987). Nearest neighbor classification rule for mixtures of discrete and continuous random variables. *Biometrical Journal*, **29**, 953–959.
Wold, H. O. (1975). *Modelling in Complex Situations with Soft Information*. Third World Congress of the Economic Society, 21–26 August, Toronto, Canada.
Wold, H. O. (1985). Partial least squares. In *Encyclopedia of Statistical Sciences, Volume 6*, (eds S. Kotz and N. L. Johnson), pp. 581–591. Wiley, New York, USA.
Wold, S., Martens, H. and Wold, H. O. (1983). The multivariate calibration problem in chemistry solved by the PLS method. *Lecture Notes in Mathematics: Procedings of the Conference on Matrix Pencils*. (eds A. Ruhe and B. Kågström), pp. 286–293. Springer-Verlag, Heidelberg, Germany.

Wolfe, J. H. (1965). *A Computer Program for the Maximum Likelihood Analysis of Types*. Technical Bulletin 65–15, US Naval Personnel and Training Research Activity, San Diego, USA.

Wolfe, J. H. (1967). *Normix: Computational Methods for Estimating the Parameters of Multivariate Normal Mixtures of Distributions*. Technical Bulletin SRM 68–2, US Naval Personnel and Training Research Activity, San Diego, USA.

Wolfe, J. H. (1970). Pattern clustering by multivariate mixture analysis. *Multivariate Behavioral Research*, **5**, 329–350.

Wolfe, J. H. (1971). *A Monte-Carlo Study of the Sampling Distribution of the Likelihood Ratio for Mixtures of Multinormal Distributions*. Technical Bulletin STB 72–2, US Naval Personnel and Training Research Activity, San Diego, USA.

Wong, M. A. and Lane, T. (1983). A *k*th nearest neighbour clustering procedure. *Journal of the Royal Statistical Society, Series B*, **45**, 362–368.

Wright, S. (1921). Correlation and causation. *Journal of Agricultural Research*, **20**, 557–585.

Wright, S. (1934). The method of path coefficients. *Annals of Mathematical Statistics*, **5**, 161–215.

Wright, S. (1960). Path coefficients and path regression: alternative or complementary concepts? *Biometrics*, **16**, 189–202.

Wright, S. (1968). *Evolution and the Genetics of Population. Volume 1: Genetic and Biometric Foundations*. University of Chicago Press, Chicago, USA.

Wyman, F. J., Young, D. M. and Turner, D. W. (1990). A comparison of asymptotic error rate expansions for the sample linear discriminant function. *Pattern Recognition*, **23**, 775–783.

Yates, F. (1937). *The Design and Analysis of Factorial Experiments*. Imperial Bureau of Soil Science, Harpenden, England.

Young, D. M. and Odell, P. L. (1986). Feature-subset selection for statistical classification problems involving unequal covariance matrices. *Communications in Statistics, Theory and Methods*, **15**, 137–157.

Young, D. M., Marco, V. R. and Odell, P. L. (1987). Quadratic discrimination: some results on optimal low-dimensional representation. *Journal of Statistical Inference and Planning*, **17**, 307–319.

Young, G. (1941). Maximum likelihood estimation and factor analysis. *Psychometrika*, **6**, 49–53.

Yule, G. U. (1900). On the association of attributes in statistics. *Philosophical Transactions of the Royal Society of London, Series A*, **194**, 257-319.

Yule, G. U. (1912). On the methods of measuring the association between two attributes. *Journal of the Royal Statistical Society*, **75**, 579–642.

Zadeh, L. A. (1965). Fuzzy sets. *Information and Control*, **8**, 338–353.

Zadeh, L. A. (1977). Fuzzy sets and their application to pattern classification and clustering analysis. In *Classification and Clustering* (ed J. van Ryzin), pp. 251–299. Academic Press, New York, USA.

Author Index

Subject Index